The Processes of Fossilization

The Processes of Fossilization

edited by
Stephen K. Donovan

Columbia University Press
New York

Columbia University Press
New York   Oxford

Copyright © 1991 Columbia University Press

Library of Congress Cataloging-in-Publication Data

The Processes of Fossilization / edited by Stephen K. Donovan
    p.   cm.
  'First published in Great Britain in 1991 by Belhaven Press (a
division of Pinter Publishers)' – t.p. verso
  Includes bibliographical references and index.
  ISBN 0-231-07674-6
  1. Fossilization.   2. Taphonomy.   I. Donovan, S. K.
QE721.2.F6P76   1991
560—dc20   91-16327
                                CIP

Printed and bound in Great Britain

CONTENTS

LIST OF FIGURES

LIST OF TABLES

LIST OF CONTRIBUTORS

Stephen K. Donovan, Department of Geology, University of the West Indies, Mona, Kingston 7, Jamaica

Gerhard C. Cadée, Netherlands Institute for Undersea Research (NIOZ), P.O. Box 59, 1790 AB Den Burg, Texel, The Netherlands

Karla M. Parsons, West Indies Laboratory, Fairleigh Dickinson University, Teague Bay, Christiansted, St. Croix, U.S. Virgin Islands 00820–4542

Carlton E. Brett, Department of Geological Sciences, University of Rochester, Rochester, New York 14627, USA

Michael L. McKinney, Department of Geological Sciences, University of Tennessee, Knoxville, Tennessee 37916–1410, USA

Maurice E. Tucker, Department of Geological Sciences, University of Durham, Science Laboratories, South Road, Durham, DH1 3LE, England

Neville T.J. Hollingworth, Department of Earth Sciences, The Open University, Walton Hall, Milton Keynes MK7 6AA, England

Michael J. Barker, Department of Geology, Portsmouth Polytechnic, Burnaby Road, Portsmouth, PO1 3QL, England

Peter A. Allison, Postgraduate Research Institute for Sedimentology, University of Reading, Whiteknights, Reading, RG6 2AH, England

Derek E.G. Briggs, Department of Geology, University of Bristol, Wills Memorial Building, Queen's Road, Bristol, BS8 1RJ, England

David R. Greenwood, Department of Geology, University of Saskatchewan, Saskatoon, Saskatchewan, Canada

Ronald E. Martin, Department of Geology, University of Delaware, Newark, Delaware 19716–6000, USA

W. David Liddell, Department of Geology, Utah State University, Logan, Utah 84322–4505, USA

Stephen E. Speyer, Department of Geology, Arizona State University, Tempe, Arizona 85287–1404, USA

William B. Boston, Department of Geological Sciences, Wright Geological Laboratory, Rutgers University, New Brunswick, New Jersey 08903, USA

Royal H. Mapes, Department of Geological Sciences, Ohio University, Athens, Ohio 45701, USA

David M. Martill, Department of Earth Sciences, The Open University, Milton Keynes, MK7 6AA, England

INTRODUCTION

Stephen K. Donovan

If not a change in direction, then certainly a widening of interests has been apparent in the literature of the broad field of palaeoecology and related studies over the past fifteen years or so. This diversification has been catalysed by an increasing awareness that the post-mortem processes which have affected fossil organisms are an important factor in both sedimentological and palaeontological research, whereas relatively little attention was paid to them previously except to generate a series of 'just so' stories of mainly curiosity value. Post-mortem processes occur in two broad environments. Biostratinomic processes are those which occur between death and final burial, while diagenetic changes take place during burial. The science that unites these two sub-disciplines is taphonomy (Efremov, 1940). It is taphonomy, in the broadest sense, that is the subject of the present volume.

Archaeologists and palaeontologists studying terrestrial vertebrates have long been aware of the importance of taphonomic (particularly biostratinomic) processes in moulding the fossil record and a diverse literature is available on their subject (see, for example, Behrensmeyer and Hill, 1980; Brothwell, 1981; Brain, 1981; Klein and Cruz-Uribe, 1984; Weigelt, 1989). However, the only general text on the biostratinomy of marine environments that has hitherto been available is Schäfer's excellent volume on the *Aktuo-Paläontologie* of the southern North Sea basin (1972). Although the original German text is now nearly 30 years old, Schäfer's observations are still a primary data source for the student of death, decay and disarticulation. For the palaeontologist, this book represents a double-edged blade that must be used with discretion. The importance of Schäfer's commentary is that it is based on primary observations of dead and dying organisms, providing data on the demise and decay of a spectrum of extant shallow marine animals.

Direct comparisons can therefore be made between many extant groups and their fossil progenitors. However, many groups that are important as fossils were not, and could not, be considered in Schäfer's study, either because they are not at present living in the shallow southern North Sea (for example, the stalked crinoids) or because they are extinct (such as the trilobites).

Fossilization was conceived as a separate study from previous texts on taphonomy, being mainly concerned with selected groups and topics that are important in the fossil record, and which were in need of detailed review. This book is divided into two slightly unequal parts. The shorter first half considers the history of taphonomic studies, diagenesis and those factors which affect our understanding of the fossil record. The second section considers the response to fossilization of a broad range of groups that followed a variety of lifestyles. Although this list is not exhaustive, the diversity of fossils discussed will make this section useful when considering a much wider range of organisms than is described herein.

My thanks go to Iain Stevenson of Belhaven Press for his help and encouragement in producing this book. To the contributing authors, whom I assailed by letter, telephone, cable and fax, thank you all for your excellent contributions and for putting up with such an impatient editor. The Department of Geology of the University of the West Indies provided invaluable logistic support, particularly in paying for my endless stream of mail. I thank Gerhard Cadée for his comments on an early version of this introduction.

REFERENCES

Behrensmeyer, A.K. and Hill, A.P., 1980, *Fossils in the making: vertebrate taphonomy and paleoecology*, University of Chicago Press, Chicago.

Brain, C.K., 1981, *The hunters or the hunted?*, University of Chicago Press, Chicago.

Brothwell, D.R., 1981, *Digging up bones* (3rd edn), British Museum (Natural History) and Oxford University Press, London and Oxford.

Efremov, J.A., 1940, Taphonomy: new branch of paleontology, *Pan-American Geologist*, **74** (2): 81–93.

Klein, R.G. and Cruz-Uribe, K., 1984, *The analysis of animal bones from archaeological sites*, University of Chicago Press, Chicago.

Schäfer, W., 1972, *Ecology and palaeoecology of marine environments* (English edn, originally published 1962), University of Chicago Press, Chicago.

Weigelt, J., 1989, *Recent vertebrate carcasses and their paleobiological implications* (English edn, originally published 1927), University of Chicago Press, Chicago.

Chapter 1

THE HISTORY OF TAPHONOMY

Gerhard C. Cadée

As palaeontology now prepares for a great leap forward into a computerised age there is perhaps a danger that it may lose sight of its historic origins in the 'steam age' of science and before (Rudwick, 1972, p.266).

INTRODUCTION

When, in a paper in an obscure American journal, Efremov (1940) coined the word 'taphonomy' for the study of the 'laws' governing the transition of organic remains from the biosphere to the lithosphere, he was aware of the fact that work had already been devoted to this topic, albeit under a different name: 'Neither the problems nor methods are new' (Efremov, 1940, p.92). In a later publication (Efremov, 1953, but written in 1943) he gave much of the background missing in his earlier paper. Strictly speaking a search for 'laws' is inappropriate (Wilson, 1988), but it is obvious that Efremov was aiming at principles, a word he used later (Efremov, 1958).

Viewpoints differ regarding the definition and scope of taphonomy (Behrensmeyer and Kidwell, 1985; Wilson, 1988). It comprises the study of the processes between death and burial of an organism, including cause and manner of death, decay, decomposition, transportation and burial. Efremov (1940; 1953) also included fossil diagenesis. In its broadest definition it is synonymous with the study of the fossilization process (Müller, 1976; 1979), and one may wonder: why this new word? Taphonomy can be divided into 'necrolysis' (necrosis, necrobiosis), the death and decomposition of an organism; 'biostratonomy' (biostratinomy), the sedimentary history of the fossil until burial; and 'fossil diagenesis', the chemical and mechanical alterations within the sediment (Seilacher, 1976). Overlap with existing terminology is obvious. 'Palaeobiology' (the relation between fossil organisms and their environment) as defined by Abel (1912) included necrolysis, but now encompasses evolution, palaeoecology and taphonomy (Briggs and Crowther, 1990). 'Biostratonomy', as used by Weigelt (1927a; 1927b; 1930b), involves necrolysis and biostratonomy as defined above. 'Aktuopalaeontology' (Richter, 1928) includes the study of taphonomy of recent organisms. Palaeontologists apparently have never been very strict in the use of definitions and many words have evolved to have meanings quite different from the original. Taphonomy is rather new in this terminology and it took some 30 years before it became accepted and widely used. It certainly proved to be a useful (and euphonious) addition to our terminology.

In this chapter I will trace the roots of taphonomy, although limitations on space prevent me from being exhaustive in my analysis. I will follow the contributions of a few palaeontologists to taphonomy, and, biased by my own interests, omit certain important works. Deecke (1923), in his review of fossilization, complained that the subject was too many-sided almost 70 years ago.

The history of taphonomic research is closely linked with that of palaeontology. Olson (1980) wrote a history of taphonomy, but mainly based on literature in English that appeared after Efremov's (1940) paper. I will focus on earlier and non-English literature. Behrensmeyer and Kidwell (1985) and Wilson (1988) also include brief histories. Many references to earlier papers, also in other languages, can be found in Deecke (1923) and Müller (1951; 1976; 1979). For discussion of the history of palaeontology the reader is referred to, for example, Lyell (1830–3), Zittel (1899), Rudwick (1972), Ellenberger (1988) and Hölder (1989).

THE EARLY HISTORY

Studies of the fossilization process could not start before the true nature of fossils was ascertained. A few Greek scientists had already grasped that fossils were the remains of organisms that once lived, but on the authority of Aristotle and his ideas on the spontaneous generation of living organisms, fossils were for almost 2000 years considered to be the unsuccessful efforts of a *vis plastica*, which was continually striving to produce organic bodies. To cite Ellenberger (1988, p.36): 'Au total, il est clair, qu'il n'y a pas de paléontologie antique'.

Fracastoro (1483–1553) was one of the first to deny the Aristotelian view. He accepted fossils as the remains of animals drowned during the biblical Deluge, a very popular concept until the middle of the nineteenth century.

Taphonomic reasoning in the truest sense was used by da Vinci (1452–1519) (translation in MacCurdy, 1938; and partly in Mather and Mason, 1939) to prove that fossil shells found in Monferrato in Lombardy were not the remains of the Deluge. Using his observations on Recent shells, he stated that living bivalves could not travel fast enough to keep pace with the rising waters of the Deluge and traverse a distance of 250 miles (from the Adriatic to Monferrato) in 40 days (the duration of the Deluge). Nor could waves carry empty shells over this distance; he observed that dead shells were never found far away from the living ones. In the Monferrato mountains da Vinci observed layers with articulated bivalves *in situ*, demonstrating they had lived there, and a little higher he found layers of the disarticulated shells, interpreted as separated valves cast up by the waves. Elsewhere in his notebooks da Vinci explained how fishes are embedded in mud, where the organic matter decays, and only the skeleton is left. He also recognized that marine shells become embedded under mud transported by rivers to the sea, and, after decay of the organic matter, are filled with mud. Da Vinci's influence on the development of palaeontology is difficult to trace; his notes circulated widely before and after his death (Duhem, 1906–13), but they were transliterated only much later (by MacCurdy, 1938).

Steno (Niels Stensen, 1638–86), according to many textbooks in geology, founded geology as a science. Scherz and Pollock (1969) gave an English translation of his works. Steno argued that (Neogene) deposits of Tuscany had formed gradually, layer by layer, as sediments under water. The organic remains found in these rocks are very similar to those now living in the sea. Animals on the bottom, already dead or still alive but unable to escape because they were fixed to the bottom, were smothered by a new sediment layer. Those that were able to escape repopulated the new surface. Alterations in composition of the organic remains occur after embedding and organic matter is lost, so it is the shells that are best preserved. The much debated 'Glossopterae' (tongue-stones) had not grown *in situ*; inasmuch as they often showed signs of decay and thus were interpreted as relics of an earlier period. These were shark teeth comparable to those of living sharks. In his almost mathematical reasoning, Steno makes a clear distinction between his 'observations' and 'conjectures' (leaving open other possible explanations). For further discussion of the important role of Steno, see Rudwick (1972) and Ellenberger (1988, pp. 232–316). Steno used taphonomic arguments, but he was not the first 'taphonomist', as Plotnick and Speyer (1989) have incorrectly stated; da Vinci was earlier.

THE BIRTH OF PALAEONTOLOGY

Palaeontology as a science became firmly established in the early nineteenth century with Cuvier (vertebrates) and Lamarck (invertebrates) as its 'founders'. The biblical Deluge as an explanation for the presence of marine fossils far from the sea became replaced by the theory of a series of 'revolutions' (Cuvier, 1812; 1825; d'Orbigny, 1849).

Actualism (the continental European term for uniformitarianism, that is, the methodology of inferring the nature of past events by analogy with processes observable at present) has been of prime importance in understanding the fossilization process. Actualism is usually thought to have been originally postulated by Hutton (1788) and advocated by Lyell (1830–3), but it certainly had earlier roots (Hooykaas, 1959), as it was hinted at by, for example, Moro (1740; see Hooykaas, 1959) and Hoff (1822–41), among others. Its roots can even be traced to Heraclites (544–483 BC), who postulated continuity in processes (see Schäfer, 1980, p.74), and da Vinci (see above). However, Lyell's *Principles* (1830–3) were certainly most influential in the spread of acceptance of actualism.

Cuvier, in the 'discours préliminaire' to his *Recherches sur les ossemens fossiles* (1812), enlarged and published in several editions separately as *Discours sur les révolutions* (1825), apart from giving a description of his theory of 'revolutions' (catastrophes causing local extinctions), made a number of 'taphonomic' remarks. He was not an actualist and clearly stated that forces acting now are insufficient to explain the catastrophes which separate his geological periods defined by different faunas. He found fossil bones encrusted with oysters and other marine organisms. This indicated to Cuvier that terrestrial vertebrates were killed by rapid 'catastrophic' marine transgressions and buried under marine sediments. Bone-bearing sediments in the Paris Basin were confined to low-lying areas, so marine incursions had been local. Bones were well preserved, showing little or no signs of transport, so the vertebrates had lived near where they were found. Mammoths discovered frozen in Siberia were victims of the last catastrophe; they could never have lived in this cold environment.

Based on his own fruitful collaboration with Brongniart, Cuvier stressed the importance of combining the knowledge gained by the taxonomically orientated palaeontologists working in museums, with that of field geologists, who usually lacked detailed palaeontological knowledge. Cuvier posed questions, such as whether fossils had lived where they are found as fossils, which could not be answered from the armchair. He particularly used terrestrial vertebrates to illustrate his theories, because they were far better known than, for example, marine molluscs, of which many unknown species might live in the still unknown deep sea (the *Porcupine, Lightning* and *Challenger* expeditions started unlocking the secrets of the deep sea in the second half of the nineteenth century). This argument he probably borrowed from Buffon (1749). The process of diagenesis was still poorly understood; Cuvier had difficulties in understanding how soft sediments deposited on the sea floor became the hardened sandstones and chalks that he studied in the Paris basin.

Early 'taphonomic' observations were presented by Buckland (1823) with regard to the bones of Kirkdale Cave. This cave, he concluded, had been the den of antediluvial hyenas, and the dietary habits of these animals had produced the seemingly haphazard assemblage of teeth and bones. He observed that the fossil bones were mostly broken and gnawed, and proved his conclusion with an actualistic experiment. Buckland gave the bone of an ox to a Cape hyena in a travelling collection that happened to pass through Oxford, and observed that this hyena broke and gnawed the bone to produce a pattern of 'preservation' very similar to those found in Kirkdale Cave. The assemblage of bones in the cave was not due to a mass-mortality, but was the result of gradual accumulation over a long period (see also Rupke, 1983). In his famous 'Bridgewater Treatise', Buckland (1836) again derived conclusions from the state of preservation of fossils. The presence of complete skeletons with intact skin of fishes and marine reptiles, and of cephalopods with intact ink-bags (Figure 1.1), in Lias deposits indicates sudden death and rapid burial in the absence of scavengers. Buckland recognized that widespread concentration of bones and coprolites represent hiatusses in sedimentation. Buckland further noted that distortion of fishes in the German Kupferschiefer was not due to agonies of death, but to unequal contractions of muscles after death (studied in more detail later by Weigelt, 1927a).

D'Orbigny's *Cours élémentaire* (1849) illustrates the rapid progress that was being made in understanding the fossilization process. He gave probably the first definition of fossilization and a review of the state of the art. Fossilization is:

Tout ce qui se rattache, plus ou moins directement, aux changements par lesquels un corps vivant et jadis animé a passé d'un époque, alors actuelle, à un autre époque qui n'est plus, en laissant dans les couches terrestres des traces impérissables de sa forme caracteristique.

Remineralization of organic remains during diagenesis was thought to be partly due to electrochemical processes, with the remains acting as an electrode on which ions collect. It was the heyday of experiments with electricity and d'Orbigny's suggestions nicely adopted the new insights gained.

D'Orbigny argued that to become fossilized, organic remains must be buried, preferably under water, and recognized that a difference exists between the style of preservation shown by floating and non-floating remains. Carcasses float because of gases produced during putrefaction. They are transported to the coast, where they may be buried in quiet localities by rapid sedimentation, but most fall apart on exposed rocky or sandy shores. D'Orbigny's ideas concerning transport of carcasses by rivers and their consequent accumulation in estuaries was based on observations in Europe, where it was the practice to throw the carcasses of dogs and cats into the rivers. In South America, where d'Orbigny stayed for eight years, he never saw carcasses floating in rivers. Of the remains of vertebrates and terrestrial snails, nothing is preserved due to subaerial destruction. Traces (birds, raindrops) indicative of shallow water cannot be produced in deeper water. D'Orbigny also determined that transported shells can be recognized as such, because they are chaotically distributed, broken and abraded in shallow-water deposits. In deeper-water deposits shells remain entire and without abrasion.

D'Orbigny extended the theory of revolutions of Cuvier and supposed that 27 revolutions had occurred during the history of the Earth, which destroyed life all over the world. This catastrophic killing was due to rapid sedimentation, which killed not only benthic, but also pelagic organisms, a possibility that d'Orbigny proved experimentally to be feasible.

Gressly (1814–65) was another early (and self-taught) geologist who studied the recent environment as a clue to the past (Martin, 1965). He is best known as the inventor of the facies concept resulting from his studies in the Jura mountains (Wegmann, 1963). The account of his stay along the Mediterranean Sea (Gressly, 1861) contains many 'taphonomic' observations. He observed, probably as one of the first, the separation of left and right shells of bivalves by waves oblique to the beach; he noted the orientation on the beach of most valves with convex side upward; and mentioned abrasion of

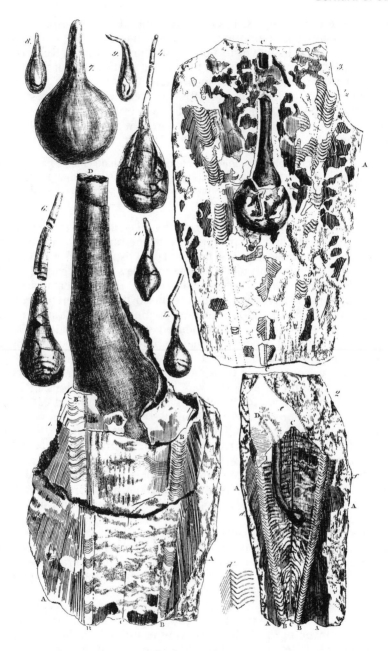

Figure 1.1   Ink-bags and traces of the internal horny shell of fossil *Loligo*-like dibranchiate cephalopods, collected by Mary Anning and others from the Lias (Lower Jurassic) of Lyme Regis, southern England; proof for Buckland of immediate and rapid burial (from Buckland, 1836, Pl. 45).

the umbones. He suggested the use of left/right valve ratios in fossil-shell layers to estimate wave and current direction. Moreover, he studied littoral and sublittoral zonation for comparison with the various facies he could discern in the Jurassic. He was one of the first to note the differences between life assemblages and death assemblages.

THE GERMAN FOUNDATIONS OF TAPHONOMY

The centre of continental palaeontological work changed from France to the German-speaking part of Europe at the end of the nineteenth century. J. Walther (1860–1937) contributed significantly to the development of the actualistic method ('ontological method' of his own terminology) for the interpretation of older deposits. Walther was trained in both biology and geology (Weigelt, 1930a). He studied recent marine sediments and their organisms in Naples from 1883 (Walther, 1910), and travelled through many parts of the world. His travels resulted in publications on both recent marine environments and fossil deposits, in addition to his many publications on deserts. His study of the famous Jurassic Solnhofener Plattenkalk (Walther, 1904) was a landmark publication in biostratonomic research, although some of his conclusions are no longer accepted (Barthel, 1978; Seilacher *et al.*, 1985). Walther considered the Solnhofen deposits as a 'Leichenfeld', in which most animals arrived dead except for a few benthic organisms which died at the end of 'death marches'. Absence of bacteria, scavengers and waves, plus rapid burial acted to keep remains, even of medusae, intact. Walther considered his *Einleitung* (1893–4) as a revised version of Hoff's *Geschichte* (1822–41). The former gives an ample discussion of the actualistic ('ontological') method, but Recent counterparts could not always be found for fossil organisms and the Recent environment is still not well known. A chapter is devoted to discussion of the incompleteness of the fossil record; there is no quantitative correspondence between fossils and the once living fauna, a topic studied in more detail later (see McKinney, this volume).

O. Abel (1875–1946) was another productive early twentieth-century palaeontologist. He published a number of books touching on taphonomy (for example, Abel, 1912; 1927; 1935). He was influenced by Dollo, whom he visited in Brussels in 1900 and who taught him to view fossils not only as documents in chronology and evolution, but also as animals that once lived (Ehrenberg, 1978). Abel became particularly interested in palaeobiology and in reconstructing palaeoenvironments. He used taphonomic data to decipher the mode of formation of what we now call 'fossil-Lagerstätten' (Seilacher, 1970), and made comparisons with recent environments to construct 'Lebensbilder' of famous fossil localities (Abel, 1927). Like Walther (1893–4) before and Schäfer (1962) after him, he stressed: 'Wir werden nie vergessen dürfen, dass wir das Leben der Vorzeit nur dann verstehen können, wenn wir uns in das Leben der Gegenwart einen Einblick verschafft haben.'

Abel made taphonomic interpretations on a broad variety of fossil associations. For example, the Miocene coastal deposits of the Vienna basin were compared with the Recent rocky coast of the Adriatic to demonstrate how poorly rocky coastal life is represented in the fossil record. Abel observed that floating carcasses of whales may reach the coast, where they normally disarticulate with the help of scavengers, but this process may already start during flotation; the first part of the skeleton to disappear to the bottom is the lower jaw (later called the 'Unterkiefergesetz' by Weigett, 1927a). Careful study of the bones of the Miocene bone beds of Pikermi (a famous fossil-hunting locality in Greece, where in 1838 the first fossil monkey was found) revealed these remains to have been quietly buried in fluvial sediments in three successive layers (Abel, 1927). Ungulate leg bones show fractures very similar to those sustained by humans during skiing accidents; marks of gnawing by predators are also present. Abel suggested periodic catastrophic mortality of herds of ungulates, which in panic (due to prairie fires?) plunged down steep river banks, breaking their legs and thus making them easy prey for predators. Embedding of their remains must have been fast because subaerial destruction of carcasses is rapid; Abel cited the rapid destruction of an African elephant, mainly due to scavengers like hyenas, jackals and vultures, with only the skull resisting weathering for a few years. This comparison anticipated similar reports in Behrensmeyer and Hill (1980).

Another famous 'fossil-Lagerstätte' is the Wealden of Bernissart, where 29 entire skeletons of the dinosaur *Iguanodon* were found in 1877. They now form the 'pièce de résistance' of the Brussels Museum of Natural History. Abel (1927) could not accept catastrophic killing in this example, because the skeletons occurred in a deposit of sediment 34 m thick, separated from each other by 2–3 m of sediment. The *Iguanodons* must have died one by one and embedding must have been rapid, because skeletons are almost intact and traces of putrefaction are absent.

Deecke (1923) nicely summarized the knowledge collected on the fossilization process in its three aspects—death, burial and diagenesis. He defined fossilization as: 'Die Summe der Vorgänge, welche zur Erhaltung der Reste geführt haben'. He already at that time complained: 'Der Stoff ist ungemein vielseitig, er schwoll mir unter der Feder fast unheimlich an und eine Hauptaufgabe war, ihn übersichtlich zu ordnen'.

Plant taphonomy was by no means neglected during this period. Potonié (1910) reviewed research on peat, lignite and coal deposits. Discussions on the organic versus inorganic origin of these deposits continued from Anaximenes (588–525 BC) to *c*.1830. W. Hutton (1833; see Buckland, 1836) proved their botanical origin by microscopical studies. The next questions to be solved were whether they were allochthonous or autochthonous (for which the discovery of *in situ* trees and roots was decisive), and marine or terrestrial, in origin. A Recent analogue for the environment of coal formation was found in the *Taxodium* swamps of delta systems in the southeastern USA (Potonié, 1910).

TAPHONOMY AS A SCIENCE

Weigelt's monograph (1927a, translated 1989) was the first full-scale taphonomic ('biostratonomic') research effort to document vertebrate death, decay, decomposition, putrefaction, disarticulation, transport and burial in Recent environments, and to determine their relevance to fossil preservation. Although biostratonomy is now often defined as dealing only with the process of embedding, it is clear from his papers (1927a; 1927b; 1930a; 1930b) that Weigelt included necrotic processes. Weigelt (1927a; with additions in 1930b) integrated numerous isolated studies by others, but the centrepiece of his book is formed by the thorough description of the Smither's Lake 'Leichenfeld' which he came across during a stay of 16 months in the US Gulf Coast area.

Weigelt spent several months studying the fate of carcasses of alligators, turtles, fishes, birds and cattle which had drifted together at the southern margin of Smither's Lake. In December 1924 a sudden sharp drop in temperature during a northern wind caused the death of many vertebrates, including over a million cattle. A large series of photographs documents his research. In November 1925 most of the carcasses were decayed or hidden under sediment and vegetation. Sediment coverage might preserve such carcass concentrations. Weigelt realized that such mass-mortalities may occur regularly. Under suitable conditions, including rapid sedimentation, several Leichenfelder may be preserved in a sedimentary sequence.

With his colleagues he indeed found a succession of such carcass concentrations in the Eocene lignites of Geiseltal (Weigelt, 1933; Voigt, 1962). His main purpose in studying Recent examples was to understand better the biostratonomy of fossil-Lagerstätten, particularly the Leichenfelder excavated under his supervision in the Geiseltal. Weigelt came to the conclusion that many fossil-Lagerstätten represent catastrophic (better anastrophic = local) mass-mortalities.

Many of the themes of Weigelt's research (for example, the role of insects in decomposition; burial of land vertebrates in marine strata; carcasses at environmental boundaries) have been discovered independently by modern investigators unaware of Weigelt's pioneer work. Weigelt's studies clearly focused on the extra information biostratonomy can give for understanding how fossil assemblages are formed.

Wasmund (1926) is usually cited as having introduced the community concept in taphonomy, but in palynology this idea has been present since c.1900 (Faegri and Iversen, 1964). Wasmund compared the biocoenosis with the 'thanatocoenosis' (a term which he did not define strictly, unfortunately, but which he used for all assemblages of organic remains whether allochthonous, autochthonous or a mixture). From his study of recent thanatocoenoses in lakes, he drew conclusions on the transport of organic remains and compared his results with those obtained in marine habitats, seeing no principal differences. Quenstedt (1927) introduced the term 'taphocoenosis' for organic remains embedded in the sediment, to replace Wasmund's thanatocoenosis.

Particularly due to the activities of R. Richter, a special institute was founded in 1928, at Senckenberg am Meer in Wilhelmshaven along the border of the Wadden Sea, to study recent processes of sedimentation and fossilization in the sea. Richter (1928) defined 'Actuopalaeontology', the main field of research for this new institute, as the study of the life and death of organisms in their Recent habitat, including the post-mortem history of their remains, ichnology and the study of facies. It is clear that this concept partly overlaps biostratonomy and taphonomy, but deals only with Recent organisms. Häntzschel (1956) reviewed the work of the first 30 years.

Schäfer (1962) superbly summarized the actuopalaeontological work from the North Sea, mainly, but not only based on work at Wilhelmshaven. Only after translation in 1972 did the German work in this field become better known. Death, disintegration and embedding of representatives of all palaeontologically interesting groups of organisms were studied in the field and in experiments (Figure 1.2). The largest part of the book is devoted to traces left in the sediment by more than 350 North Sea organisms, using a classification following Seilacher (1953) into dwelling, food-collecting, escape and resting traces. Schäfer's classification of sediments with their fossil content into five first-order biofacies, based on the presence or absence of an autochthonous benthic fauna (indicating water oxygenation) and the bedding type of the sediment (indicating energy), was an early attempt to bring some order to this field.

Taphonomy of macroscopic plant remains in this period (c.1925–70) was somewhat neglected. Weigelt (1928) studied biostratonomy of plant remains in the Upper Permian Kupferschiefer (Zechstein, Germany). Orientation, fragmentation and maceration of the plant remains made it possible to discern various grades of plant transport and sorting. Taxonomic information was lost, but taphonomic and environmental information was gained. No comparison was made with Recent analogues except by Chaney (1924), who tried to solve the question of whether dominant plant fossils reflect what were dominant plants in the original vegetation. He compared the Tertiary Bridge Creek flora with plant accumulations in stream basins in its nearest Recent Redwood analogue. The four dominant tree species made up 85 per cent of the plant remains in the Recent Redwood, but ferns and herbs were underrepresented. Only recently has similar research on plant taphonomy been picked up again (for example, Ferguson, 1985; Greenwood, this volume).

In palynology such quantitative comparisons between vegetation and pollen deposition started with Hesselman (1919; see Faegri and Iversen, 1964). The importance of this type of research was well understood, because it is the investigation of changes in the vegetation (due to climatic changes), particularly during the Quaternary, that is the purpose of much palynological research. Little attention, however, was paid to preservation of pollen grains. Usually pollen analysis was only applied to such sediments as gyttja, dy or peat, in which good preservation was generally assumed to be self-evident. With the extension of pollen analysis to nearly every type of sediment, pollen corrosion could no longer be overlooked. Havinga (1967), among others, started experiments in pollen corrosion, both in the laboratory and by burial of pollen in various types of soil. He found strong differences in corrosion susceptibility for pollen of different plant species, which could easily lead to misinterpretation of pollen diagrams even of some types of peat.

THE PROLIFERATION OF TAPHONOMY

The German taphonomic tradition (Walther, Abel, Weigelt, Wasmund, Schäfer) did not find wide support outside Germany. This was partly due to language, but anti-German sentiments probably also played a role. The growing influence of national socialism under Hitler in Germany in the 1930s made many palaeontologists elsewhere reluctant to read German papers and follow German research lines. One understands these anti-German sentiments by reading, for example, Richter (1934). The university library at Groningen in the Netherlands went so far as to cancel subscriptions to some German journals in 1933. Moreover, a few German taphonomists showed pro-Nazi sympathies, which are usually not mentioned in their obituaries (except, for example, by Gripp, 1964, on Wasmund).

The taphonomy that became popular in North America rather followed the Russian tradition of Efremov (1940), particularly due to its introduction by Olson. The incompleteness of the fossil record was the main reason for Efremov to advocate a better study of processes and principles governing the transition of organic remains from the biosphere to the lithosphere (one of his definitions of taphonomy), rather than the elucidation of palaeoenvironments, which was the primary thrust of the German school. Efremov's main research interest concerned terrestrial vertebrates and he supposed taphonomical studies to be easier in the marine habitat because the biocoenosis should be better preserved here; a marine fauna can be preserved *in situ*, whereas a terrestrial vertebrate fauna is almost always transported. The methods that Efremov proposed for taphonomical studies, like the study of spatial distribution of organic remains in the sediment and experiments on artificial fossilization and transport, are not significantly different from those of Weigelt and Richter. His new ideas concerned the quantification of the incompleteness of the (terrestrial) fossil record, by estimating the content of fossil localities and estimating for each period the area of continental sedimentation in comparison with the entire continent, in order to calculate the ratio of the embedded to the total fauna.

Quantification of this incompleteness ('information loss') later became a major goal in taphonomic studies in both the marine and terrestrial environments. Lawrence (1968) concluded that 7–67 per cent of the species in marine communities were soft-bodied and thus had little potential for preservation. He studied a Recent oyster community as an analogue for an Oligocene oyster community and concluded that 75 per cent of the macroscopic species in the Oligocene oyster community had not been preserved. Along comparable lines, a number of Recent, mainly shallow marine environments were studied to ascertain the fidelity of the fossil record by comparing living and fossil faunas (reviewed in Scott and West, 1976; Kidwell and Bosence, in press). General models for the formation of fossil assemblages, stressing the importance of biasing effects like skeleton destruction, transport and time-averaging, were formulated by Johnson (1960) and Fagerstrom (1964).

One of the taphonomic aspects studied in both terrestrial vertebrates and marine shell-bearing organisms became size-frequency distribution, which was used as an indicator of transport (Boucot, 1953) and, in untransported

Figure 1.2 An actualistic experiment of the first steps in fish fossilization in quiet water, under anoxic conditions (that is, without scavengers); five successive stages in the decomposition of the sea-scorpion *Cottus scorpius*. The fish sinks upside down to the bottom (A), but soon starts to float due to gas production (B). Gas is released during further decomposition and the fish sinks again (C), reaching the bottom (D). Stages 1–4 take four days at 18°C. (E) Three months later all the organic matter is decomposed, leaving a largely disarticulated skeleton on the bottom of the aquarium (from Schäfer, 1962, Figures 20 and 21, reproduced with the permission of Verlag Waldemar Kramer, Frankfurt am Main).

assemblages, as a tool to interpret the population dynamics of the fossil population (Kurtén, 1953). Hallam (1972) gave a review and Cadée (1982) mentioned some pitfalls of such studies.

Seilacher (1976) gave new impetus to the German tradition in palaeoecology, by the foundation of a 'Sonderforschungsbereich Palökologie' in Tübingen in 1970. This research project aimed at a more systematic development to palaeoecology (including taphonomy) by giving it the 'framework' it still lacked. This enabled the integration of many diverse lines of research into a system, permitting a better understanding of their interrelationships. Post-mortem changes were first seen as a taphonomic overprint causing information loss, but soon embedding and preservation proved to give additional information on sediment genesis that inorganic particles did not provide (Seilacher, 1976). Of the four sub-projects (fossil-Lagerstätten, fossil diagenesis, constructional morphology and fossil assemblages), the first two have already produced a wealth of taphonomic information.

The rationale behind Seilacher's (1970) concept of fossil-Lagerstätten was a sedimentological one. Lagerstätten are considered end members of ordinary sedimentary facies. The study of organic remains gives information on biotope conditions, the fate of the soft parts, transport, burial and diagenesis, allowing a better understanding of sediment genesis (Seilacher *et al.*, 1985). A first step was a genetic classification of fossil-Lagerstätten (Seilacher, 1970), discriminating between concentration deposits (accumulations of disarticulated organic remains) and conservation deposits (where, due to incomplete decay of proteins, entire organisms are preserved). Processes of concentration and mode of preservation have been used in order further to subdivide conservation and concentration deposits.

A further step presented a conceptual, heuristic framework for the classification of conservation deposits (Seilacher *et al.* 1985). Based on three factors—stagnation (the hydrographical regime), obrution (the sedimentological regime) and bacterial sealing (the early diagenetic regime)—conservation deposits were plotted in a triangular diagram. Fossil-Lagerstätten are no longer seen (only) as unique occurrences, but also as part of a continuum, facilitating comparisons. Such a comparative analysis, although tentative, stimulates research better than a statement that each taphocoenosis presents a singular historical event (Dodson, 1980; *contra* Olson, 1980).

The 'taphofacies' concept articulated by Speyer and Brett (1986) also aims at 'comparative taphonomy' (Brett and Baird, 1986); details of fossil preservation reflect sedimentary conditions and may aid in interpretation of depositional environments. Kidwell *et al.* (1986) proposed a standardised descriptive terminology and a genetic classification of fossil concentrations, improving comparative taphonomic analysis.

Increasing evidence shows that rigorous application of uniformitarianism is not always possible. Walther (1893–4, p.1003) already stated that organisms have changed during geologic history, and thus processes influenced by organisms have changed as well. Preservation potential may have changed during geologic history; rate of bioturbation, and the diversity of scavengers and predators, have undoubtedly changed (Conway Morris, 1985). Vermeij

(1977) and Taylor (1980) documented these changes in predators: fishes, now the most important benthic predators, radiated at the end of the Cretaceous; whereas predatory gastropods increased exponentially from Late Cretaceous to early Tertiary; crustaceans diversified in the early Tertiary. Walker and Diehl (1985) suggested that marine cementation may have been more common in the early Palaeozoic, due to a different water chemistry. This might have resulted in better preservation of early Palaeozoic carbonate shelf communities, because the fossils were more often locked in place by very early cementation.

After 1970 there was an exponential increase in papers on taphonomy. Taphonomic processes were no longer seen only as causing an information loss, but also as a feedback mechanism (Kidwell and Jablonski, 1983), or, sedimentologically, as an information gain (Einsele and Seilacher, 1982; Thomas, 1986; Wilson, 1988) *sensu* Weigelt's use of biostratonomy, hence the belated translation of his major work (Weigelt, 1927a) in 1989. Special conferences have been devoted to, or dealt with, particular aspects of taphonomy: in 1976, taphonomy and vertebrate palaeoecology (Behrensmeyer and Hill, 1980); in 1984, the positive aspects of taphonomy (see Thomas, 1986); in 1985, extraordinary fossil biotas (Whittington and Conway Morris, 1985); in 1986, ecological and evolutionary implications of taphonomic processes (Behrensmeyer and Kidwell, 1988); again in 1986, plant taphonomy (Gastaldo, 1989); *etc.* A newsletter was started by Plotnick and Speyer (1989).

With this exponential increase in papers, we really need the synthesis Efremov was after. Efremov (1958) enumerated three general 'principles' in taphonomy, Wilson (1988) came to nine 'generalizations', but after some 60 years of research this seems rather meagre. The more promising lines of research are the genetic classifications of fossil deposits. We are now able to classify fossil-Lagerstätten, fossil concentrations and taphofacies according to their genesis, using taphonomic criteria. This should provide the framework to bring order in the overwhelming amount of taphonomic observations.

Unfortunately, taphonomy has developed rather independently in vertebrates, invertebrates and plants, whereas a taphonomy in pollen analysis and marine micropalaeontology developed rather late. Research on the quantitative correspondence of living and dead fauna and flora in modern environments has developed largely independently: for leaf assemblages (for example, Chaney, 1924); ostracods (for example, Kornicker, 1958); palynology (for example, Hartman, 1968); a marine oyster community (Lawrence, 1968); marine molluscs (for example, Cadée, 1968); Foraminifera (for example, Murray, 1968; Martin and Liddell, this volume); terrestrial vertebrates (for example, Behrensmeyer and Boaz, 1980); and diatoms (for example, Kosugi, 1989). Such uniformitarian investigations in the marine environment indicate that the compositions of death assemblages are in many respects faithful reflections of long-term community structure (Kidwell and Bosence, in press).

Taphonomists working in different fields should be brought together, as their goals are similar. Language barriers and political differences have hampered good scientific exchange. This has caused a deplorable information loss in taphonomy!

SUMMARY

Taphonomic observations are as old as palaeontology. Taphonomy's goals show a cyclic pattern: in its early history information on the depositional environment was stressed (Weigelt); then (Efremov) insight in the incompleteness of the fossil record (information loss); and most recently sedimentologic and palaeoenvironmental information gain from taphonomic studies has been appreciated. The uniformitarian method has been used from the beginning in understanding taphonomic processes, but the marine environment, particularly its biological factors, has changed during geologic history and so strict actualism must have inherent limitations. The development of taphonomy has suffered from language barriers and from independent growth of taxon-specific subdisciplines. The search for general 'laws' has not been very successful, but synthesis in taphonomy is in progress, promoted by genetic classifications of fossil-Lagerstätten, fossil concentrations and taphofacies.

ACKNOWLEDGEMENTS

I thank Verlag Waldemar Kramer for permission to reproduce Figures 20 and 21 (Figure 1.2 herein) from Schäfer (1962). The author is very grateful to Drs G.J. Boekschoten, A. Brouwer, S.M. Kidwell and S.K. Donovan for critically reading the typescript.

REFERENCES

Abel, O., 1912, *Grundzüge der Paläobiologie der Wirbeltiere*, Schweitzerbart'sche Verlagsbuchhandlung, Stuttgart.
Abel, O., 1927, *Lebensbilder aus der Tierwelt der Vorzeit* (2nd edn), G. Fischer, Jena.
Abel, O., 1935, *Vorzeitliche Lebensspuren*, Verlag G. Fischer, Jena.
Barthel, K.W., 1978, *Fossilien aus Solnhofen*, Ott Verlag, Thun.
Behrensmeyer, A.K. and Boaz, E.D., 1980, The Recent bones of Amboseli park, Kenya, in relation to east African paleoecology. In A.K. Behrensmeyer and A.P. Hill (eds), *Fossils in the making*, University of Chicago Press, Chicago: 72–92
Behrensmeyer, A.K. and Hill, A.P., 1980, *Fossils in the making*, University of Chicago Press, Chicago.
Behrensmeyer, A.K. and Kidwell, S.M., 1985, Taphonomy's contribution to paleobiology, *Paleobiology*, **11**(1): 105–19.
Behrensmeyer, A.K. and Kidwell, S.M. (eds), 1988, Ecological and evolutionary implications of taphonomic processes, *Palaeogeography, Palaeoclimatology, Palaeoecology*, **63**(1/3): 1–291.
Boucot, A.J., 1953, Life and death assemblages among fossils, *American Journal of Science*, **251**(1): 25–40.
Brett, C.E. and Baird, G.C., 1986, Comparative taphonomy: a key to paleoenvironment interpretation based on fossil preservation, *Palaios*, **1**(3): 207–27.
Briggs, D.E.G. and Crowther, P.R., 1990, *Palaeobiology, a synthesis*, Blackwell, Oxford.

Buckland, W., 1823, *Reliquiae diluvianae; or, observations on the organic remains contained in caves, fissures, and diluvial gravel, and on other geological phenomena, attesting the action of an universal deluge*, John Murray, London.

Buckland, W., 1836, *Geology and mineralogy considered with reference to natural theology*, Bridgewater Treatise, Pickering, London.

Buffon, Comte de, 1749, *Histoire naturelle, générale et particulière, avec la description du cabinet du Roy, 1, Théorie de la terre*, Imprimerie Royale, Paris.

Cadée, G.C., 1968, Molluscan biocoenoses and thanatocoenoses in the Ría de Arosa, Galicia, Spain, *Zoologische Verhandelingen, Rijksmuseum van Natuurlijke Historie, Leiden*, **95**: 1–121.

Cadée, G.C., 1982, Low juvenile mortality in fossil brachiopods, some comments, *Interne Verslagen Nederlands Instituut voor Onderzoek der Zee*, **1982**(3): 1–29.

Chaney, R.W., 1924, Quantitative studies of the Bridge Creek flora, *American Journal of Science*, 5th series, **8**(44): 126–44.

Conway Morris, S., 1985, Concluding remarks, *Philosophical Transactions of the Royal Society*, London, **B311**: 187–91.

Cuvier, G., 1812, *Recherches sur les ossemens fossiles de quadrupèdes, ou l'on rétablit les caractères de plusieurs espèces d'animaux que les révolutions du globe paroisent avoir détruites*, Deterville, Paris.

Cuvier, G., 1825, *Discours sur les révolutions de la surface du globe*, Dufour et d'Ocagne, Paris.

Deecke, W., 1923, *Die Fossilisation*, Borntraeger, Berlin.

Dodson, P., 1980, The progress of taphonomy, *Science*, **210**(4470): 631–2.

Duhem, P., 1906–13, *Études sur Léonard de Vinci, ceux qu'il a lus et ceux qui l'ont lu* (3 vols), A. Hermann, Paris.

Efremov, J.A., 1940, Taphonomy: new branch of paleontology, *Pan-American Geologist*, **74**(2): 81–93.

Efremov, I.A., 1953, Taphonomie et annales géologiques, *Annales du Centre d'Études et de Documentation Paléontologiques* (translation by S. Ketchian and J. Roger), **1953**(4): 1–164.

Efremov, I.A., 1958, Some considerations on biological bases of paleozoology, *Vertebrata Palasiatica*, **2**(2/3): 83–98.

Ehrenberg, K., 1978, Othenio Abels Werden und Wirken, *Mitteilungen Gesellschaft geologische Bergbaustudien Österreich*, **25**(1): 1–23.

Einsele, G. and Seilacher, A. (eds), 1982, *Cyclic and event sedimentation*, Springer-Verlag, Berlin.

Ellenberger, F., 1988, *Histoire de la géologie, tome 1. Des anciens à la première moitié du XVIIe siècle*, Lavoisier, Paris.

Faegri, K. and Iversen, J., 1964, *Textbook of pollen analysis* (2nd edn), Blackwell, Oxford.

Fagerstrom, J.A., 1964, Fossil communities in paleoecology: their recognition and significance, *Geological Society of America Bulletin*, **75**(12): 1197–216.

Ferguson, D.K., 1985, The origin of leaf-assemblages—new light on an old problem, *Review of Palaeobotany and Palynology*, **46**(1): 117–88.

Gastaldo, R.A. (ed.), 1989, Plant taphonomy—organic sedimentary processes, *Review of Palaeobotany and Palynology*, **58**(1): 1–94.

Gressly, A., 1861, Erinnerungen eines Naturforschers aus Südfrankreich, *Album Combe-Varin*, **1861**: 210–96.

Gripp, K., 1964, *Erdgeschichte von Schleswig-Holstein*, Wachholtz, Neumünster.

Hallam, A., 1972, Models involving population dynamics. In T.J.M. Schopf (ed.),

Models in paleobiology, Freeman, Cooper and Company, San Francisco: 62–80.

Häntzschel, W., 1956, Rückschau auf die paläontologischen und neontologischen Ergebnisse der Forschungsanstalt 'Senckenberg am Meer', *Senckenbergiana Lethaea*, **37**(3/4): 319–30.

Hartman, A.A., 1968, A study on pollen dispersal and sedimentation in the western part of the Netherlands, *Acta Botanica Neerlandica*, **17**(6): 506–49.

Havinga, A.J., 1967, Palynology and pollen preservation, *Review of Palaeobotany and Palynology*, **2**(1): 81–98.

Hesselman, H., 1919, Iakttagelser över skogsträdpollens spridningsförmåga, *Meddelanden från Statens Skogsförsöksanstalt*, **16**(1): 27–39.

Hoff, K.E.A. von, 1822–41, *Geschichte der durch Überlieferung nachgewiesenen natürlichen Veränderungen der Erdoberfläche*, J. Perthes, Gotha.

Hölder, H., 1989, *Kurze Geschichte der Geologie und Paläontologie*, Springer-Verlag, Berlin.

Hooykaas, R., 1959, *Natural law and divine miracle*, Brill, Leiden.

Hutton, J., 1788, Theory of the Earth; or an investigation of the laws discernable in the composition, dissolution and restoration of land upon the globe, *Transactions of the Royal Society of Edinburgh*, **1**(2): 209–304.

Hutton. W., 1833, Observations on coal, *Edinburgh Philosophical Magazine*, 3rd series, **2**: 302–4.

Johnson, R.G., 1960, Models and methods for analysis of the mode of formation on fossil assemblages, *Geological Society of America Bulletin*, **71**(7): 1075–86.

Kidwell, S.M. and Bosence, D.J.W., in press, Taphonomy and time-averaging of marine shelly faunas. In D.E.G. Briggs and P.A. Allison (eds), *Taphonomy, releasing information from the fossil record*, Plenum Press, New York.

Kidwell, S.M., Fürsich, F.T. and Aigner, T., 1986, Conceptual framework for the analysis and classification of fossil concentrations, *Palaios*, **1**(3): 228–38.

Kidwell, S.M. and Jablonski, D., 1983, Taphonomic feedback: ecological consequences of shell accumulation. In M.J.S. Tevesz and P.L. McCall (eds), *Biotic interactions in Recent and fossil benthic communities*, Plenum, New York: 195–248.

Kornicker, L.S., 1958, Ecology and taxonomy of Recent marine ostracodes in the Bimini area, Great Bahama Bank, *Bulletin of the Institute of Marine Science*, **5**: 194–300.

Kosugi, M., 1989, Process of formation in fossil diatom assemblages and the paleoecological analysis, *Benthos Research*, **35/36**: 29–40.

Kurtén, B., 1953, On the variation and population dynamics of fossil and recent mammal populations, *Acta Zoologica Fennica*, **76**(1): 1–122.

Lawrence, D.R., 1968, Taphonomy and information losses in fossil communities, *Geological Society of America Bulletin*, **79**(10): 1315–30.

Lyell, C., 1830–3, *Principles of geology*, (3 vols), John Murray, London.

MacCurdy, E., 1938, *The notebooks of Leonardo da Vinci* (2 vols), Jonathan Cape, London.

Martin, G.P.R., 1965, Zum einhundertsten Todestag des Wegbereiters von Paläobiologie und Paläökologie am 13. April 1965, *Natur und Museum*, **95**(5): 217–26.

Mather, K.F. and Mason, S.L., 1939, *A source book in geology*, McGraw-Hill, New York.

Moro, A.L., 1740, *De' crostacei e degli altri marini corpi che si truovano su' monti*, S. Monti, Venice.

Müller, A.H., 1951, Grundlagen der Biostratonomie, *Abhandlungen der deutschen*

Akademie der Wissenschaften zu Berlin, Klasse für Mathematik und allgemeine Naturwissenschaften, **1950**(3): 1–147.

Müller, A.H., 1976, *Lehrbuch der Paläozoologie*, G. Fischer Verlag, Jena.

Müller, A.H., 1979, Fossilization (taphonomy). In R.C. Moore, R.A. Robinson and C. Teichert (eds), *Treatise on invertebrate paleontology, Part A, Introduction*, Geological Society of America and University of Kansas Press, Boulder, Colorado, and Lawrence: A2–A78.

Murray, J.W., 1968, The living Foraminiferida of Christchurch Harbour, England, *Micropaleontology*, **14**(1): 83–96.

Olson, E.C., 1980, Taphonomy: its history and role in community evolution. In A.K. Behrensmeyer and A.P. Hill (eds), *Fossils in the making*, University of Chicago Press, Chicago: 5–19.

d'Orbigny, M.A., 1849, *Cours élémentaire de paléontologie et de géologie stratigraphiques, Vol. 1*, V. Masson, Paris.

Plotnick, R.E. and Speyer, S.E., 1989, *Death, decay and disintegration, the newsletter for research in taphonomy*, University of Illinois and University of Arizona, Chicago and Tucson, **1**: 1–33.

Potonié, H., 1910, *Die Enstehung der Steinkohle und der Kaustobiolithe überhaupt*, Borntraeger, Berlin.

Quenstedt, W., 1927, Beiträge zum Kapitel Fossil und Sediment vor und bei der Einbettung, *Neues Jahrbuch für Mineralogie, Geologie und Paläontologie, Abteilung B, Geologie und Paläontologie*, **58**: 353–432.

Richter, R., 1928, Aktuopaläontologie und Paläobiologie, eine Abgrenzung, *Senckenbergiana*, **10**(6): 285–92.

Richter, R., 1934, 'Natur und Volk', *Natur und Volk*, **64**(1): 1–3.

Rudwick, M.J.S., 1972, *The meaning of fossils, episodes in the history of palaeontology*, Macdonald, London.

Rupke, N.A., 1983, *The great chain of history, William Buckland and the English school of geology (1814–1849)*, Clarendon Press, Oxford.

Schäfer, W., 1962, *Aktuo-Paläontologie nach Studien in der Nordsee*, Verlag W. Kramer, Frankfurt am Main.

Schäfer, W., 1972, *Ecology and paleoecology of marine environments* (translation by I. Oertel and G.Y. Craig), Oliver and Boyd, Edinburgh.

Schäfer, W., 1980, *Fossilien, Bilder und Gedanken zur Paläontologischen Wissenschaft*, Verlag W. Kramer, Frankfurt am Main.

Scherz, G. and Pollock, A.J., 1969, Steno geological papers, *Acta Historica Scientarum Naturalium et Medicinalium*, **20**: 1–370.

Scott, R.W. and West, R.R., (eds), 1976, *Structure and classification of paleocommunities*, Dowden, Hutchinson and Ross, Stroudsburg, Pennsylvania.

Seilacher, A., 1953, Studien zur Palichnologie I. Über die Methoden der Palichnologie, *Neues Jahrbuch für Geologie und Paläontologie, Abhandlungen*, **96**: 421–52.

Seilacher, A. 1970, Begriff und Bedeutung der Fossil-Lagerstätten, *Neues Jahrbuch für Geologie und Paläontologie, Monatshefte*, **1970**(1): 34–9.

Seilacher, A., 1976, Sonderforschungsbereich 53, 'Palökologie': Arbeitsbericht 1970–1975, *Zentralblatt für Geologie und Mineralogie*, II, **1976**(5/6): 203–494.

Seilacher, A., Reif, W.-E. and Westphal, F., 1985, Sedimentological, ecological and temporal patterns of fossil Lagerstätten, *Philosophical Transactions of the Royal Society, London*, **B311**: 5–24.

Speyer, S.E. and Brett, C.E., 1986, Trilobite taphonomy and middle Devonian taphofacies, *Palaios*, **1**(3): 312–27.

Taylor, J.D., 1980, The evolution of predators in the Late Cretaceous and their ecological significance. In P.L. Forey and P.H. Greenwood (eds), *The evolving biosphere*, British Museum (Natural History), London, and Cambridge University Press, Cambridge: 229–40.

Thomas, R.D.K. (ed.), 1986, Taphonomy: ecology's loss is sedimentology's gain, *Palaios*, **1**(3): 206–338.

Vermeij, G., 1977, The Mesozoic marine revolution: evidence from snails, predators and grazers, *Paleobiology*, **3**(3): 245–58.

Voigt, E., 1962, Johannes Weigelt als Paläontologe, *Mitteilungen aus dem Geologischen Staatsinstitut in Hamburg*, **31**(1): 26–50.

Walker, K.R. and Diehl, W.W., 1985, The role of marine cementation in the preservation of Lower Palaeozoic assemblages, *Philosophical Transactions of the Royal Society, London*, **B311**: 143–53.

Walther, J., 1893–4, *Einleitung in die Geologie als historische Wissenschaft*, Verlag G. Fischer, Jena.

Walther, J., 1904, Die Fauna der Solnhofener Plattenkalk, bionomisch betrachtet, *Jenaische Denkschriften*, **11**(1): 133–214.

Walther, J., 1910, Die Sedimente der Taubenbank im Golfe von Neapel, *Abhandlungen der koeniglich Preussischen Akademie der Wissenschaften, Philosophisch-Historische Classe*, **1910**(3): 1–49.

Wasmund, E., 1926, Biocoenose und Thanatocoenose, *Archiv für Hydrobiologie*, **17**(1): 1–116.

Wegmann, E., 1963, L'Exposé original de la notion de faciès par A. Gressly (1814-1865), *Sciences de la Terre*, **9**(1): 85–119.

Weigelt, J., 1927a, *Rezente Wirbeltierleichen und ihre paläobiologische Bedeutung*, Verlag M. Weg, Leipzig.

Weigelt, J., 1927b, Über Biostratonomie. Eine Betrachtung zu Dollos siebzigstem Geburtstag, *Der Geologe*, **1927**(42): 1069–76.

Weigelt, J., 1928, Die Pflanzenreste des mitteldeutschen Kupferschiefers und ihre Einschaltung ins Sediment, *Fortschritte der Geologie und Paläontologie*, **6**(19): 395–592.

Weigelt, J., 1930a, Der Lebensgang von Johannes Walther, *Walther Festschrift, Leopoldina*, **6**: 3–10.

Weigelt, J., 1930b, Vom Sterben der Wirbeltiere, *Walther Festschrift, Leopoldina*, **6**: 281–340.

Weigelt, J., 1933, Die Biostratonomie der 1932 auf der Grube Cecilie im mittleren Geiseltal ausgegrabenen Leichenfelder, *Nova Acta Leopoldina, (N.F.)*, **1**: 157–74.

Weigelt, J., 1989, *Recent vertebrate carcasses and their paleobiological implications* (translation by J. Schaefer), University of Chicago Press, Chicago.

Whittington, H.B. and Conway Morris, S., (eds), 1985, Extraordinary fossil biotas, *Philosophical Transactions of the Royal Society, London*, **B311**: 1–192.

Wilson, M.V.H., 1988, Taphonomic processes: information loss and information gain, *Geoscience Canada*, **15**(2): 131–48.

Zittel, K.A. von, 1899, *Geschichte der Geologie und Paläontologie bis Ende des 19. Jahrhunderts*, Oldenbourg Verlag, Munich.

Chapter 2

TAPHONOMIC PROCESSES AND BIASES IN MODERN MARINE ENVIRONMENTS:
An actualistic perspective on fossil assemblage preservation

Karla M. Parsons and Carlton E. Brett

INTRODUCTION

A major goal of palaeoecological studies is the accurate reconstruction of ancient environments and communities. Palaeontologists yearn to restore the flesh and blood of fossil remains, and thereby to depict in detail the appearance, and analyse the interactions and dynamics, of ancient communities. Unfortunately, in most normal fossil assemblages, this reconstructional process remains qualitative, imprecise and utterly incomplete. Conversely, the reciprocal operation—reduction of living communities and their death assemblages to a potential fossil sample—appears, at least on first reflection, to be feasible. Understanding of the processes of *Aktuo-Paläontologie*, and rates of organism degradation, burial and early diagenesis, are also keys to better reconstruction of ancient ecosystems.

Detailed comparative studies of taphonomy in modern marine and non-marine environments were initiated by the seminal research of the German school of *Aktuo-Paläontologie* (Weigelt, 1927; Richter, 1929; Schäfer, 1972). Recently, many palaeontologists have also focused at least a portion of their research programmes on the study of actualistic taphonomic processes (for recent summaries, see Powell *et al.*, 1982; Behrensmeyer and Kidwell, 1985; Kidwell and Behrensmeyer, 1988; Kidwell and Bosence, in press). Certain of these studies have dealt with the specifics of decay processes and rates in particular taxonomic groups; see Allison (1986; 1988) and Plotnick (1986) for arthropods; Meyer (1971), Meyer and Meyer (1986), Liddell (1975), Greenstein (1989), and Kidwell and Baumiller (1990) for echinoderms. Schäfer (1972) summarized qualitative observations on the decay of many marine vertebrates and invertebrates. All of these studies point to the rapidity of degradation, especially in normal environments.

A second category of studies comprises more general analyses of the relationship of living assemblages (biocoenoses) to death assemblages accumulating in the same environments; these 'live/dead' studies include works by Boucot (1953), Jones (1969), Warme (1969), Warme *et al.* (1976), Schopf (1978), Dörjes (1972), Carthew and Bosence (1986), and, more recently, those of the Texas A & M researchers (Cummins *et al.*, 1986a; 1986b; Davies *et al.*, 1989a; 1989b; see also the review paper by Kidwell and Bosence, in press). In general, many of these studies deal with the 'negative aspects' of taphonomy, in underscoring the high degree of taphonomic bias present under even the best of conditions.

A third, more recently developed line of investigation makes use of quantifiable and qualitative aspects of preservation in one or more categories of preservable remains to delineate associations of taphonomic properties, called taphofacies (*sensu* Speyer and Brett, 1986). These studies include works by Fürsich and Flessa (1987), Henderson and Frey (1986), and Parsons (1989). Such studies underscore the close relationship between taphonomic signatures (measured in terms of indices of breakage, abrasion, dissolution, encrustation, bioerosion, etc.) and habitat, and, thus, the feasibility of taphofacies analysis for reconstructing ancient environments. Furthermore, a few studies are beginning to discern taphonomic factors (such as orientation, size frequency and, perhaps, articulation) which may be useful in distinguishing processes occurring during final transport and burial from background taphonomic processes (see particularly Davies *et al.*, 1989a; Powell *et al.*, 1990).

In turn, this last approach is just beginning to define a fourth major line of research in actualistic taphonomy, that is, the relationship between predicted or potential death assemblages and the actual accumulated fossil record. In particular, it is now evident that short-term skeletal accumulations of the type normally studied by actualistic taphonomists are distinct from, and may actually bear little resemblance to, the long-term accumulation that occurs below a taphonomically active zone (TAZ; see Davies *et al.*, 1989a; 1989b). If there is one unifying theme in all of these studies it is that of the ephemeral nature of all potential fossil remains and the absolute importance of residence time/burial in generating the final record. Thus, simple study of inferred life to death assemblage relationship is not sufficient to predict the actual assemblage of most environments. That requires understanding of the additional step which commonly may involve burial processes that are of such a very low frequency as rarely to be observed by researchers. Only by coring beneath the TAZ can workers begin to restore more accurately the original history of the buried assemblage.

These aspects of taphonomic research outline the major objectives of the neontological study of preservation. They are: to determine processes and rates of organism destruction and burial; to understand the preservational biases involved in conversion of living to dead potential fossil assemblages; to define modern taphofacies and their relationship to environmental processes; and to determine the burial processes and dynamics which are involved in producing the stratinomy of actual fossil assemblages.

LIVING/DEAD ASSEMBLAGE COMPARISONS

How faithful a representation of the original community is present in its dead assemblage? The degree of resemblance between the live and potential fossil assemblage, referred to as 'fidelity' (Kidwell and Bosence, in press), can only be assessed by detailed comparison of modern habitats and their accumulating sedimentary records. By sampling extant communities from a taphonomic perspective, it is possible to discern the extent of bias that exists in the conversion of the living biota into a residue of geochemically resistant skeletal elements and trace fossils. In actuality a series of questions are answered by comparing life (biocoenosis) and death (taphocoenosis) assemblages, including: what proportion of the original living assemblage is potentially preservable; what proportion of the potentially preservable or predicted death assemblage is actually to be found accumulating in the surficial sediments beneath a community; to what extent skeletal remains of various communities are transported laterally and commingled; how biased are the proportions of different species, in terms of numbers of individuals, biomass or other metrics, and in terms of proportions of trophic and life habit groupings; and finally, to what extent the long-term accumulations (deeper within sediment) resemble predicted and/or surficial, short-term accumulations? These questions have been dealt with in a variety of comparative studies of modern marine environments. In the following sections we briefly summarize the results of these studies. Most of this research has involved sampling of shallow marine (especially lagoon, bay, tidal flat and shallow shelf) environments with unconsolidated substrates. Typically, sampling involves suction dredging and analysis of shells extracted from the upper centimetres of sediment. Surveys of live versus dead assemblages on hard substrates are more limited, but have been attempted in a few areas of shallow shelf and reefal hardgrounds, and on skeletal hard substrates.

Proportions of marine fauna preservable

Numerous comparative studies of live versus dead assemblages have established the extreme degree of bias present in most fossil assemblages of marine environments. Non-skeletonized organisms comprise 50 per cent to over 95 per cent of the total number of species in most marine communities. Obviously, these organisms normally leave no record of body fossils. A second category of lightly skeletonized organisms, such as arthropods with non-mineralized chitinous cuticles or echinoderms with thin plates, may potentially leave some remains in the death assemblage, although they are only rarely preserved and are clearly underrepresented. The third category, the well-skeletonized organisms, including many molluscs, barnacles, some echinoderms, bryozoans, corals, and the teeth and bones of some vertebrates, should be preservable if buried relatively quickly.

Even among the 'well-skeletonized' category, however, differences exist in preservability. For example, aragonitic molluscan shells typically undergo much more rapid dissolution in the upper sediments than do calcitic shells (Koch and Sohl, 1983). Differential thickness of shells as well as the proportion of contained organic matrix may also play significant roles in the rate of skeletal degradation.

Soft substrates
Using these criteria, many workers have examined various marine biotas to assess the percentage of taxa that should be represented in the fossil record. The percentage of well-skeletonized species in an assemblage varies widely from near zero to over 70 per cent, but tends to average close to 30 per cent (Kidwell and Bosence, in press). High-diversity communities may be dominated by soft-bodied organisms and would yield low-diversity fossil assemblages. For example, in an early study, Jones (1969) analysed 176 dredge samples from the southern California shelf yielding 1473 macroinvertebrate species. Of these, 523 were polychaete worms with no preservation potential (except possibly for chitinous jaw parts and traces), 419 were crustaceans which Jones argued would leave a poor record at best, 408 were molluscs with good preservation potential, and 64 were echinoderms with variably moderate to poor preservation potential. All told, Jones (1969) found that on the order of 30 per cent of species from this diverse biota should be represented in a potential fossil record.

In a similar study Dörjes (1972) used dredge samples from 0–15 m depths taken from near the shore to 20 km offshore on the Georgia shelf. He, too, found a high proportion of non-preservable or very poorly preservable taxa among 268 macroinvertebrate species. In terms of species, again about a third (90 species) were readily preservable molluscs, while the rest were polychaetes (76 species) and crustaceans (67 species) with little to no preservation potential. However, when the sample was examined in terms of numbers of specimens, the potential death assemblage appeared much worse, as only 3.5 per cent of total individuals were molluscs, whereas polychaetes and crustaceans comprised 56 per cent and 15 per cent respectively, of the individuals recovered.

Schopf (1978) made a comparative study of fossilization potential among three broadly defined environments in Friday Harbour, Washington. He examined the samples for systematic habitat-related differences in proportions of preservable organisms. He classified all organisms into one of three categories based on *a priori* criteria: those that would probably be well-represented (such as clams); those poorly represented (for example, starfishes); and those not represented in the fossil record (such as worms). On this basis he predicted that 29 per cent of the rock-dwelling fauna, 32 per cent of sandy shelf biota, and 30 per cent of mud-bottom species would yield many recognizable fossils; 41 per cent of the sand fauna, 42 per cent of the rock fauna and 38 per cent of the mud fauna were predicted to yield a few recognizable fossils; the remainder in each environment would have no record. He then sampled the sediments accumulating in Friday Harbour in order to determine the proportion of organisms actually represented and to test his predictions; about 29 per cent of the total fauna was represented. Thus prediction and actual sampling of the short-term death assemblage both indicated a similar and environment-independent proportion (29–30 per cent) of preservable species richness.

Schopf also examined the actual record of fossils of the various taxa as indicated by the *Treatise on invertebrate paleontology*. The results were slightly more encouraging. He found that 32 per cent of the sand fauna, 39 per cent of the rock-dwelling fauna, and 44 per cent of the mud fauna had actually been preserved in the record. Hence, in the long term the mud fauna was actually best represented. Schopf attributed these slight discrepancies to differences in the proportions of various trophic groups. He noted, for example, that sessile filter feeders and herbivores were much more apt to be well skeletonized (67 per cent) and, therefore, preserved, than infaunal deposit feeders such as polychaetes or vagrant detritus feeders such as crabs or asteroids (16–27 per cent of species preservable). Thus, depending on the constraints of a given environment, the preservation of total richness might be better or worse in proportion to the numbers of various trophic groups present.

Trace fossil comparisons

One possible means of retrieving some of the information lost with the decay of soft-bodied organisms is through study of distinctive trace fossils. In some instances the trace-makers can be identified at least at a general level (for example, clionid sponge borings). Such traces may be used to determine the presence and relative frequency of the producers, and/or their interactions with other organisms (such as predation intensity; see Stanton *et al.*, 1981).

Soft-substrate traces are generally difficult or impossible to correlate with specific producing organisms, although they may be discrete and distinctive. In such cases, these ichnofossils might at least serve as a rough guide as to the diversity of soft-bodied infaunal organisms. Hertweck (1972) re-examined the 268 species of macroinvertebrates on the Georgia Shelf from the standpoint of trace production. Of these, only 40 (mainly soft-bodied) produced a distinctive trace. Unfortunately only about half of these traces were actually preservable in the sediments. The production of multiple different trace-fossil types by single species of trace-makers, and of nearly identical traces by distantly related species, further limits the utility of ichnofossils as a guide to the richness and relative frequency of soft-bodied organisms.

Hard substrates

Relatively few studies have considered the preservation potential of communities encrusting hard substrates. Rasmussen and Brett (1985) studied and compared the proportions of skeletonized and soft-bodied organisms encrusting platy corals and cave roofs within a tropical reef setting from St Croix, US Virgin Islands. Removal of soft-bodied organisms by mechanical means (brushing) or soaking in chlorine bleach solution revealed the skeletons of numerous early encrusters beneath later-stage forms, thus somewhat increasing total species richness and proportions of preservable species. Again, they found a similar percentage of preservable species (about 30 per cent for cave-roof encrusting communities and slightly higher, 35–40 per cent, for undersurfaces of platy corals).

In contrast, Liddell and Ohlhorst (1988) found a high proportion (about 70 per cent) of preservable skeletonized species on shallow reef surfaces; however, only about 1.8 per cent of deeper-reef (greater than 120 m) encrusting communities were found to be preservable. When encrusting communities were examined from the standpoint of relative areas covered by colonial and solitary organisms, a much lower percent preservability was found (about 10 per cent of total encrusted area); the remainder of the surface area, formerly nearly totally covered by sponges, tunicates and various algae, would appear as blank unoccupied spaces in the fossil record.

Summary
Overall it appears that in most communities a relatively low proportion of species and a lower proportion of individuals could be represented in a potential death assemblage. Other studies of potential death assemblages include those of Warme (1969; 1971), Warme *et al.* (1976), Stanton (1976), Peterson (1976), Staff *et al.* (1986), and Cadée (1968). Carthew and Bosence (1986) recorded variable proportions of preservable species (5–66 per cent) and individuals (1–87 per cent) in British shallow shelf communities. However, in general, preservable species average about 30 per cent and preservable individuals less than 10 per cent; the actual sample is also highly biased toward sedentary epifaunal species. No strong correlation is evident between percent preservable taxa and substrate or water depth (Staff *et al.*, 1985), although Liddell and Ohlhorst's (1988) data suggest a decrease in preservability of reef taxa with depth.

Transport of potential fossil skeletons

A persistent concern of palaeoecological studies is the degree to which skeletal elements are transported and, therefore, to which spatially separate community remains are admixed. This is best tested by examining the distribution of living and dead organism remains across a gradient of closely associated environments. Many studies provide substantial data that indicate marginal post-mortem transport except in very nearshore environments, and even here most exotic shells are derived from immediately adjacent habitats (Kidwell and Bosence, in press). Furthermore, although 'exotic taxa' comprise a high proportion of species, they generally account for less than 10 per cent of individuals (Henderson and Frey, 1986).

In his detailed study of molluscan biocoenoses and taphocoenoses from Ría de Arosa, a deep bay along the west coast of Galicia, Spain, Cadée (1968) observed little or no transport of shells outward from the central bay although there was minor transport of outer bay and shelf species into the central bay. Overall, however, he argued that transport was a minor factor in biasing shell assemblages.

Detailed study of beach and tidal inlet shells from Sapelo Island, Georgia, USA (Henderson and Frey, 1986), indicated substantial differential transport in these areas. Left–right valve sorting is a major factor in bivalve shells produced by oblique swash of waves onto the shoreface. The observed bias in

valve proportions proves the importance of selective transport in this environment. Henderson and Frey (1986) observed major faunal redistribution and
mixing in a tidal inlet channel from Sapelo Island. Moreover, they observed a
scour hole within the channel that served as a stratigraphic leak from relict
Holocene and Pleistocene shells that became mixed with the modern assemblage. Henderson and Frey predicted that reworked assemblages should be
common in tidal inlet/shoal/barrier island facies. This prediction appears to be
borne out by the work of Davies *et al.* (1989a) in the San Luis Pass, a
microtidal inlet on the Texas coast. These workers found relatively little shell
material that was previously *in situ* and in fact observed mixtures of shells
from adjacent bay and beach environments.

Conversely, most studies of lagoonal and shallow subtidal shell assemblages reveal little or no mixing even of closely adjacent environments,
despite minor within-habitat transport (Cummins *et al.*, 1986a). For example,
Warme (1971) noted closely associated mud and sand facies in Mugu Lagoon,
California, with distinctive communities. Cluster analysis of shells from sediments in these areas revealed little lateral mixing between them. Similarly,
Miller (1988) found little evidence for mixing at Tague Bay, St Croix. Mollusc
samples for this study were collected across environmental gradients in a
grassy lagoon. Using a combination of cluster analysis, polar ordination,
moving averages of species and correlation analyses of species and vegetation, he found excellent correspondence between density of sea grass (that is,
environment) and its contained molluscan death assemblage. Dense patches
of sea grass produced faunas dominated by epifaunal gastropods and lucinid
bivalves, while bare areas were dominated by other bivalves and some
infaunal gastropods. The transition areas showed transitional faunas reflecting environment type rather than mixing of end-member faunas (that is,
transport). These patterns were only evident in the dead assemblages. Live
specimens were scarce and therefore not as useful.

Surprisingly, Fürsich and Flessa (1987) found relatively little shell transport
even on tidal flat environments in the Gulf of California. Shell assemblages
from the inner, outer and mid-flat environments appeared to faithfully replicate the patterns of the living communities. Similarly, Walker *et al.* (1989)
were able to correctly distinguish inner, middle and outer shelf habitats on the
Texas continental margin on the basis of dead shell assemblages.

In summary, most offshore and low-energy environments display little or
no transport of benthic organism remains, but transport may be significant in
high-energy settings. Obviously pelagic and epiplanktonic organisms cannot
be preserved in their habitats and some (such as cephalopods) may be rafted
for vast distances after death (see Boston and Mapes, this volume).

Potential death assemblage versus actual accumulation

Given that only a small proportion (typically 30 per cent or less) of the living
benthic community is capable of being fossilized, we can next turn to a more
detailed consideration of just that potentially preservable fraction. How well

does the skeletonized assemblage actually accumulating (in a short-term sense) in the sediments reflect the potential or predicted death assemblage based on *a priori* assumptions? This question has been examined in many studies (Valentine, 1961; Johnson, 1965; Cadée, 1968; Warme, 1971; Warme *et al.*, 1976; Peterson, 1976; Stanton, 1976). Kidwell and Bosence (in press) noted that dead/live fidelity can be measured in three different ways: percentage of live species in death assemblages; percentage of species in death assemblages found in live assemblages; and percentage of dead individuals represented in the live community. In an excellent summary of 18 live/dead studies in intertidal (5), shallow subtidal (5), and open shelf (8) settings, Kidwell and Bosence (in press) demonstrated high fidelity for the first of these indices (54–100 per cent, with means 82–99 per cent), considerably lower fidelity for the second (10–64 per cent, with means 31–49 per cent for different settings) and similarly low and variable for the third (2–99 per cent, with means 60–86 per cent). The general result of these studies is that although most species predicted to occur in the death assemblage are, indeed, present, the proportions and rank abundances of species are not faithfully reproduced. Moreover, the match between the death assemblage and the skeletonized fraction of the living community censused at a particular point in time is typically rather poor, in that the actual death assemblage contains many species that are not living in the area at the time of the census. For example, Valentine (1961), working in the Santa Monica shelf in California, found that only 14–19 per cent of the species of dead shells were represented by living molluscs in the same samples. Johnson (1965) found only 50 per cent of dead shells represented in the living community at Tomales Bay, California. Shifflett (1961) and Murray (1965) found a similarly poor correlation between living and dead foraminifera in single samples. In each case, the source of the discrepancy appears to be short-term time-averaging of patchy communities which enriches the death assemblage.

Cadée (1968) made a very detailed comparison of biocoenoses and taphocoenoses from Ría de Arosa. He recognized, perhaps for the first time, the importance of time-averaging, noting that many species of molluscs occurred in local clumps, and pointed out that because of short-term migration of these clumps no close correspondence could be expected between the living community and skeletal debris beneath it in any one sample. To correct for this he made an assessment of the potentially preservable component of a broader region. He also integrated observations made over a period of three years. When this integrated sample was then compared with observed death assemblages, a relatively close correspondence could be found. All of the major species found in the death assemblage were found among the living community of the general region in the middle and outer portion of the bay.

Several later studies done along similar lines (Warme, 1969; 1971; Warme *et al.*, 1976; Ekdale, 1974; Stanton, 1976) also showed a rather close correspondence in terms of presence/absence data between life and death assemblages. These studies made use of cluster analyses to discern recurrent groupings of living taxa (communities) that characterized given environments; they then used the same clustering procedures to define recurrent groupings based on shells obtained from sediment samples in the same areas.

In most cases, clusters based on dead molluscan shells matched rather closely those based on live communities and with depositional environment. In nearly all cases the actual death assemblage was more diverse than the potentially preservable portion of the living community and in several instances the death assemblage was a better predictor of the environment than the living assemblage. These observations confirmed the importance of short-term temporal fluctuations and the phenomenon of time-averaging was emphasized. That is, the actual death assemblages represent integrations of minor fluctuating biocoenoses averaged over some period of time.

In a recent study of living and dead molluscs from Bahia la Choya, in the Gulf of California, Fürsich and Flessa (1987) used a similar approach to recognize molluscan assemblages across a gradient of inner tidal flat to shallow subtidal environments. Results of cluster analysis based on relative abundance data (cosine theta similarity indices) for the molluscs produced very good overlap between living and dead assemblages. Hence, in this instance, even the relative proportions of preservable species appear to be preservable, although with minor bias in some taxa.

None the less, despite these encouraging results, most studies reveal numerous discrepancies between the proportions of preservable taxa in living communities and their dead counterparts. This problem remains even when data on the live communities are integrated over relatively long sampling intervals and broader regions, although longer observation times for the living community substantially improve all fidelity measures (Kidwell and Bosence, in press). For example, Cadée (1968) noted substantial over-representation of certain species and underrepresentation of others in death assemblages of Ría de Arosa. In a study on living and dead molluscs in two California lagoons, Peterson (1976) noted similar discrepancies in relative abundance. Kidwell and Bosence (in press) noted that shells which are most subject to rapid destruction in marine environments tend to be 'short-lived, small, or opportunistic forms' because they are often weakly skeletonized; the more persistent 'core' of many communities is more apt to be preserved.

Possible causes of the bias between living and dead assemblages were thoroughly reviewed by Cadée (1968) and include: selective transport and mixing of shells between habitats; mixing of modern shells with reworked older, including fossil, remains; selective removal of shells by predators; differences in rates of shell production; differential mechanical destruction; and differential solubility; to which could be added biogenic sorting.

In most cases the first factor has been ruled out by the relatively good correspondence between species composition of living and dead assemblages or by direct observation; selective transport might be a factor in high-energy nearshore environments (see above). Neither Cadée (1968) nor Peterson (1976) considered this to be a major factor. Major temporal mixing of shells (as opposed to short-term time-averaging) is probably only a factor in areas of relatively slow sediment accumulation or erosion.

Cadée (1968) underscored the possible role of selective predation in biasing samples, although this factor has been largely ignored by later workers.

Obviously, species with higher rates of productivity (or more rapid turn-over) may produce larger numbers of skeletons than those which have lower rates. Cadée argued that this might be important in explaining the overrep-resentation of some opportunistic nuculid bivalves.

Differential mechanical strength of shells would appear to be a factor primarily in high-energy environments where at least local transport of shells occurs. Nonetheless, Noble and Logan (1981) observed biased valve ratios in living brachiopod assemblages from environments in which no major trans-port is taking place. They attributed this bias to preferential destruction of more fragile brachial valves, a phenomenon common in fossil assemblages. This raises the possibility that mechanical agents may act upon shells without producing major transport. Storm-generated waves and currents, and the consequent impacts of shells with one another, provide a possible explanation for selective fragmentation even in normally quiet water environments. The bacterial degradation of organic matrix within shells may increase their susceptibility to fragmentation. Differences in amounts of organic matrix may therefore make shells more or less preservable.

Differential dissolution of shells is an important factor that may bias samples. Variation in shell mineralogy and microarchitecture may influence solubility (see Koch and Sohl, 1983). Peterson (1976) examined this pos-sibility quantitatively in shells from Mugu Lagoon, California. He experi-mentally determined weight loss in shells due to solution within the upper 50 cm of sediment. He buried shells and exhumed them to determine dry weight loss. He extrapolated to average residence times for various taxa (time before complete shell loss by solution) and determined highly variable persist-ences ranging from 4 years to nearly 900. He was then able to make adjust-ments from the potential death assemblage to predict a longer-term assemblage that more closely resembled the proportion of species. Obviously, however, since many of these same types of bivalve occur as fossils there must be some threshold level within sediment at which dissolution stops altogether.

Tanabe *et al.* (1986) have noted an additional bias in live/dead comparisons based on life habits. Deep infaunal bivalves are much more likely to die in the sediment and thus to be preserved *in situ* than are shallow infaunal or epifaunal species. Consequently, assemblages may be biased with respect to infauna. In any event, Tanabe *et al.* reasoned, as did Staff *et al.* (1986), that whole articulated shells can give a better sense of the living community than do disarticulated remains.

A final bias recorded is biogenic sorting. Hermit crabs may selectively remove gastropod shells from certain environments (Frey, 1987; Walker, 1988; 1989). Additionally, tidal flat molluscs are often scavenged by gulls and this may sort out particular species, or sizes, of taxa.

More detailed comparative studies of live and dead assemblages of total biotas have been carried out for shallow (about 1 m) marine communities in Laguna Madre and upper Copano Bay, Texas (Staff *et al.*, 1985; 1986; Powell *et al.*, 1990). These researchers used a variety of metrics to compare the composition of living, potential (predicted) death assemblages and actual (short-term) accumulations from box cores taken over a two-year period. Taxonomic composition of the samples reflected a relatively high proportion of potentially preservable species; 20 of 38 taxa (53 per cent) in Copano Bay and 31 of 58 taxa (53 per cent) in Laguna Madre. Most of the potentially preservable forms were found in the actual dead assemblages, along with a large number of other taxa not present in the living community. The total richness of the death assemblage—88 taxa for Copano Bay and 65 taxa for Laguna Madre—exceeds the total richness of the living community in each case. This latter counterintuitive observation must reflect admixture of temporally and/or spatially separate communities (see below). Following Stanton *et al.* (1981), the researchers also compared the richness represented by adult skeletons in the death assemblage to the living sample and found a much better correspondence (20 of 25 taxa in the Laguna Madre death assemblage sample also appeared in the living assemblage). This presumably reflects the occurrence of a large number of opportunistic juveniles and ephemeral forms in the death assemblage that are not viable in the environment and are absent in the actual living assemblage at any one time. Transport and short-term environmental change may also account for presence of taxa in death assemblages not found in living communities.

The order of abundance of organisms was not faithfully recorded in the death assemblages of Laguna Madre or Copano Bay. Indeed, most of the dominant taxa in the death assemblage were not common among the skeletonized (preservable) fractions of the living communities and vice versa. The most dominant organisms in the two death assemblages ranked 10th and 17th in the potential (predicted) death assemblage of the extant community. Again, the authors argued that differences in rank abundance probably reflect one or more of the following factors: importation of many dead shells into the area; differential preservability of the skeletons in different species; and inadequate temporal sampling of rapidly fluctuating community structure.

Strangely, the researchers found that ranking of living species by biomass (dry weight for soft-bodied organisms estimated from shell size using equations developed by Powell and Stanton, 1985) was more accurately mirrored in biomass proportions (estimated similarly) of the death assemblage. Thus, four of the top six species accounting for greatest biomass in the live community were also in the top six for the potential death assemblage. Moreover, these were also found in almost the same biomass ranking in the actual death assemblage.

Why biomass proportions should be more accurately preserved than rank ordering is not entirely clear. Nor is it clear how general this result is. Presumably it reflects the fact that, at least in the Texas lagoons, the largest and heaviest (in part due to large shells) organisms are readily preservable.

Fürsich and Flessa (1987) made a comparably detailed study of live/dead relationships of molluscs in the Gulf of California at Bahia la Choya, Mexico. In this study Fürsich and Flessa compared grab-sample live and dead assemblages along a transect of environments ranging from salt marsh, tidal channel, inner, middle and outer flats, and shallow subtidal in a macrotidal regime. They reported better correspondence between living and dead species assemblages using adult shells only. Size-frequency surveys across tidal flats indicated the likelihood of juvenile mortality and/or transport in some inshore areas.

Rather than comparing proportions of various species in a single environment as in the case of Staff *et al.* (1985), Fürsich and Flessa plotted the absolute frequency (number of specimens) for each species along the subtidal to inner tidal flats transect, and compared frequency histograms for the living and dead shell assemblages along the gradient. They noted a close corre-spondence in peaks and troughs of distribution between the two samples, indicating, in contrast to the Staff *et al.* study, that relative frequency of living forms was mirrored in the potential fossil assemblage, at least for many species.

Again, Kidwell and Bosence (in press) provided a useful summary of fidelity of rank and relative abundance for death assemblages derived from nine nearshore and shallow shelf communities. They concluded that in gen-eral only about half of the six most common taxa of the living community are among the top six in the death assemblage, and relative and rank abundance of these common organisms were very poorly reflected in the death assemblage.

Trophic groups and life habit group comparison

One of the important potentially preservable aspects of community structure is the proportion of various trophic, or feeding, groups such as the percentage of filter feeders, deposit feeders, herbivores and carnivores. Also closely related is the proportion of life habit groups such as epifauna and infauna; the two types of information are commonly lumped to give, for example, the proportion of infaunal deposit feeders, infaunal suspension feeders, etc. Trophic/life-habit proportions are known to vary consistently among different physical environments and, therefore, may potentially give very useful palaeoecological information (Walker, 1972; Walker and Bambach, 1974), but how well do the apparent ratios in death assemblages represent trophic/life-habit groupings in living communities? Unfortunately, several studies point to relatively poor preservation of trophic/life-habit proportions.

As already noted, Schopf (1978) argued that epifaunal suspension feeders are inherently more likely to be well skeletonized and thus more preservable than are infaunal organisms. The former are exposed to physical and biologi-cal stresses that favour possession of a protective skeleton, whereas in the latter, a soft, flexible body probably favours wriggling or crawling through sediments and skeletal protection is less critical. Indeed, most numerically based studies have noted a strong preservational bias toward both epifauna and suspension-feeding modes of life.

In their detailed studies of life/death assemblages in Laguna Madre and Copano Bay, Texas, Staff *et al.* (1985; 1986) assigned various species to deposit feeders (including herbivores and detritivores), suspension feeders and parasites/predators, and also to infauna and epifauna. Using numerical abundance data for 29 data sets published in the literature, including their own (numbers of specimens belonging to particular trophic/life-habit groups), they observed, as did previous authors, a slight bias toward suspension feeders, though not as strong as expected. In fact, only in 15 cases was the difference between live and dead assemblage greater than 10 percentage points. Use of biomass as a proxy for relative frequency accentuated the bias towards suspension feeders (21 cases with greater than 10 percentage points' difference). A similar enhancement of bias occurred in comparing epifaunal with infaunal proportions. This apparently reflected the fact that the epifaunal suspension feeders are not only well skeletonized, but also more bulky (weighty) than the numerically more dominant infaunal species. Biomass measurements favour suspension feeders in both the living and dead assemblages.

On the whole, trophic group proportions are rather poorly preserved in most fossil samples. The effect may be more or less strong depending upon substrate conditions. Soft mud or sands may initially carry much higher proportions of deposit than suspension feeders, whereas hardgrounds and skeletal hard substrates will show initially higher percentages of epifaunal suspension feeders and, presumably, less biased representation.

Death assemblages versus actual fossil record

Actualistic studies have provided a general view of the degree of bias to be expected in fossil samples from a few environments. However, these studies have not fully dealt with the problem of longer-term accumulation of organism remains as parts of the permanent fossil record. How well do the skeletal assemblages in the upper few centimetres of unconsolidated sediment, as typically sampled in actualistic studies, resemble the permanent record? At first reflection this might appear to be an unreasonable question. After all, it might appear that these remains are part of the permanent record. However, as will be noted, there may be considerable discrepancy between the two types of sample.

One approach to this problem is to make comparative studies of Holocene subfossil or Late Pleistocene fossil assemblages and comparable Recent environments. For example, Fürsich and Flessa (1987) compared molluscan assemblages in Pleistocene limestones with Recent tidal flat skeletal accumulations in the Gulf of California. Similarly, we have examined or compared potential fossil assemblages of the St Croix shelf with actual fossil assemblages containing identical species in Late Pleistocene carbonates on the island. Two salient observations must be explained in these cases and, we suspect, as a general rule.

The first significant difference observed by Fürsich and Flessa (1987) and by ourselves is that the fossils are commonly better preserved than would be predicted from recent sediment sampling. This is manifest in the higher proportions of closed, articulated shells, the relatively high frequency of unbroken and uncorroded valves, and the presence of *in situ* fossils. In the case of St Croix limestones we find certain relatively continuous horizons that display preservation remarkably better than predicted. Large colonies of relatively fragile ramose scleractinian corals (*Porites, Acropora*) and relatively large numbers of articulated, tightly closed bivalves are present in the limestone at several locations. In contrast, intact branching coral skeletons and closed articulated shells were only rarely observed in detailed sampling by Parsons of apparently similar patch reef to lagoonal sediments on the modern St Croix shelf. Indeed, some aspects of the actual fossil assemblages more closely resembled features of the living communities than of death assemblages. Why does this discrepancy exist? At present we can only offer a tentative explanation for this phenomenon. The extraordinary preservation of many fossil assemblages runs counter to evidence of relatively rapid degradation of skeletal remains on the sea floor. We surmise that such preservation can only result from rapid entombment of organisms in sediment below the TAZ. In turn, this evidence, together with the existence of disjunct fossil beds (see below), suggests a highly episodic mode of accumulation for the permanent fossil record. Without doubt, many bioclasts preserved in particular beds represent earlier generations of organisms that had accumulated gradually as parts of time-averaged assemblages. However, the final processes resulting in the permanent stratigraphic accumulation of a particular bed were both abrupt and associated with unusual, probably disequilibrium, conditions.

The second observation is that fossils tend to occur in discrete layers, beds or biostromes, which are sharply set off from adjacent, more sparsely fossiliferous sediment layers. This contrasts with the common appearance of homogeneity in upper sediment samples obtained by box coring in comparable modern environments. Many thicker shell beds undoubtedly record multiple episodes of burial and exhumation. As in the Jeram model of Seilacher (1985), it is probable that once a particularly thick shell pavement is formed it may serve as a 'reference horizon' or armour, below which little or no erosion takes place. Instead, later generations of skeletal material may accumulate upon the original pavement of shells by winnowing of intervening, surrounding, fine-grained matrix sediments in periodic storms. The skeletal pavement may also serve as a foundation for hard-substrate communities which utilize skeletal parts as attachment areas and, therefore, a skeleton-dominated build-up may result, a process referred to as 'taphonomic feedback' (Kidwell and Jablonski, 1983). In this way a biostromal layer, including such organisms as ramose corals and possibly even small patch reefs, might develop on the pavement.

It remains somewhat unclear what processes initiate and terminate such skeletal accumulations. Thicker shell beds commonly display sharply defined soles with evidence of piping of shell material into firm-ground burrows that are cut into over-compacted muds. These resemble the tubular tempestites of Wanless (1986), that is, burrows that act as traps for storm-concentrated shell debris. This observation suggests that shell-bed initiation may be associated with minor discontinuities. The initial pavement accumulates during a period of minimal sediment deposition. Aggregation of skeletal debris may reflect a combination of erosion and repeated winnowing. The shell armour, together with the firmness of the underlying sediments, may inhibit erosion beyond a certain point and allow formation of the reference horizon. The minor discontinuities may commonly be associated with sediment-starved time intervals, perhaps during marine transgressive events. This would imply that many thicker shell beds and skeletal accumulations are manifestations of longer-term, perhaps cyclic, processes encompassing hundreds to a few thousand years. The upper surfaces of skeletal beds commonly display well-preserved, even *in situ*, fossils that suggest rapid smothering of terminal communities by fine-grained sediments (Parsons *et al.*, 1988). Again, it is not fully understood why these terminal sediment blankets were never winnowed, as we suppose most previously deposited sediment bundles have been. Presumably, shell-bed generation is terminated by particularly thick sediment deposition events. If, during minor sea-level rise or climatic cycles, storm-wave base was also rising it could reach a threshold level which would bring the sea bottom below the zone of frequent winnowing and aid in preservation of the terminal sediment blanket on top of the shell bed. In any case, the process of shell-bed accumulation appears to be episodic rather than steady-state in many stratigraphic successions.

Given their episodic mode of accumulation, the composition at least of the terminal shell-rich beds and thin isolated shell layers may reflect far greater fidelity to an original single community and less time-averaging than is commonly assumed. Conversely, within many thicker fossil beds there may be a strong degree of time-averaging. Despite reports of 'indefinitely long half lives' for larger and more robust shells (Kidwell and Bosence, in press), in most skeletal remains there is evidence of a limited residence time on the sea floor (Davies *et al.*, 1989b). The excellent condition of many skeletons in ancient shell beds further implies that accumulation does not occur simply by shell production without burial. Rather, most skeletal aggregations reflect nearly continuous, but interrupted, burial of remains in thin sediment layers, punctuated by very brief periods of exhumation, reworking and skeletal concentration.

Absolute dating of shells from modern tidal flats in the Gulf of California indicates a poor correlation between stratigraphic position and the age of the shells (Flessa *et al.*, 1990). Shells in the upper metre of sediment may range in age from a few tens of years to over 3500 years in age and shells observed in cores may occur out of their proper temporal order with respect to depth in the core. This 'stratigraphic disorder' of potential fossils implies a strong degree of time-averaging in most modern sediments. Shells up to about 4000 years old have been commingled within the upper metre of sediment. Such disorder implies either mixing of shells by bioturbation or reworking via exhumation and reburial cycles. In computer simulations of disorder, Cutler and Flessa (1990) found that bioturbational mixing probably plays a much weaker role in stratigraphic disordering of shells than does reworking. The latter process is highly efficient in scrambling the temporal relationships of shells. Again, this study underscores the importance of physical reworking and amalgamation mainly by storm processes in the formation of the skeletal aggregations.

It is noteworthy that physical reworking does not necessarily produce its own taphonomic signature, provided that it acts only briefly. Furthermore, skeletal condition has been found to be a poor indicator of shell age, as some geologically young shells are highly corroded while other apparently ancient shells are nearly pristine. The taphonomic condition of shells in mixed assemblages may provide clues to their original habitat and/or that area or site in which the shells have the longest post-mortem residence time. Hence, mixed taphonomic signatures may permit identification of substantial time-averaging, and even the recognition of laterally transported and mixed skeletal accumulations. If environments imprint diagnostic taphonomic signatures on shells, then these may be used to diagnose the site of original skeletal accumulation, whether or not the shells remained in that site.

TAPHONOMIC PROCESSES AND PROPERTIES

The study of modern taphonomic processes allows workers to interpret death assemblages based on the knowledge gained by observing remains in the context of contemporary physical and biological processes. With this information, modern assemblages and the remains of individual organisms can be described in terms of their taphonomic history with greater confidence than for fossil assemblages. Armed with quantitative information on how a skeleton obtains its particular taphonomic signature, extrapolation to fossils with similar modes of preservation can be made.

Taphonomic processes can be subdivided into two categories. The first are processes that cause modification to the skeleton, including encrustation, fragmentation, abrasion, dissolution and bioerosion/corrosion (Figure 2.1). Second, there are processes which affect the relationship among shells in an assemblage such as orientation and articulation (see Table 2.1 for a summary). In modern settings, these features are most commonly measured on mollusc remains because they are readily preservable, easily collected and cross-cut many environments. Less work has been done on foraminifera, although they also have favourable attributes (see Martin and Liddell, this volume). Less useful taxa include those that are environment-specific (for example, corals and vertebrates) or less readily preservable (such as arthropods and echinoderms).

Taphonomic properties

Encrustation
Encrustation is a ubiquitous phenomenon in marine settings. Any exposed, stable substrate is likely to be settled upon by a variety of epifauna. Encrusting organisms settle on firm substrates such as shells (live and dead), rocks, echinoderm tests, vertebrate remains, and even sea grasses and algae. Common encrusting fauna which leave preservable remains include coralline algae, foraminifera, coelenterates, serpulid worms (Figure 2.1A), bryozoans, barnacles (Figure 2.1A) and some molluscs.

Encrustation is a good indicator of exposure at the sediment–water interface. Most hard parts are encrusted after death, but some are encrusted during life, so caution must be used in estimating time and exposure since death. Processes such as burial, surface irregularities (that is, ornamentation) or possibly mineralogical differences in substrate may control the amount and kind of infestation.

Table 2.1 *A summary of taphonomic indices and their implications for the physical, chemical and biological conditions under which they affect skeletal remains*

Taphonomic feature	Implications	References
Abrasion	The wearing-down of skeletons due to their differential movement with respect to sediments is an indicator of environmental energy. Significant abrasion is most often found on skeletal material collected from beaches, or areas of strong currents or wave action.	Driscoll and Weltin (1973), Driscoll (1967b).
Articulation	Multi-element skeletons are soon disarticulated after death. Articulated skeletons, then, indicate rapid burial or otherwise removing the skeleton from the TAZ.	Allison (1986; 1988), Plotnick (1986).
Bioerosion	Bioerosion encompasses many different corrasive processes by organisms. The most pervasive causes of degradation are boring and grazing. Bioerosion erases a large amount of information from the fossil record, but it also leaves identifiable traces made by organisms on remaining hard skeletons. Therefore, bioerosion adds information on the diversity of ancient assemblages. Also, patterns and processes of bioerosion vary among environments due to the distribution of bioeroders, energy levels and other habitat differences.	Odum and Odum (1955), Warme (1977), Pleydell and Jones (1988), Boekschoten (1966), Perkins and Tsentas (1976), Fütterer (1974).
Dissolution	Skeletal remains are often in equilibrium with surrounding waters, but changes in chemical conditions can cause skeletons to dissolve. Dissolution represents fluctuations in temperature, pH or pCO_2 in calcium carbonate skeletons. Silicious skeletons dissolve more readily because normal sea water is usually undersaturated with respect to silica.	Davies et al. (1989b), Flessa and Brown (1983), Alexandersson (1978).
Edge rounding	Broken edges of skeletons become rounded due to either dissolution or abrasion of the exposed surface. The processes that control edge rounding are not fully known, but are probably a combination of dissolution, abrasion and bioerosion. Rounding gives an estimate of time since breakage.	Davies et al. (1990).
Encrustation	The overgrowth of hard skeletal substrates by other organisms is a common occurrence. Besides indicating exposure of the skeleton above the sediment–water interface, encrustation can specify environment. Different patterns of encrustation as well as different biota occur in different environments.	Rasmussen and Brett (1985), Driscoll (1967a), Driscoll (1968).
Fragmentation	Breakage of skeletons is usually an indication of high energy resulting from wave action, currents, tides or winds. Fragmentation can also be caused by other organisms through either predation or bioturbation.	Müller (1979).
Orientation	After death, skeletal remains are moved by the transporting medium and orientated relative to their hydrodynamic properties. Fossil skeletons in life position indicate rapid burial, attachment to a firm substrate or death of in-place infauna. Hard parts tend to orientate long-axis parallel to unidirectional flow in current-dominated areas and perpendicular to wave crests on wave-dominated bottoms.	Nagle (1967), Johnson (1957), Emery (1968), Salazar-Jiménez, et al. (1982), Clifton and Boggs (1970), Brenchley and Newall (1970).
Size	After death, a skeleton behaves as a sedimentary particle and is moved and sorted with respect to the carrying capacity of the flow due to currents, waves or tides. Size can, therefore, be an effective indicator of flow capacity in a hydraulic or wind-driven system.	

The rate of infestation, and the type, diversity and relative abundance of encrusting organisms, can vary between environments. These differences may be due to disturbance frequency (for example, rolling shells are unfavourable for larval settlement of some taxa, but favourable for others), degree of exposure (buried shells are not encrusted as readily) and light intensity (upper surfaces of substrates collect different assemblages from undersides and surfaces inside caves).

Epibiont encrusters include both mineralized organisms that are preservable and non-preservable forms. The non-preservable fraction (soft algae, non-boring sponges, etc.) may have an impact on the mineralized forms and, thus, contribute to the signature of the environment and the community. Relationships between these different encrusters in submarine caves were studied by Rasmussen and Brett (1985), who documented the amount of information lost from the encrusting community when the soft-bodied forms die and leave no body-fossil remains. They found that some mineralized encrusters are dissolved by later-stage soft-bodied forms, so that the preserved community is biased toward the earlier stages of encrustation. They concluded that mineralized epibiont encrusters inside caves are commonly destroyed by the erosive action of the sponges. On more exposed surfaces where these types of sponge are less common, encrusting organisms are potentially better preserved.

Methods for quantifying epibiont coverage range from simple presence/absence to more accurate measurements of area and diversity/abundance of encrusting species. Most commonly, either presence/absence of epibionts is recorded (Frey and Howard, 1986; Meldahl and Flessa, 1990) or abundance is estimated on a scale of absent, low, moderate and high encrustation (Fürsich and Flessa, 1987). In a somewhat more detailed study, Driscoll (1970) counted species of encrusters on bivalves and compared these numbers among shells using normalized surface areas (Driscoll, 1968). Parsons (1989) estimated percentage coverage of shells by epibionts using a comparison chart (Pichon, 1978) similar to those used in petrology for estimating percentages of minerals in thin-sections. Parsons also noted presence/absence of individual epibiont taxa (for example, four types of foraminifers, two serpulids, algae, etc.) on shells. Rasmussen and Brett (1985) obtained detailed information on coverage by photographing the substrate surface and then digitizing the area covered by epibionts. This method resulted in an accurate, yet time-consuming, measure of surface area encrusted.

The information gained by these methods has been interpreted in relation to a number of environmental variables. Meldahl and Flessa (1990) found that low-energy environments favoured encrustation, while in high-energy settings (that is, in shallow subtidal and lower intertidal settings) shells displayed fewer incidences of encrustation. Parsons (unpublished data) found that encrustation was correlated more with exposure than with energy in reef-related carbonate systems. Shells found on hardgrounds (reefs, patch reefs, pavements) were highly encrusted, but the high-energy beach environment showed less encrustation due either to the inability of encrusters to settle on shells that are constantly being buried and exhumed, or to high abrasion that could wear epibionts off the shells very quickly. Low-energy grass beds,

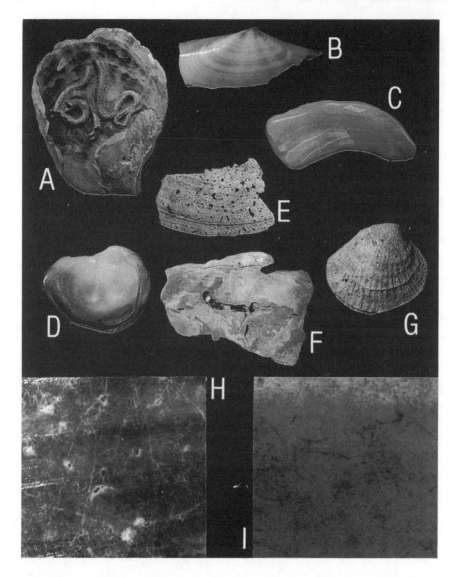

Figure 2.1 Examples of the taphonomic characteristics of molluscs. (A) Encrustation of *Isognomon alatus* by serpulid worms and a barnacle. (B) Fragmentation; fresh and sharp breaks on *Tellina radiata*. (C) Fragmentation; worn breaks on *Codakia orbicularis*. (D) Abrasion on *Chama sarda* collected from a beach. (E) Pitting, corrosion and borings on an unidentified fragment. (F) Dissolution; boreholes enlarged by dissolution on the oyster *Isognomon alatus*. (G) Bioerosion, corrosion and abrasion on *Chione cancellata*; note pitted surface and loss of sculpture. (H) Bioerosion; unidentified microborings on *Bulla striata*. (I) Bioerosion; unidentified microborings on interior of *Chione cancellata*. A is × 1, B–D × 1·5, E–G × 2, and H and I × 18.

however, display high levels of encrustation on epifaunal gastropods (exposed) and low encrustation on infaunal bivalves (not exposed), while the environment as a whole is low-energy. Driscoll (1968) related the amount of encrustation to differential burial rates of shells placed in 10 m of water off of Buzzards Bay, Massachusetts. Often the degree of encrustation is controlled by an interaction between energy and exposure. For example, hardgrounds are areas of high rates of encrustation and are often high-energy areas with waves or currents keeping them swept clean of sand. Likewise, the experimental situation of Driscoll was both exposure- and energy-related due to the effects of migrating sand ripples burying the shells.

Fragmentation
The degree to which a skeleton is broken reflects the amount of physical energy in any given environment. Breakage usually occurs as a result of hard impacts with other skeletons or pebbles, although biological factors may also cause fragmentation (for example, crabs, Trewin and Welsh, 1976; birds, Cadée, 1989; Meldahl and Flessa, 1990). Fragmentation tends to be highest on beaches, in active tidal channels and on reef crests where water turbulence is high and grain sizes are large. Less fragmentation occurs in subtidal environments and in restricted lagoons.

Breaks tend to follow lines of weakness such as changes in skeletal structure and mineralogy, lines of ornamentation such as ribs and growth lines on molluscs, or sharp changes in convexity (see Müller, 1979). Gastropods tend to break at the aperture and in high-spired forms the body whorl separates from the spire. Colonial organisms, such as corals, break at branch points and at lobe intersections depending on colony shape and construction. Size, thickness, ornamentation and mineralogy of the skeleton can all contribute to the resistance to breakage. Because of the differential thresholds of fragmentation in different kinds of skeleton, it is preferable to use a single species or genus that is found across all environments for taphofacies analysis.

Methods for measuring fragmentation vary widely from simply counting numbers of broken and whole skeletons to much more elaborate techniques. Davies *et al.* (1989a) subdivided breakage into major (more than 20 per cent of skeleton broken away) and minor; a similar method was used by Parsons (1989). Fürsich and Flessa (1987) used a ratio of complete shells to fragments, which is valid if encrustation and/or abrasion is either equal for whole shells and fragments or else is an infrequent occurrence.

Fragmentation is clearly a measure of environmental energy, with the exception of biologically mediated breakage (predation and bioturbation). The condition of broken surfaces (Figures 2.1B and 2.1C) as they are exposed to wear, corrosion and encrustation (Davies *et al.*, 1990) can also give a rough estimate of the time elapsed since breakage. Breakage followed by immediate burial, however, stops the clock and preserves the sharp breaks.

Abrasion
Abrasion of skeletons occurs when they are exposed to moving particles or when the skeletons themselves are moved relative to other particles. In

general, intense abrasion is associated with nearshore wave (Figure 2.1D), current or tidal action. Vigorous movement of skeletons, or sands past skeletons, requires a high-energy setting. The most common effect of abrasive action on mollusc shells is a loss of surface ornamentation. Fine ornamentation is lost in the first few hours of exposure to strong abrasive action, while coarse ornamentation may be only moderately affected for long periods of time. In high-energy areas (for example, in the surf zone) abrasive smoothing occurs within a short span of time after death of the molluscs. In low-energy areas (such as in subtidal zones) the process occurs much more slowly.

In molluscs, other factors have also been shown to be important. For example, the rapidity and intensity of abrasion within any given energy regime is dependent on the surface area per unit weight of the shell (Driscoll, 1967b). Shells with a small surface area to weight ratio are more durable than those with larger surface area per unit weight. Grain size of surrounding sediments is also extremely important in determining shell durability with respect to abrasion. The larger the grain size, or the lower the sorting, the faster and more intense the abrasive effect (Driscoll and Weltin, 1973). Driscoll also noted that abrasion differentially affects various areas on valve surfaces. The umbo is most strongly affected, followed by the postero-dorsal margin and the post-umbonal slope.

Müller (1979) discussed abrasion in terms of ambient polishing (general surface abrasion) and faceting, which he described as localized abrasion due to differential wear on the skeleton. Ambient polishing affects the periostracum and fine details of the shell first, after which the valves of clams become separated and hinge teeth are worn down, as are the edges of the shell. For the polishing medium, angular grains are more effective than rounded grains and fine grains are more effective than coarse, which tend to move more slowly in the water.

Faceting is subdivided into anchor facets, roll facets and glide facets. Anchor facets occur when the skeleton is firmly anchored to the substrate, causing selective abrasion of the exposed surface. Roll facets are caused by strong fluctuating currents in coarse-grained sands such as tides, strong longshore currents or swash zones. Roll facets in bivalves usually begin with abrasion of the umbo ('umbo-faceting'). As abrasion continues the worn area works its way from the hinge line out following growth lines. This eventually results in the shell having a horseshoe-shaped appearance. Glide facets occur when skeletons move over abrasive substrates. Cone-shaped shells will glide with the wider, apertural end toward the substrate which eventually wears away the lower side, leaving only the apex or spire. A special form of glide faceting occurs in hermited (pagurized) gastropod shells. Dragging of the shell over the substrate leaves distinctive marks on or near the aperture (Walker, 1988; 1989).

Abrasion has been measured qualitatively in modern specimens using presence/absence (Meldahl and Flessa, 1990) or a scale of no, minor and major abrasion (Davies et al., 1989a; Parsons, 1989). To assess abrasion quantitatively would require detailed microscope work or preferably scanning electron microscope examination in order clearly to separate mechanical abrasion from the effects of chemical dissolution and microborings by algae.

Macroscopically, cues such as smoothing of ornamentation and localized faceting help to separate abrasion from dissolution and bioerosion.

Dissolution
Skeletal hard parts are often subjected to dissolution both prior to and after burial. Dissolution usually appears as general pitting and corrosion of the skeletal surface (Figure 2.1E). Evidence for dissolution in exposed samples generally implies conditions of high salinity, low temperature or active bioturbation. Tropical marine waters are normally supersaturated with respect to calcium carbonate, which facilitates its precipitation by organisms. Alexandersson (1978) has found areas in the North Sea where waters are undersaturated with calcium carbonate; similar areas are known in the Baltic Sea. Open-marine systems in tropical seas are oversaturated with respect to calcium carbonate and therefore other factors must be controlling dissolution. Fluctuations in pH, pCO_2 and temperature can all cause dissolution. These changes probably occur in the microhabitat surrounding skeletons on the sea floor or within surrounding sediments, although this has been under debate (Alexandersson, 1978). In the TAZ, interstitial water is presumably undersaturated with respect to calcite and aragonite. Sulphuric acid can be produced by the degradation of organic carbon and the oxidation of hydrogen sulphide at the oxic–anoxic interface. Bioturbation of the upper 10–20 cm mixes the acid into the sediment, and even enhances acid production by increasing sulphate supply and interstitial water exchange (Aller, 1982; Davies *et al.*, 1989b).

Siliceous skeletons (for example, radiolarians and sponge spicules) are easily dissolved because the oceans are undersaturated with respect to silica. As temperatures decrease (that is, with depth) silica dissolves more slowly, allowing the preservation of siliceous skeletons in deep-oceanic sediments. Little is known about the diagenesis of phosphatic skeletons in marine environments. The oceans may be saturated with respect to dahlite (carbonate apatite) and phosphate minerals are common, but there have been few studies of their susceptibility to dissolution (Dodd and Stanton, 1990). Fossil evidence points to the resistance of phosphate minerals to degradation because phosphate pebbles and steinkerns are often the only evidence remaining in highly condensed and reworked lag deposits on ancient bedding planes (see Martill, this volume).

Flessa and Brown (1983) experimentally addressed the controls of skeletal mineralogy on dissolution. They exposed different skeletons to acid baths and found that calcitic hard parts with a high magnesium carbonate content (common in echinoderms) were the most soluble, followed in descending order by aragonitic (for example, many molluscs) and low magnesium calcitic hard parts (for example, oysters and brachiopods). In addition, they found that skeletons with high surface area to weight ratios dissolve more rapidly than those with low surface area to weight ratios such as thick and compact clam shells. Peterson (1976) came to similar conclusions using field burial experiments.

Skeletal porosity and architecture can also affect the rate of dissolution because they increase the effective surface area. Although barnacles are made of calcite and have a low surface area to weight ratio, they dissolve quickly because of their porosity. Secondarily bored skeletons would also be expected to dissolve more rapidly due to increased surface area created by the borings (Figure 2.1F).

Chemical dissolution is often recognized on skeletons by a generally chalky appearance which occurs very early in the process of dissolution. Many aragonitic shells lose their colour and lustre first, and then become chalky. In contrast, however, oysters become more lustrous due to progressive slough-ing of foliar calcitic layers to expose fresh calcite.

Surface etching during dissolution can result in accentuation of ornamen-tation, whereas abrasion subdues ornament. This contrast may prove to be a good guide for separating abrasion from dissolution in hand specimens. As dissolution advances, skeletons become thinner and more fragile, followed by perforation of the skeleton. In the laboratory experiments of Flessa and Brown (1983), perforations appeared first in the muscle scar region on the interior of bivalve shells. Under field conditions, however, Davies *et al.* (1989b) found that dissolution attacked the outer surfaces first. In bored shells, holes were made larger and previously unnoticed holes showed up as a result of dissolution.

Like other taphonomic indices, dissolution has been measured in different ways by different investigators in various settings. Davies *et al.* (1989b) measured the amount of dissolution on their specimens by noting whether they were chalky, had minor or major pitting, or if other evidence of major corrosion (that is, dissolved holes) was present. They then calculated a ratio of minor to major dissolution for whole shells and fragments, and for different localities. Parsons (1989) and Meldahl and Flessa (1990) simply noted the presence/absence of evidence for dissolution (pitting, chalkiness, etc.). Parsons also noted the amount of colour loss, which is a useful parameter in modern samples to describe the early stages of dissolution.

In all but the clearest examples (usually in modern environments), it is extremely difficult to differentiate between physical corrosion and microbor-ings. For example, many carbonate skeletons are constructed with organic matter in the crystal lattice that acts as a binding agent for carbonate crystals. As the organic material chemically decays, often with the help of bacteria, the crystals are released from the skeleton (Swinchatt, 1965). The result of this activity closely resembles dissolution, but the process actually involves bac-terial breakdown of the skeleton.

Corrosion and abrasion can also be difficult to distinguish. Nevertheless, there are several clues that can be used in some cases to distinguish dissolu-tion from abrasion. The accentuation of shell sculpture during dissolution mentioned above is important, but is most useful when dissolution is fairly active. The surface appearance of an abraded shell will appear glossy and smooth while a dissolving shell will be pitted and chalky. The shell margin will be rounded by abrasion, but sharpened and thinned by dissolution.

Bioerosion
Given the productivity of modern marine systems, we would expect thick deposits of recognizable skeletal remains. Instead of thick, continuous shell

deposits, we find numerous, thin shell beds, or widely scattered shells on the sea floor. In reefs, instead of thick accumulations of upright coral heads, we find degraded rubble piles (Hubbard *et al.*, 1990). This is related to the fact that the great amount of carbonate produced is continually being destroyed by other organisms. The term for this process, bioerosion, was introduced fairly recently by Neuman (1966), although the concept has been recognized for some time. The most recent reviews of the subject in reefs can be found in Hutchings (1986). The controlling factors limiting the intensity of bioerosion in any habitat are the presence of bioeroding organisms and suitable substrates. Bioerosion has also been shown to be proportional to nutrients (Moore and Shedd, 1977).

Bioerosion is the result of a search for food (grazers and predatory boring molluscs) and shelter (endoliths; Figure 2.1I). Grazing organisms are most often feeding on algae that grow on the surfaces of skeletons, just underneath the outer surface (microboring algae; Figure 2.1H) or within the living tissues of organisms such as corals. Other foods the grazers feed on are sponges and fungi. Grazers include various fishes (scarids and acanthurids), sea urchins, molluscs (gastropods and polyplacophorans) and polychaetes. The fishes graze primarily on larger skeletons such as corals. They grind the surface of both living and dead coral surfaces with their boney mouth plates leaving characteristic scrape marks behind. They ingest chunks of the skeleton as well as the algae and living tissue. While the plant matter, and to a lesser extent the coral tissue, are digested, the carbonate is passed through the digestive system and excreted as sand-sized particles. Sea urchins scrape algae from primarily dead skeletal surfaces and other carbonate substrates, such as beach rock and submerged limestones. The sea urchins scrape using the Aristotle's lantern which has a set of five teeth made of calcite. Some sea urchins are mobile grazers (for example, *Diadema*) while others (such as *Echinometra*) graze in one place, resulting in a protective cave that could be considered a borehole. Grazing molluscs include many archeogastropods and chitons. Molluscan grazers use a radula to rasp at almost any hard substrate on which algae has grown. Radular scrape marks are difficult to distinguish macroscopically, but they can be recognized under SEM. Chitons scrape at algae-covered substrates and leave a visible trace. The 'tooth' marks of chitons consist of bundles of parallel scrape marks about 5 mm in length and about 0.06 mm apart (Boekschoten, 1966).

Unfortunately, the action of most grazing organisms is to grind away the surface of a skeleton. Many grazers leave no recognizable traces and, more importantly, they erase evidence of earlier bioeroders and other taphonomic features. There is no way to estimate the amount of material lost in many skeletal remains, especially colonial skeletons which have no predictable size or shape. However, when grazing is recognized, it points to the presence of a food source for the grazer, which is often algae, and thus can indicate presence of sufficient light. Sometimes the grazing organism can be identified from the traces it leaves on the substrate.

While boring organisms are responsible for lesser amounts of bioerosion than grazers, they are nevertheless very important in the process of degrading original skeletal material. Unlike grazers, the record left behind is well preserved and easily distinguished. The principle borers (volumetrically) are the demosponges. Algal microborings are also common and possibly more widespread, although they do less damage. Members of the class Polychaeta and the phylum Mollusca also bore into other skeletons, albeit on a lesser scale.

The boring sponges (family Clionidae) are responsible for 95 per cent of boring in corals (MacGreachy, 1977) and are destructive agents on molluscs, barnacles and other carbonate skeletons (Neumann, 1966). The sponge begins the boring process by dissolving around a characteristically shaped chip that is undercut and freed, and eventually passed out through the excurrent canal (Fütterer, 1974). Clionid boreholes are 0.06–1.6 mm in diameter and the holes can be aligned in rows, but most often are randomly spaced on the substrate. Single sponges can overtake large coral heads as well as very small mollusc shells. The sponge establishes itself first by forming subsurface galleries, then it overgrows the surface of the skeletal substrate, finally extending beyond the original substrate and attacking surrounding surfaces (Driscoll, 1967a). *Cliona* are found in world-wide tropical seas and their borings have been identified at least as far back as the Cretaceous.

Polychaetes are important boring organisms in larger skeletons such as corals, as are boring bivalves such as *Lithophaga* and *Gastrochaena*. *Lithophaga* leaves a characteristic borehole and often their shells are left in the borings to be preserved. These borers are important mainly because of their size. The large-diameter holes cause rapid weakening of the host skeleton (usually corals) and contribute to the toppling of large colonies. Polychaetes, bivalves and clionid sponge borings together exert a major control on reef-accretion rates in tropical seas.

On a much smaller scale, the algae bore into carbonate substrates in many environments. Endothilic algae made up the highest proportion of biomass of any other group on the reefs at Eniwetok Atoll (Odum and Odum, 1955) and this probably holds true for many tropical carbonate systems. Algal borings range from fine, branched networks to single, pinpoint holes. Blue-greens are the most prevalent boring algae followed by the greens (Perkins and Tsentas, 1976). They are commonly found boring into substrates in high-energy areas where the substrate will collect less sediment (Boekschoten, 1966). Boring algae can so degrade the outer surface of molluscs that they create a zone of mud-sized matrix surrounding fossil shells viewed in thin-section (the 'micrite envelope'). This is a useful indicator of exposure of the shell to light at the sediment–water interface prior to final burial.

Perkins and Tsentas (1976) experimented with rates of algal infestation and found the substrates (both organic and inorganic calcium carbonate) that they had planted at shelf sites on St Croix, US Virgin Islands, were bored within nine days. They were heavily infested within two to four months. Thus, algal borings are a good indicator of the photic zone (algae photosynthesize), orientation (more common on 'up' sides exposed to light) and energy. They are quick to inhabit substrates, so these signatures can be found on many skeletal remains. Their only drawback is that they require SEM analysis for proper identification. They also are easily confused with fungal borings which do not require light for growth.

Several organisms erode skeletal substrates by chemical means. For example, some non-boring demosponges corrode the underlying substrate, as mentioned above, and the sea grass *Thalassia* will leave white linear traces of dissolution where the rhizomes have come into contact with buried shells.

Bioerosion leaves a useful palaeoecologic mark in the form of trace fossils on skeletal remains. Many non-preservable organisms can be shown to have been present in a fossil assemblage. There is much valuable palaeoecological information to be gained by identifying the traces and knowing something about the habits of similar modern taxa. Unfortunately, the net result of bioerosion is a wholesale loss of skeletal material, and therefore much potentially useful palaeoecological information is lost as well.

Methods of documenting bioerosion in taphofacies analysis include noting the presence/absence of traces (*Cliona* traces, gastropod boreholes, rhizome etchings and scrape marks made by polychaetes, echinoderms and fishes). Algal microborings are handled by noting no, minor (1–25 per cent), moderate (25–75 per cent) and major (75–100 per cent) infestation (Fürsich and Flessa, 1987). Davies *et al.* (1990) measured 'biological interaction' on their shells which includes bioaddition (including encrustation) as well as bioerosion. They counted shells as either biologically altered or unaltered. Several other authors have also used presence/absence of evidence of bioerosion as an index (Meldahl and Flessa, 1990; Frey and Howard, 1986).

Corrasion
The term 'corrasion' was introduced by Brett and Baird (1986) to describe general degradation of skeletal surfaces when it is too difficult or time-consuming to attempt to separate out the many possible contributors to dissolution, bioerosion (collectively considered as 'corrosion') and abrasion. Without SEM analysis it is often difficult or impossible to separate dissolution from extensive algal borings, for example. Also, in most settings all three of these degrading activities occur together, thus confusing the matter even further.

Whenever possible, an attempt should be made to separate the genetic origin of the markings because the taphonomic information can be very valuable. However, when samples are too degraded to permit specific identification of abrasion or corrosion, general surface degradation can still be recognized and thus corrasion can be a useful indicator of the overall condition of preservation. The term 'corrasion' is useful in fossil specimens where time and contact with pore waters may have made detailed surface analysis even more difficult.

Corrasion is a fairly recent addition to the pool of taphonomic indices and thus has not been extensively used in modern studies. A corrasion index would best be measured semi-quantitatively (that is, 0, 1–25, 25–50, 50–75 and over 75 per cent).

Articulation
Articulation of multi-element skeletons is a valuable indicator of the time of exposure since death or of the energy of the depositional environment. Multi-element skeletons that are found articulated as fossils almost always

indicate rapid burial. Echinoderms and arthropods are easily disarticulated after death. They have been shown experimentally to degrade within hours after death of the organism (Greenstein, 1989; Allison, 1988; Kidwell and Baumiller, 1990; Plotnick, 1986). In general, bivalves, with their tough ligament, can hold together much longer, but they tend to disarticulate under high-energy conditions. However, observations of thick assemblages of articulated shells on the beaches of southern St Croix demonstrate that rapid burial can preserve significant numbers of articulated shells even in high-energy settings.

Once disarticulated, bivalves may undergo valve sorting. The two valves often have different hydrodynamic properties and thus are transported differentially (Frey and Henderson, 1987). Therefore, it can be useful to note right/left valve ratios in the fossil assemblage. A significant concentration of one valve type indicates uneven hydrodynamic properties such as beaches (the swash zone) or areas with tidal influence.

Articulated remains are uncommon in most modern environments studied by the authors, but they can be very important locally. For example, in our own studies we have noted whether or not bivalves were articulated with a ligament. Davies *et al.* (1990) went further. They subdivided articulation into articulated with ligament, interlocked by the dentition, and paired (in articulated position, but not attached). The latter two categories require examination in place or the valves will become separated.

Orientation

Skeletons behave as sedimentary particles as soon as the organism dies. Waves and currents will transport and reorientate particles relative to their hydrodynamic properties. Nagle (1967) determined that mollusc shells orientate differently under current-dominated systems than under wave-influenced systems. He observed shells in flumes and wave tanks, and found that shells generally orientate long-axis parallel to current directions and perpendicular to waves. Concave-up and -down positions of bivalved shells (Brachiopoda and Bivalvia) record relative current velocity. Shells deposited in areas with strong currents, wave or migrating ripples (Clifton and Boggs, 1970) tend to rest concave-down, while those in quiet water (deep ocean, protected bays) tend to rest concave-up (Emery, 1968).

Other measures of orientation are also useful. For example, recognizing skeletons in life position almost always signals rapid burial. Measurements of absolute azimuthal orientation can help reconstruct direction of transport, but it must be remembered that only the last current direction before burial is recorded, which may or may not be an average current direction. Nevertheless, shell orientation can be an important tool for reconstructing wave and current properties of ancient environments and thus should be a part of modern taphonomic studies.

Orientation can either be the life position of the shells, or it can be due to post-mortem reorientation by fluid flow or biological reworking. Skeletons should be noted as being in life position, their inclination to bedding or sediment–water interface and the azimuthal orientation of a long axis or other appropriate axis depending on the organism. These measurements can only be taken *in situ* and thus subsurface measurements must be taken from carefully extracted cores (see Davies *et al.*, 1989a).

TAPHOFACIES ANALYSIS

Taphofacies, or groupings of assemblages based on similar taphonomic histories (Speyer and Brett, 1986), can be valuable in the interpretation of ancient depositional environments. Several workers have begun studying taphofacies in modern environments. Such studies include those of Parsons (1989) in reef-related carbonate environments on St Croix; Meldahl and Flessa (1990) in intertidal to shallow shelf siliciclastic environments from Cape Cod, Massachusetts; Fürsich and Flessa (1987) in a tidal flat in Bahia la Choya (Sonora, Mexico); Frey and Howard (1986) in an offshore, shelf setting on Sapelo Island, Georgia; Davies *et al.* (1989a) in a hurricane-influenced inlet on the Texas coast; and Staff and Powell (1990a) comparing the above inlet (Davies *et al.*, 1989a) to the inner continental shelf on the Texas coast. These studies provide insights into the origin of taphonomic features of individual skeletons such as abrasion, fragmentation, bioerosion, dissolution/corrosion, articulation and encrustation, as well as assemblage-level properties including size frequency, live/dead ratios and orientation. When these measurable characteristics are analysed for modern assemblages, a taphofacies characteristic of particular depositional environments can be defined semi-quantitatively.

These studies use a variety of sampling methods, including surficial box cores, scrape samples and suction dredging (airlifting) to obtain samples of only living or recently dead and little-transported material. This is important because there should be confidence that the specimens collected from a particular environment have actually acquired a taphonomic signature for that environment and have not been transported from another locality or been dredged from a deeper level that may have accumulated at a time when local conditions were significantly different. It has been demonstrated (Davies *et al.*, 1989a; Staff and Powell 1990a; 1990b) that skeletal fragments as well as whole and identifiable skeletons should be included in taphonomic studies. The fragments may be more sensitive to taphonomic processes and thus record useful, and often different, information than whole or only slightly broken skeletons.

Associated measurements and observations also provide valuable information about the environment from which the samples were collected. Sediment samples should be taken from the same sampling sites to gain possible insight into current and wave conditions, and possibly provenance of sediments. Ambient oceanographic conditions are measured to define the environments based on criteria independent of faunal and taphonomic properties of associated biota. Measurements include current, tidal range, depth, wave heights, wave period, wave direction and distance from shoreline. It is also useful to know storm frequency and, when possible, storm effects on bottom conditions (that is, dominant transport direction, burial depths, etc.). Associated organisms that are not preservable or rarely preserved are also measured as they can influence the preservable fraction in many ways. For example, density of surface vegetation can influence infaunal density and diversity. Further, the soft-bodied fraction of a marine community can leave recognizable traces in the surrounding sediments and also on hard-part remains (for example, sponge borings, rhizome traces and algal microborings).

One of the most important aspects of modern studies is that they can establish a reference signature or taphofacies for given well-characterized environments if it can be demonstrated that skeletons are long-term residents of that environment. Furthermore, the physical and biological processes responsible for various taphonomic features are potentially observable directly and can be used to develop process-response models of taphofacies comparable to models established for lithofacies.

Modern taphofacies studies

There has been increasing interest over the past decade in quantifying taphonomic signatures in modern depositional systems. While each study has added its own insight, they have all shared the common objective of identifying subtle cues or gradients that would be difficult or impossible to notice if approached solely from an ancient (fossil) perspective. The luxury of readily observable two- or three-dimensional patterns in a present-day system has allowed investigators to quantify not only subtle taphonomic variations, but also the energy regime, degree of exposure, substrate stability, water chemistry, light, and associated flora and fauna that have influenced them.

In a study on St Croix, Parsons (1989) and Parsons and Brett (research in progress) established that taphonomic features of mollusc shells can be reliable indicators of environmental energy and substrate in a tropical reef and lagoon system (Figure 2.2). This study measured all of the previously discussed taphonomic properties of shells and shell fragments, whether identifiable to species or not, and showed that different environments impart specific taphonomic signatures to the molluscs. While environments with similar hardground substrates (forereef, backreef and patch reef) were not distinguishable from one another, they were easily separated as a group from lagoons, mud bottoms, beaches, etc. Lagoon *Thalassia* beds, lagoon *Callianassa* mounds, mud bottom, beach and bare sand areas all had distinct taphonomic signatures. These signatures indicate essentially *in situ* acquisition of taphonomic characteristics, because transport in this reef and lagoon system appears to be minimal (except during major storms). In the St Croix study (Parsons, 1989) sandy environments seaward of the forereef slope, in the lagoon and on the beach differed in type and density of sea grass, intensity of bioturbation and energy. These different sites are sedimentologically similar, so they would be difficult to distinguish in the fossil record. Because the molluscs showed significantly different taphonomic signatures due to differences in energy, bioturbation and algal or grass cover, preservational characteristics could provide clues to distinguish between these otherwise similar environments in fossil samples where sedimentologic or taxonomic information could not.

This study showed that substrate type, sedimentation rate and energy are important factors controlling the taphonomy of molluscs in carbonate environments. The substrate, in part, dictates the length of time of exposure a shell will have from the death of the organism until it becomes buried or destroyed. Sedimentation rate also influences the time of exposure, and energy levels affect abrasion levels and burial rates. This type of study acts as a basic reference for taphofacies in a modern system. Furthermore, because taphofacies cross-cut taxonomic and physical boundaries, taphonomic characteristics provide an indicator of physical and associated biological environments independent of sedimentology and systematics. This identification of a suite of taphonomic characteristics from related environments potentially provides information on the physical and biological properties of a physical system that could not be otherwise obtained.

Studies which also analyse assemblages according to biofacies, sedimentological facies, and ecological zones provide a comparison of taphofacies to other physical and biological zonation systems. Meldahl and Flessa (1990; Figure 2.3 herein) collected samples from eight different shallow shelf environments in Provincetown Harbour, Cape Cod, Massachusetts, and defined six different biofacies using cluster analysis. Diversity and trophic data provided additional information on environmental zonation. Meldahl and Flessa concluded that their biofacies were controlled primarily by tidal exposure time. They also noted that taphonomic effects on shells in the low-energy areas were dominated by encrustation and bioerosion, while abrasion was the dominant factor in high-energy zones. They further concluded that the order of acquisition of taphonomic features (their 'taphonomic pathways') varied between high-energy and low-energy environments. In general, this study concluded that the taphofacies pattern observed is due primarily to energy variations which control the amount of shell stability and reworking and, secondarily, that facies patterns are influenced by substrate conditions, exposure time and shell transport. Most importantly, however, it points out that taphofacies can provide information separate from biofacies or ecology because there are different controlling factors for each type of zonation.

Fürsich and Flessa (1987; Figure 2.4 herein) also used an integrated approach to taphofacies analysis of molluscan assemblages from a tidal flat in the northern Gulf of California. They measured species composition, relative abundance, species richness, specimen size, percentage articulation (bivalves), types and percentages of bored and encrusted shells, and preservation quality. They established standards for shells ranging from pristine to very degraded for each of seven taphonomic features: abrasion; loss of lustre; loss of colour; fragmentation; bioerosion; encrustation; and dissolution. Specimens were assigned points according to their preservation class which were summed to give a measure of the quality of preservation for each shell.

Fürsich and Flessa found that larger shells were more often found in their own life habitat than were small ones. The distribution of live versus dead taxa showed very good agreement from environment to environment among most species. Species zonation closely followed substrate patterns and only

	ABRASION	BIOEROSION	ENCRUSTATION	FRAGMENTATION	EDGE ROUNDING	COLOR LOSS	RHIZOME ETCHINGS	ARTICULATION
FOREREEF	HIGH	HIGH	HIGH	HIGH	HIGH	HIGH	NONE	NONE
BACKREEF	HIGH	LOW	HIGH	HIGH	MOD	HIGH	NONE	NONE
PATCH REEF	HIGH	LOW	HIGH	HIGH	MOD	MOD	NONE	NONE
GRASS BED	MOD	LOW	MOD	MOD	MOD	MOD	HIGH	LOW
BIOTURBATED SAND	LOW	NONE	LOW	MOD	LOW	LOW	LOW	NONE
MUD	NONE	NONE	NONE	MOD	NONE	LOW	NONE	LOW
BEACH	LOW	HIGH	LOW	HIGH	HIGH	MOD	NONE	NONE

Figure 2.2 Summary of the taphonomic characteristics of molluscs of the carbonate reef-related environments of St Croix, US Virgin Islands. Note that no two environments (read horizontally) have the same taphonomic signature. Abrasion, bioerosion, fragmentation and colour loss were measured on a scale of high, medium, low and absent. A designation of 'high' means that the average of the shells in the forereef samples was high for those characteristics. Encrustation was measured as a percentage of surface area covered, estimated to the nearest 10 per cent. The remainder were noted as present or absent.

rarely were the distributions of dead fauna obscured by post-mortem transport. The distribution of the taphocoenosis was very close to the distribution of the living communities.

Boring and encrustation was found to correlate with the substrate and to the rate of burial. A general increase in the quality of preservation across the tidal flats was related to residence time on the sea floor and frequency of reworking events. As energy decreased towards shore, the probability that shells would be exhumed and reworked went down. Thus, better preservation inferred lower-energy regimes.

Frey and Howard (1986) studied mollusc assemblages on the offshore shelf near Sapelo Island, Georgia, USA. They designated shells as either 'new' or 'old' on the basis of loss of colour, gloss and ligament, and having a bleached or chalky appearance. They also quantified the degree of boring and encrustation.

From these data they were able to assess the provenance of the shell assemblages and concluded that in general their samples were not transported. Both Holocene and recently exhumed Pleistocene shells became encrusted when they were exposed for at least one season of larval settlement and metamorphosis of encrusting species. Therefore, intensity of boring and encrustation was related more closely to the length of time a shell was exposed at the sediment–water interface than to its age. However, older shells tended to be more bored and encrusted than younger ones because they would have had more chances to be exposed for at least one season of larval settlement which initiates the process of boring and encrusting. Old shells were most important at the landward and seaward ends of the transect, presumably because the sedimentation rate is slower there, leading to a greater exposure time.

The study of assemblages buried *in situ* can provide information on highly preservable material that is not too far removed from surface conditions as would be true for fossil assemblages. Davies *et al.* (1989a) documented the taphonomic signature of shells in 75 mm diameter vibrocores (1–2 m long) taken from a microtidal inlet on the Texas coast. The majority of skeletons were bivalves and the researchers ignored the occasional arthropod or echinoderm remains. This study encompassed a large number of taphonomic features as well as taxonomy, sedimentology (grain size and orientation of skeletal particles), shell provenance (habitat of origin) and size-frequency distributions.

The cores were divided into shell beds and intervening non-shelly intervals. The different grades of accumulation (shell-poor sands, slightly shelly sands, shell-rich sands and shell gravels) were felt to represent the differences in physical processes that formed the accumulations. Shells were assigned to original habitat groups with a few species falling into the 'unrestricted habitat' category (for example, *Mulinia lateralis*, which was the most common shell in all the cores regardless of environment). By comparing the taphonomic signature of the shells in shell beds to the shells found scattered in the 'background' sediments, Davies *et al.* identified taphonomic differences between skeletons found within shell beds and intervening, and presumably non-transported, skeletons. The underlying goal was to determine the taphonomic signature, if any, of transport.

In their study, the investigators examined breakage, abrasion, dissolution, size sorting, edge rounding, biotic interactions and articulations of shells. They concluded that the habitat of origin was the most important factor in determining the taphonomic signature of the shells (Figure 2.5). Depth in core and sediment type had lesser effects on the amount of dissolution, edge rounding, abrasion and shell size.

Davies *et al.* further concluded that the displaced shells do not obtain a depositional overprint through transport and accumulation in their final resting place. A storm would leave its signature in shell orientation and size-frequency distribution instead. Therefore, shells moved by a major storm would retain their original taphonomic characteristics and attempts to relate these to the final environment of deposition would yield misleading results.

Thus, care must be taken when analysing the taphonomy of fossil assemblages. The taphonomic signature on shells can be a reliable provenance indicator, but not a depositional indicator (excluding articulation and orientation, which are affected by transport). The treatment of fossil assemblages

	ABRASION	ALGAL BIOEROSION	CLIONA BIOEROSION	ALGAL ENCRUSTATION	BRYOZOAN ENCRUSTATION	BARNACLE ENCRUSTATION	BREAKAGE	CORROSION
SALT MARSH	LOW	MOD	LOW	LOW	NONE	MOD	MOD	HIGH
INNER FLATS	LOW	LOW	NONE	LOW	NONE	MOD	HIGH	HIGH
MIDDLE FLATS	MOD	LOW	NONE	LOW	NONE	NONE	HIGH	HIGH
OUTER FLATS	MOD	LOW	LOW	LOW	NONE	LOW	MOD	HIGH
SUBTIDAL (1m)	LOW	LOW	NONE	LOW	LOW	LOW	LOW	MOD
SUBTIDAL (4-10m)	LOW	MOD	LOW	LOW	LOW	MOD	MOD	HIGH
BEACH	HIGH	LOW	LOW	NONE	NONE	LOW	HIGH	MOD

Figure 2.3 Summary of the taphonomic characteristics of molluscs from Provincetown Harbour, Cape Cod. The environments include salt marsh through tidal flats to subtidal and beach on a terrigenous embayment. Note that taphonomic signatures are distinct for the different environments. Measurements represent percentages of valves (bivalves) or individuals (gastropods) having evidence of each of the characteristics. Fragmentation was based on 10 per cent or more of the shell missing. Data from Meldahl and Flessa (1990).

should, therefore, include a preliminary determination of transport or *in situ* preservation. Orientation of specimens should be determined before removal from the surrounding matrix. When major transport has been ruled out, then a taphofacies study can be confidently used to define biologic zones. If transport is evident, taphonomy could be recording the provenance of an assemblage, or multiple environments of origin for the skeletons.

Staff and Powell (1990a) compared the analysis of a microtidal inlet (Davies *et al.*, 1989a) to shells collected on the inner continental shelf (Figure 2.6). The purpose was first to establish that the different environments imparted unique taphonomic signatures for the different benthic habitats and, second, to determine if these signatures would survive transport, therefore allowing for determination of provenance for assemblages containing mixtures of local and transported shells.

Staff and Powell found that the inner shelf assemblages contained a mixture of well-preserved fresh and taphonomically altered shells. The inlet, on the other hand, contained almost exclusively taphonomically altered shells. They attributed the preservational differences between the two habitats to variations in water energy and to the temporal relationship between mortality and final burial. The inner shelf environment is dominated by burial due to storms and death within the sediment because of the infaunal habitat of many of the organisms that live there. Staff and Powell found that dissolution and unaltered breakage dominate the taphonomic signature of the inner shelf. In contrast, the inlet is a higher-energy environment with frequent reworking. The signature here is dominated by abrasion and edge rounding, but this environment is best described as a mixture of taphonomic signatures because of the large proportion of transported shells that make up the inlet assemblage.

The conclusion of this study is that the taphonomic signatures of shells from different environments were distinct. Also, the signature of the shelf-derived shells in the inlet was conservative in that the signature of the original habitat survived transport to the inlet. Processes in the inlet imparted their effects (for example, abrasion) onto the allochthonous shells, but the original signal was still recognizable. This result is encouraging because it implies that taphonomic signatures can be used to determine the provenance of transported shells and possibly will also prove useful in palaeoenvironmental studies.

The studies summarized above compare taphofacies characteristics among clearly defined environments. The assumption for comparative taphonomy is that skeletons acquire a signature that reflects the physical, chemical and biological characteristics of the environment, and that the controlling parameters are relatively uniform in any one environment. To test the validity of this assumption, Staff and Powell (1990b) examined within-environment variability of taphonomic signatures on molluscs from the shallow continental shelf of Texas. The sample sites differed little in water depth (15–22 m) or sedimentology and have been combined as replicates of one environment in an earlier study (Staff and Powell, 1990a).

The results showed that some taphonomic characteristics are as variable within the habitat as they were between habitats in the earlier Staff and Powell (1990a) study. For example, articulation in the inlet/beach (5 per cent) was not significantly different from the shelf environment (0 per cent), but it showed significant variation among the shelf sites (2–9 per cent). The within-habitat variations could possibly be linked to small changes in depth, sediment texture and perhaps frequency of storm reworking. An interesting result of their analysis was that large shells appeared to be more conservative in their taphonomic signature and thus more indicative of the overall geographic environment (shelf, in this case). On the other hand, small shells were more sensitive to taphonomic processes and thus recorded subtle changes within the shelf environment.

The significant result of these many varied studies of taphofacies in modern settings is that the taphonomic signature was unique for each subenvironment of the overall study area. Figures 2.2–2.6 summarize the overall findings of these studies in terms of taphonomic features. The methods, and often the goals, of these studies were different, but the underlying theme suggests that taphofacies may become a useful tool in palaeoenvironmental studies. Davies *et al.* (1990) have attempted to standardize the methods for determining taphofacies which would render the results of future studies more broadly applicable.

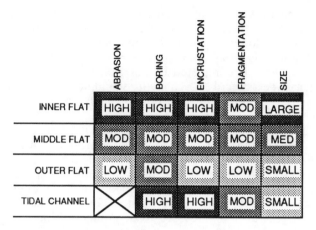

Figure 2.4 Summary of the taphonomic characteristics of tidal flat environments from the Gulf of California. Taphonomic signatures are distinct for each environment. Measurements of abrasion, boring intensity, encrustation and fragmentation were assigned points ranging from pristine to fully degraded. Data from Fürsich and Flessa (1987).

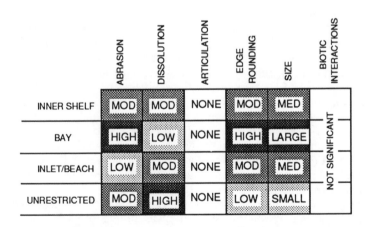

Figure 2.5 Summary of the taphonomic characteristics of molluscs derived from terrigenous environments off of the Texas coast. Note that no two signatures are alike. Abrasion was measured as none, minor or major. Dissolution was measured as none, minor, major or extreme. Edge rounding was classified as sharp, chipped, slightly worn or smooth. The summaries were based on the order of the least squares means. Data from Davies et al. (1989a).

A comparison of the findings of these studies allows some generalizations to be made about the relationship of the environment to the resulting taphonomic signature. High-energy areas such as the forereef and beach tend to be high in abrasion, fragmentation and edge rounding (Figures 2.2, 2.3, and 2.6) due to stronger wave energy and coarser grain size. Davies *et al.* (1989a; Figure 2.5 herein) demonstrated low abrasion in the inlet/beach environment, but this may be a low-energy beach, or perhaps the beach signature was overpowered by the general inlet taphonomic signature. Zones dominated by exposure (hardgrounds and reefs) were primarily influenced by biological processes (bioerosion, encrustation). Encrustation, bioerosion and borings dominate on the reef subenvironments (Figure 2.2). Algal bioerosion and encrustation by barnacles dominated the taphonomic signatures in areas of rocky exposure on the tidal flats of Cape Cod (Meldahl and Flessa, 1990; Figure 2.3). Degrees of corrosion and dissolution are caused by factors such as sediment type and interstitial pore-water chemistry. These vary greatly across environments and can only be useful locally; thus, the differences in dissolution in the above modern studies cannot be generalized.

The most useful indices from these taphofacies studies include size, abrasion, dissolution, edge rounding and encrustation (Figure 2.7). Of these characteristics, only dissolution is less preservable. Surface pitting, etching and chalkiness can be caused by later diagenetic processes, so using dissolution in a taphonomic analysis of fossil assemblages must be done with great care.

Modern taphofacies studies provide examples of how taphonomic indices or gradients can be quantitatively addressed in modern environments where: environment of deposition can be readily observed; subtleties of taphonomy can be identified and measured; and the two can be effectively related. In these ideal instances, quantitative information about modern depositional systems, biotic assemblages and taphonomic overprints can be applied in detail to their ancient counterparts. In areas where logistical considerations or poor outcrop exposure are limiting factors, however, more generalized schemes of taphonomic characterization can still be effectively employed.

Brandt (1989) developed a system of taphonomic 'grades' whereby a high-grade assemblage is characterized by a very well-preserved set of fossils and a low-grade assemblage represents poorly preserved fossils. She demonstrated this method on fossil assemblages, but it could prove useful for delineating modern taphofacies as well. In ancient systems, taphonomic grades are applied to individual beds or assemblages of fossils based on the observation of taphonomic features (for example, breakage, articulation, corrasion and orientation) as well as sedimentologic characteristics (such as percentage matrix, size sorting and percentage size grading). In a modern setting, taphonomic grades could be assigned to surface assemblages along an environmental gradient and then grouped by similarity to delineate taphofacies. All of the features listed above are as easily assessed in modern settings with the possible exception of percent matrix, which could be difficult to measure in unconsolidated sediments. The method is not as detailed as some of the studies discussed above, but it could serve as an excellent field method for preliminary examination of a modern area, or in instances where time is limited.

	ABRASION	BIOLOGICAL ALTERATION	FRAGMENTATION	EDGE ROUNDING	DISSOLUTION	SIZE	ARTICULATION	PERIOSTRACUM OR LIGAMENT
MICROTIDAL INLET	MOD	LOW	MOD	HIGH	LOW	SMALL	NONE	NONE
SHALLOW SHELF	NONE	LOW	MOD	LOW	LOW	SMALL	LOW	LOW

Figure 2.6 Summary of the taphonomic characteristics of molluscs from Texas coast environments. Note that the adjacent environments produce distinct taphonomic signatures. Abrasion and dissolution were measured as none, minor or major. Biological alteration was noted as altered or unaltered. Edge rounding was measured as in Davies *et al.* (1989a; see Figure 2.5). Data from Staff and Powell (1990).

SUMMARY

The concept of taphofacies has evolved as a result of the increased use by palaeontologists of the more positive aspects of taphonomy. It has long been understood that the processes of abrasion, dissolution, fragmentation, etc., gradually remove information from any fossil assemblage. In addition, however, these same processes leave evidence of their actions on remaining hard skeletons. Thus, the resulting assemblage contains a record of physical, chemical and often biological characteristics of their depositional environment or, as is evident from the above discussions, of the environment in which the skeletons remained exposed for the longest time. This information is invaluable to palaeoecological interpretations.

The existing methods of fossil-assemblage reconstruction are no less useful to palaeontologists. Sedimentary analysis is a proven method for establishing physical characteristics of ancient environments including wave, current and substrate characteristics. None the less, many facies, especially fine-grained sediments, contain few or no sedimentary structures or textural heterogeneities upon which to base interpretations. Biofacies are reliable indicators of trophic structure and ecology, as well as energy, chemical conditions and substrate types. Nevertheless, there are dangers in extrapolating the habitat preferences of modern species to their ancient ancestral counterparts (taxonomic uniformitarianism). Taphonomic characteristics add another method of analysis that is less dependent on taxonomic uniformitarianism and may also help enhance sedimentary analyses by gaining insight into the characteristics of the sediment, the presence of other non-preservable organisms (known by their boreholes, etch marks and other traces left on skeletal remains), water chemistry (dissolution), and by serving as an independent check on wave and current action (fragmentation and abrasion). There is also valuable information gained from the orientation of skeletons and from size sorting of biological particles. Therefore, the taphonomic information

	Carbonate reef-related facies[1]	Terrigenous inter-tidal and shallow shelf facies[2]	Terrigenous tidal flat facies[3]	Terrigenous off-shore shelf facies *[4]	Terrigenous storm-influenced inlet facies[5]	Terrigenous inner shelf & microtidal inlet facies[6]
SIZE	⊗	—	◉	—	●	◉
LIVE/DEAD	○	—	◉	—	—	—
ARTICULATION	⊗	—	○	○	○	⊗
Left/Right valve ratios	○	—	—	—	○	○
SURFACE ALTERATIONS	◉	⊗	◉	—	◉	●
Abrasion	◉	◉	↑ (combined)	—	◉	●
Dissolution	◉	○	(combined)	—	●	●
Color loss	◉	○	(combined)	●	—	—
Ligament/Periostracum	◉	—	—	●	⊗	◉
BREAKAGE	○	◉	○	—	⊗	⊗
Edge Rounding	●	—	—	—	●	●
BIOLOGICAL INTERACTIONS	◉	◉	◉	◉	◉	○
Encrustation	●	◉	●	◉	—	—
Bioerosion	◉	◉	⊗	◉	—	—
Cliona and other borings	◉	◉	⊗	◉	—	—
Gastropod Borings	⊗	◉	—	—	—	—
Rhizome Etchings	●	—	—	—	—	—
ORIENTATION	—	—	—	—	○	—

Legend:
● VERY USEFUL
◉ MODERATELY USEFUL
⊗ LESS USEFUL
○ NOT USEFUL
— INDEX NOT USED

Figure 2.7 Summary of taphonomic characteristics and their utility for describing taphofacies as determined in several modern studies. [1]Parsons and Brett (Figure 2.2 herein); [2]Meldahl and Flessa (1987); [3]Fürsich and Flessa (1987); [4]Frey and Howard (1986); [5]Davies *et al.* (1989); and [6]Staff and Powell (1990). *The study by Frey and Howard (1986) combined several indices to categorize the shells as 'new' and 'old'. Extracting the indices from these broad classifications may not be entirely accurate.

preserved in fossil assemblages within sedimentary facies provides the opportunity to subdivide zones more finely than could be accomplished based on sedimentology or palaeontology alone.

REFERENCES

Alexandersson, E.T., 1978, Destructive diagenesis of carbonate sediments in the eastern Skagerrak, North Sea, *Geology*, **6** (6): 324–7.

Aller, R.C., 1982, Carbonate dissolution in nearshore terrigenous muds: the role of physical and biological reworking, *Journal of Geology*, **90**(1): 79–95.

Allison, P.A., 1986, Soft-bodied animals in the fossil record: the role of decay in fragmentation during transport, *Geology*, **14**(12): 979–81.

Allison, P.A., 1988, The role of anoxia in the decay and mineralization of proteinaceous macro-fossils, *Paleobiology*, **14**(2): 139–54.

Behrensmeyer, A.K. and Kidwell, S.M., 1985, Taphonomy's contribution to paleobiology, *Paleobiology*, **11**(1): 105–19.

Boekschoten, G.J., 1966, Shell borings of sessile epibiontic organisms as palaeoecological guides (with examples from the Dutch coast), *Palaeogeography, Palaeoclimatology, Palaeoecology*, **2**(2): 333–79.

Boucot, A.J., 1953, Life and death assemblages among fossils, *American Journal of Science*, **251**(1): 25–40.

Brandt, D.S., 1989, Taphonomic grades as a classification for fossiliferous assemblages and implications for paleoecology, *Palaios*, **4**(4): 303–9.

Brenchley, P.J. and Newall, G., 1970, Flume experiments on the orientation and transport of models and shell valves, *Palaeogeography, Palaeoclimatology, Palaeoecology*, **7**(3): 185–220.

Brett, C.E. and Baird, G.C., 1986, Comparative taphonomy: a key to paleoenvironmental interpretation based on fossil preservation, *Palaios*, **1**(3): 207–27.

Cadée, G.C., 1968, Molluscan biocoenoses and thanatocoenoses in the Ría de Arosa, Galicia, Spain, *Zoologische Verhandelingen* (Leiden), **95**: 1–121.

Cadée, G.C., 1989, Size-selective transport of shells by birds and its palaeoecological implications, *Palaeontology*, **32**(2): 429–37.

Carthew, R. and Bosence, D.W.J., 1986, Community preservation in Recent shell-gravels, English Channel, *Palaeontology*, **29**(1): 243–68.

Clifton, H.E. and Boggs, S., Jr., 1970, Concave-up pelecypod (Psephidia) shells in shallow marine sand, Elk River beds, southwestern Oregon, *Journal of Sedimentary Petrology*, **40**(3): 888–97.

Cummins, H., Powell, E.N., Newton, H.J., Stanton, R.J., Jr. and Staff, G., 1986a, Assessing transportation by the covariance of species with comments on contagious and random distributions, *Lethaia*, **19**(1): 1–22.

Cummins, H., Powell, E.N., Stanton, R.J., Jr. and Staff, G., 1986b, The rate of taphonomic loss in modern benthic habitats: how much of the potentially preservable community is preserved?, *Palaeogeography, Palaeoclimatology, Palaeoecology*, **52**(1): 291–320.

Cutler, A.H. and Flessa, K.W., 1990, Fossils out of sequence: computer simulations and strategies for dealing with stratigraphic disorder, *Palaios*, **5**(3): 227–35.

Davies, D.J., Powell, E.N. and Stanton, R.J., Jr., 1989a, Taphonomic signature as a function of environmental process: shells and shell beds in a hurricane-influenced inlet on the Texas coast, *Palaeogeography, Paleoclimatology, Palaeoecology*, **72**(2): 317–56.

Davies, D.J., Powell, E.N. and Stanton, R.J., Jr., 1989b, Relative rates of shell dissolution and net sediment accumulation—a commentary: can shell beds form by the gradual accumulation of biogenic debris on the sea floor?, *Lethaia*, **22**(2): 207–12.

Davies, D.J., Staff, G.M., Callender, W.R. and Powell, E.N., 1990, Description of a quantitative approach to taphonomy and taphofacies analysis: all dead things are not created equal. In W. Miller (ed.), *Paleocommunity temporal dynamics: the long-term development of multispecies assemblies. Paleontological Society Special Publication*, **5**: 328–50.

Dodd, R.J. and Stanton, R.J., 1990, *Paleoecology: concepts and applications* (2nd edn), Wiley, New York.

Dörjes, J., 1972, Georgia coastal region, Sapelo Island, U.S.A.: sedimentology and biology, VII. Distribution and zonation of macrobenthic animals, *Senckenbergiana Maritima*, **4**(1): 183–216.

Driscoll, E.G., 1967a, Attached epifauna–substrate relations, *Limnology and Oceanography*, **12**(4): 633–41.

Driscoll, E.G., 1967b, Experimental field study of shell abrasion, *Journal of Sedimentary Petrology*, **37**(4): 1117–23.

Driscoll, E.G., 1968, Sublittoral attached epifaunal development in Buzzards Bay, Massachusetts, *Hydrobiologia*, **32**(1): 27–32.

Driscoll, E.G., 1970, Selective bivalve destruction in marine environments, a field study, *Journal of Sedimentary Petrology*, **40**(2): 898–905.

Driscoll, E.G. and Weltin, T.P., 1973, Sedimentary parameters as factors in abrasive shell reduction, *Palaeogeography, Palaeoclimatology, Palaeoecology*, **13**(1): 275–88.

Ekdale, A.A., 1974, Marine molluscs from shallow-water environments (0 to 60 meters) off the northeast Yucatan coast, *Bulletin of Marine Science*, **24**(3): 638–68.

Emery, K.O., 1968, Positions of empty pelecypod valves on the continental shelf, *Journal of Sedimentary Petrology*, **38**(4): 1264–9.

Flessa, K.W. and Brown, T.J., 1983, Selective solution of macroinvertebrate calcareous hard parts: a laboratory study, *Lethaia*, **16**(2): 193–205.

Flessa, K.W., Meldahl, K.H. and Cutler, A.H., 1990, Quantitative estimates of stratigraphic disorder and time-averaging in a shallow marine habitat, *Geological Society of America Abstracts with Programs*, **22**(7): A83.

Frey, R.W., 1987, Hermit crabs: neglected factors in taphonomy and paleoecology, *Palaios*, **2**(4): 313–22.

Frey, R.W. and Henderson, S.W., 1987, Left–right phenomena among bivalve shells: examples from the Georgia coast, *Senckenbergiana Maritima*, **19**(3/4): 223–47.

Frey, R.W. and Howard, J.D., 1986, Taphonomic characteristics of offshore mollusk shells, Sapelo Island, Georgia, *Tulane Studies in Geology and Paleontology*, **19**(2): 51–61.

Fürsich, F.T. and Flessa, K.W., 1987, Taphonomy of tidal flat molluscs in the northern Gulf of California: paleoenvironmental analysis despite the perils of preservation, *Palaios*, **2**(6): 543–59.

Fütterer, D.K., 1974, Significance of the boring sponge *Cliona* for the origin of fine grained material of carbonate sediments, *Journal of Sedimentary Petrology*, **44**(1): 79–84.

Greenstein, B.J., 1989, Mass mortality of the West-Indian echinoid *Diadema antillarum* (Echinodermata: Echinoidea): a natural experiment in taphonomy, *Palaios*, **4**(5): 487–92.

Henderson, S.W. and Frey, R.W., 1986, Taphonomic redistribution of mollusk shells in a tidal inlet channel, Sapelo Island, Georgia, *Palaios*, **1**(1): 3–16.

Hertweck, A., 1972, Georgia coastal region, Sapelo Island, U.S.A.: sedimentology and biology V. Distribution and environmental significance of Lebensspuren and *in-situ* skeletal remains, *Senckenbergiana Maritima*, **4**(1): 125–67.

Hubbard, D.K., Miller, A.I. and Scaturo, D., 1990, Production and cycling of calcium carbonate in a shelf-edge reef system (St. Croix, U.S. Virgin Islands): applications to the nature of reef systems in the fossil record, *Journal of Sedimentary Petrology*, **60**(2): 335–60.

Hutchings, P.A., 1986, Biological destruction of coral reefs. A review, *Coral Reefs*, **4**(4): 239–52.

Johnson, R.G., 1957, Experiments on the burial of shells, *Journal of Geology*, **65**(5): 527–35.

Johnson, R.G., 1965, Pelecypod death assemblages in Tomales Bay, California, *Journal of Paleontology*, **39**(1): 80–5.

Jones, G.F., 1969, The benthic macrofauna of the mainland shelf of southern California, *Allan Hancock Monographs in Marine Biology*, **4**: 1–219.

Kidwell, S.M. and Baumiller, T.M., 1990, Experimental disintegration of regular echinoids: roles of temperature, oxygen and decay thresholds, *Paleobiology* **16**(3): 247-71..

Kidwell, S.M. and Behrensmeyer, A.K., 1988, Overview: ecological and evolutionary implications of taphonomic processes, *Palaeogeography, Palaeoclimatology, Palaeoecology*, **63**(1): 1–13.

Kidwell, S.M. and Bosence, D.W.J., in press, Taphonomy and time averaging of marine shelly faunas. In D.E.G. Briggs and P.A. Allison (eds), *Taphonomy: releasing information from the fossil record*, Plenum, New York.

Kidwell, S.M. and Jablonski, D., 1983, Taphonomic feedback, ecological consequences of shell accumulation. In J.S. Tevesz and P.L. McCall (eds), *Biotic interactions in recent and fossil benthic communities*, Plenum, New York: 195–248.

Koch, C.F. and Sohl, N.F., 1983, Preservational effects in paleoecological studies: Cretaceous mollusc examples, *Paleobiology*, **9**(1): 26–34.

Liddell, W.D., 1975, Recent crinoid biostratinomy, *Geological Society of America Abstracts with Programs*, **22**(6): 1169.

Liddell, W.D. and Ohlhorst, S.L., 1988, Hard substrata community patterns, 1–120 m, north Jamaica, *Palaios* **3**(2): 413–23.

MacGeachy, J.K., 1977, Factors controlling sponge boring in Barbados reef corals, *Proceedings of the Third International Coral Reef Symposium*, Rosenstiel School of Marine and Atmospheric Science, Miami, **2**: 477–83.

Meldahl, K.H. and Flessa, K.W., 1990, Taphonomic pathways and comparative biofacies and taphofacies in a recent intertidal/shallow shelf environment, *Lethaia*, **23**(1): 43–60.

Meyer, D.L., 1971, Post-mortem disarticulation of Recent crinoids and ophiuroids under natural conditions, *Geological Society of America Abstracts with Programs*, **3**: 645.

Meyer, D.L. and Meyer, K.B., 1986, Biostratinomy of Recent crinoids at Lizard Island, Great Barrier Reef, Australia, *Palaios*, **1**(3): 294–302.

Miller, A.I., 1988, Spatial resolution in subfossil molluscan remains: implications for paleobiological analyses, *Paleobiology*, **14**(1): 91–103.

Moore, C.H. and Shedd, W.W., 1977, Effective rates of sponge bioerosion as a function of carbonate production, *Proceedings of the Third International Coral Reef Symposium*, Rosenstiel School of Marine and Atmospheric Science, Miami, **2**: 499–505.

Müller, A.H., 1979, Fossilization (taphonomy). In R.A. Robinson and C. Teichert (eds), *Treatise on invertebrate paleontology, Part A. Introduction*, Geological Society of America and the University of Kansas, Boulder, Colorado, and Lawrence: A2–A78.

Murray, J.W., 1965, Significance of benthic foraminiferids in plankton samples, *Journal of Paleontology*, **39**(1): 156–7.

Nagle, J.S., 1967, Wave and current orientation of shells, *Journal of Sedimentary Petrology*, **37**(4): 1124–38.

Neumann, A.C., 1966, Observations on coastal erosion in Bermuda and measurements of the boring rate of the sponge *Cliona lampa*, *Limnological Oceanography*, **11**(1): 92–108.

Noble, J.P.A. and Logan, A., 1981, Size-frequency distribution and taphonomy of brachiopods: a recent model, *Palaeogeography, Palaeoclimatology, Palaeoecology*, **36**(1): 87–105.

Odum, H.T. and Odum, E.P., 1955, Trophic structure and productivity of a windward coral reef community on Eniwetok Atoll, *Ecological Monographs*, **25**(3): 291–320.

Parsons, K.M., 1989, Taphonomy as an indicator of environment: Smuggler's Cove, St. Croix, U.S.V.I. In D.K. Hubbard (ed.), *Terrestrial and marine ecology of St. Croix, U.S. Virgin Islands, Special Publication* **8**, West Indies Laboratory, St Croix, Virgin Islands: 135–43.

Parsons, K.M., Brett, C.E. and Miller, K. B., 1988, Taphonomy and depositional dynamics of Devonian shell-rich mudstones, *Palaeogeography, Palaeoclimatology, Palaeoecology*, **63** (1): 109–39.

Perkins, R.D. and Tsentas, C.J., 1976, Microbial infestations of carbonate substrates planted on the St. Croix shelf, West Indies, *Geological Society of America Bulletin*, **87**(11): 1615–28.

Peterson, C.H., 1976, Relative abundances of living and dead molluscs in two Californian lagoons, *Lethaia*, **9**(1): 137–48.

Pichon, M., 1978, Quantitative benthic ecology of Tuléar reefs. In D.R. Stoddart and R.E. Johannes (eds), *Coral reefs: research methods, Monographs on Oceanographic Methodology*, **5**, UNESCO, Paris: 163–74.

Pleydell, S.M. and Jones, B., 1988, Boring of various faunal elements in the Oligocene–Miocene Bluff Formation of Grand Cayman, British West Indies, *Journal of Paleontology*, **62**(3): 348–67.

Plotnick, R.E., 1986, Taphonomy of a modern shrimp: implications for the arthropod fossil record, *Palaios*, **1**(1): 286–93.

Powell, E.N. and Stanton, R.J., Jr., 1985, Estimating biomass and energy flow of molluscs in palaeocommunities, *Palaeontology*, **28**(1): 1–34.

Powell, E.N., Staff, G., Davies, D.J. and Callender, W.R., 1990, Macrobenthic death assemblages in modern marine environments: formation, interpretation and application, *Reviews in Aquatic Sciences*.

Powell, E.N., Stanton, R.J., Jr., Cummins, H. and Staff, G., 1982, Temporal fluctuations in bay environments—the death assemblage as a key to the past. In J.R. Davis (ed.), *Proceedings of the Symposium on Recent Benthological Investigations in Texas and Adjacent States*, Texas Academy of Science, Austin: 203.

Rasmussen, K.A. and Brett, C.E., 1985, Taphonomy of Holocene cryptic biotas from St. Criox, Virgin Islands: information loss and preservational biases, *Geology*, **13**(8): 551–3.

Richter, R., 1929, Gründung and Aufgaben der Forschungsstelle für Meeresgeologie 'Senckenberg' in Wilhelmshaven, *Natur und Museum*, **59**: 1–30.

Salazar-Jiménez, A., Frey, R.W. and Howard, J.D., 1982, Concavity orientations of bivalve shells in estuarine and nearshore shelf sediments, Georgia, *Journal of Sedimentary Petrology*, **52**(2): 565–86.

Schäfer, W., 1972, *Ecology and paleoecology of marine environments*, University of Chicago Press, Chicago.

Schopf, T.J.M., 1978, Fossilization potential of an intertidal fauna: Friday Harbor,

Washington, *Paleobiology*, **4**(3): 261–70.

Seilacher, A., 1985, The Jeram model: event condensation in a modern intertidal environment. In U. Bayer and A. Seilacher (eds), *Sedimentary and evolutionary cycles, Lecture Notes in Earth Sciences*, **1**, Springer-Verlag, Berlin: 336–41.

Shiflett, E., 1961, Living, dead and total foraminiferal faunas, Heald Bank, Gulf of Mexico, *Micropaleontology*, **7**(1): 45–54.

Speyer, S.E. and Brett, C.E., 1986, Trilobite taphonomy and Middle Devonian taphofacies, *Palaios*, **1**(3): 312–27.

Staff, G.M. and Powell, E.N., 1990a, Taphonomic signature and the imprint of taphonomic history: descriminating between taphofacies of the inner continental shelf and a microtidal inlet. In W. Miller (ed.), *Paleocommunity temporal dynamics, Paleontological Society Special Publication*, **5**: 370–90.

Staff, G.M. and Powell, E.N., 1990b, Local variability of taphonomic attributes in a parautochthonous assemblage: can taphonomic signature disinguish a heterogeneous environment?, *Journal of Paleontology*, **64**(4): 648–58.

Staff, G.M., Powell, E.N, Stanton, R.J., Jr. and Cummins, H., 1985, Biomass: is it a useful tool in paleocommunity reconstruction?, *Lethaia*, **18**(2): 209–32.

Staff, G.M., Stanton, R.J.,Jr., Powell, E.N. and Cummins, H., 1986, Time-averaging, taphonomy, and their impact on paleocommunity reconstruction: death assemblages in Texas bays, *Geological Society of America Bulletin*, **97**(4): 428–43.

Stanton, R.J., Jr., 1976, Relationships of fossil communities to original communities of living organisms. In R.W. Scott and R.R. West (eds), *Structure and classification of paleocommunities*, Dowden, Hutchinson and Ross, Stroudsburg, Pennsylvania: 107–42.

Stanton, R.J., Jr., Powell, E.N. and Nelson, P.C., 1981, The role of carnivorous gastropods in the trophic analysis of a fossil community, *Malacologia*, **20**(2): 451–69.

Swinchatt, J.P., 1965, Significance of constituent composition, texture, and skeletal breakdown in some Recent carbonate sediments, *Journal of Sedimentary Petrology*, **35**(1): 71–90.

Tanabe, K., Fujiki, T. and Katsuta, T., 1986, Comparative analysis of living and death bivalve assemblages on the Kawarazu shore, Ehime Prefecture, west Japan, *Benthos Research*, **30**(1): 17–30.

Trewin, N.H. and Welsh, W., 1976, Formation and composition of a graded estuarine shell bed, *Palaeogeography, Palaeoclimatology, Palaeoecology*, **19**(3): 219–30.

Valentine, J.W., 1961, Paleoecologic molluscan geography of the Californian Pleistocene, *University of California Publications in Geological Sciences*, **34** (7): 309–442.

Walker, K.R., 1972, Trophic analysis: a method for studying the function of ancient communities, *Journal of Paleontology*, **46**(1): 82–93.

Walker, K.R. and Bambach, R.K., 1974, Feeding by benthic invertebrates: classification and terminology for paleoecological analysis, *Lethaia*, **7**(1): 67–78.

Walker, S.E., 1988, Taphonomic significance of hermit crabs (Anomura: Paguridea): epifaunal hermit crab—infaunal gastropod example, *Palaeogeography, Palaeoclimatology, Palaeoecology*, **63**(1): 45–71.

Walker, S.E., 1989, Hermit crabs as taphonomic agents, *Palaios*, **4**(5): 439–52.

Walker, S.E., Kidwell, S.M. and Powell, E.N., 1989, Molluscan death assemblages on the Texas continental shelf are potentially local, not exotic accumulations, *Geological Society of America Abstracts with Programs*, **21**(6): A71.

Wanless, H.R., 1986, Production of subtidal tubular and surficial tempestites by Hurricane Kate, Caicos Platform, British West Indies, *Journal of Sedimentary Petrology*, **58**(3): 730–50.

Warme, J.E., 1969, Live and dead mollusks in a coastal lagoon, *Journal of Paleontology*, **43**(1): 141–50.

Warme, J.E., 1971, Paleoecological aspects of a modern coastal lagoon, *University of California Publications in Geology*, **87**: 1–133.

Warme, J.E., 1977, Carbonate borers—their role in reef ecology and preservation. In S.H. Frost, M.P. Weiss and J.P. Saunders (eds), *Reefs and related carbonates—ecology and sedimentology, American Association of Petroleum Geologists Studies in Geology*, **4**: 261–80.

Warme, J.E., Ekdale, A.A., Ekdale, S.F. and Peterson, C.H., 1976, Raw material of the fossil record. In R.W. Scott and R.R. West (eds), *Structure and classification of paleocommunities*, Dowden, Hutchinson, and Ross, Stroudsburg, Pennsylvania: 143–50.

Weigelt, J., 1927, *Rezente Wirbeltierleichen und ihre paläobiologische Bedeutung*, Verlag M. Veg, Leipzig. (1989 English translation: *Recent vertebrate carcasses and their paleobiological implications*, University of Chicago Press, Chicago.)

Chapter 3

COMPLETENESS OF THE FOSSIL RECORD: AN OVERVIEW

Michael L. McKinney

INTRODUCTION

The degree of completeness of the fossil record is often on the mind of any palaeontologist. What scientist would not be deeply concerned when the data represent only a partial, often non-random sample of what has occurred? Yet, in spite of intense scrutiny, there is no general consensus as to how completeness is viewed. Since long before Darwin, the record has traditionally been seen as highly incomplete. Lacking a quantitative measure of completeness, the main evidence for this view has been qualitative observations, such as the well-known morphological 'leaps' and rapid group turnover events of the record. Neo-Darwinian defendants continue to emphasize this incompleteness, focusing on the limitations of the record to support gradual evolution and other neo-Darwinian ideas (Levinton, 1988; Hoffman, 1989). Similarly, Signor (1990) has emphasized the difficulty of studying past extinctions imposed by the poor record. However, other palaeontologists have been more optimistic. For instance, Valentine (1989a) has discussed a number of lines of evidence indicating that a surprisingly large proportion of taxa may be preserved as fossils. Similarly, Paul (1982; 1985) presented an oft-cited defence of the record for phylogenetic reconstruction.

In this chapter, I present an overview of completeness of the fossil record. Space limitations prohibit an exhaustive summary, so my goal is to outline the major themes and participants in current thinking. I will emphasize basic principles because a main impediment to the discussion of completeness has been confusion over terms and concepts. (For instance, even 'completeness' itself has a wide range of meanings.) The chapter has three basic sections concerned with temporal (vertical) completeness; spatial (horizontal) completeness, and other factors affecting temporal completeness via sampling effects; and a brief attempt to put geologic completeness into an even broader context, of information theory in general.

TEMPORAL COMPLETENESS

What is completeness?

Traditionally, Earth historians have equated incompleteness with 'missing time'. Gaps, unconformities and a host of other terms have been used to describe the general notion that slices of time are unrepresented in the rock and fossil record. So commonsensical is this idea that very little rigorous thought went into it until recently. Sadler (1981) and Schindel (1982) presented important attempts to quantify and define completeness in a rigorous way. Both approaches were based on sedimentation rates, as discussed shortly, but the point here is the precise meaning they gave to the term 'completeness': the proportion of the total span of absolute geologic time encompassed in a sampled sequence that is represented by actual strata, instead of gaps (Schindel, 1982).

In his excellent study of palaeontological completeness, Allmon (1989) noted that Schindel's (1982) definition should be supplemented to include spatial completeness. That is, both sediment and fossil deposition have a geographic, as well as a temporal, component. For fossils (palaeontological completeness) the definition of 'completeness' is therefore 'the proportion of the total time and space inhabited by all the individuals of all the species of the group of interest that is represented by fossils' (Allmon, 1989, p. 143). A slight rewording will further generate a useful definition of sedimentological completeness, to indicate that it is the proportion of the total time and space occupied by the sediment of interest (for example, sediment produced by the erosion of a well-defined parent rock).

Besides adding the key concept of spatial completeness, Allmon's definition makes one other crucial point, represented by the phrase 'group of interest'. Geological completeness is meaningful only when there is a specific referent; if completeness is the proportion of past phenomena (events occurring or objects existing) that are recorded, then we must specify what these phenomena are if we are to avoid confusion. For example, the completeness of the fossil record will never exceed the completeness of the enclosing sedimentary record, and will often be less complete (Dingus and Sadler, 1982). Thus, sedimentological completeness must be differentiated from palaeontological completeness. This is probably obvious in that we are 'conditioned' to think of the record in terms of sediment and dead organisms, but, as discussed at the end of this chapter, we should keep in mind the huge number of phenomena (observational 'referents') that are not recorded at all, or very little: water chemistry, temperature, and ecological interactions are just a few of the many kinds of information that are unavailable.

Even when analysing completeness in such commonly used referents as fossils, confusion can arise because different scales and categories of the referent are observed. Thus, Valentine (1989a) estimated that 85 per cent of durably skeletonized living species may have been initially captured in the Recent rock record. Harking back to Allmon's definition, the referent, or 'group of interest', in this case is durably skeletonized living species, and the proportion recorded is 85 per cent. Valentine is not concerned with all the individuals, or the total time or space occupied by them. Only one individual of a species need be found for the species to be recorded and contribute to this kind of completeness. Bretsky (1979) also discussed the completeness of species diversity. Similarly, Valentine (1989b) estimated that up to 12 per cent of all durably skeletonized species may be recorded in the Phanerozoic fossil record. Thus, the 'group of interest' is all durably skeletonized marine species (an estimated 2 million; Signor, 1985) and the proportion of species preserved is up to 12 per cent.

In summary, for the sake of terminological precision, I suggest that Allmon's definition be used to define total palaeontological completeness, where one wishes to estimate the proportion of total time and space occupied by a group that was recorded. However, I would also allow for other kinds of palaeontological completeness, based upon the different kinds of referents ('group of interest'). As examples (with references where they are discussed), I suggest the following:

- *Ontogenetic completeness (intra-species):* the proportion of the ontogeny of a species that is recorded in the fossil record (McKinney and McNamara, in press).
- *Taxonomic completeness (multi-species):* the proportion of a higher taxon's component species that is recorded in the fossil record (Paul, 1982).
- *Ecosystem completeness (multi-species):* the proportion of an ecosystem's (or 'community's') component species that is recorded in the fossil record (Valentine, 1989a). This may be expanded to include preserved relative proportions of original individual abundance of those component species (Damuth, 1982).

This partial listing is only intended to illustrate some of the different ways that preserved proportions ('completeness') are often discussed in the palaeontological literature. Rather than being viewed as something distinct from total palaeontological completeness (Allmon, 1989), they should be seen as subsets of it. Operationally, they are much easier to analyse than total palaeontological completeness because it is virtually impossible to know the total time and space inhabited by all the individuals of all the species of organisms.

Measuring completeness

While numerous kinds of completeness can be defined, depending on the referents of interest (for example, sedimentological, and the various palaeontological ones just cited), the notion of proportion is clearly central to them all. More precisely, completeness is the proportion of some past phenomenon (referent) that has been preserved (or that preserved subset of it that has been observed, see below) up to the present. Thus, the record of species known in a certain group is very complete if the species are nearly all preserved, even if many are known from only one individual.

This intuitive meaning of completeness as proportion continues to be commonly used either explicitly or implicitly by many workers. However, Sadler (1981) and Schindel (1982) independently developed methods of estimating completeness that allowed it a much more precise meaning. It is often called 'stratigraphic completeness', a term that I will use to distinguish it from the various kinds of palaeontological completenesses discussed above. Also, while stratigraphic completeness is fundamentally a kind of sedimentological completeness, it is distinguished by its particular quantitative approach. By compiling sedimentation rates in a number of environments, Sadler and Schindel observed that average rates are higher for shorter time-spans. In large part, this occurs because progressively longer gaps occur as the overall period of deposition increases. Most importantly, because average sedimentation rate is not constant across all time-spans, anyone wishing to estimate how much 'time' is represented in a section of rock must incorporate the variable rates into the calculations. For example, in observing a given thickness of rock, we would expect that more very long time-spans (say, 10 million years) would be more likely to be represented by sediment in the rock than the more numerous shorter time-spans (say, 1 million years). More precisely:

$$T = (S/R)1000,$$

where T is the time (in millions of years) represented by a given thickness of strata, S (in metres) deposited at a short-term rate, R (in metres per thousand years). Values of R are determined from the compilations of Schindel (1982) and Sadler (1981). Completeness (C) is then estimated by:

$$C = T/D,$$

where D is the time (in millions of years) elapsing between the uppermost and lowermost boundaries of the strata being analysed. (This particular version of calculation is further described in McKinney, 1984, and McKinney and Schoch, 1985; it is mathematically equivalent to the calculations originated by Sadler, 1981, and Schindel, 1982.)

The main point of such estimates is that completeness will vary with the level of resolution (span of time) considered. At coarse levels (for example, 10 million years), a given thickness will be more complete than at finer levels because longer intervals are more likely to see deposition. Computationally, this is manifested in the lower average sedimentation rate of longer time-spans in the Sadler (1981) and Schindel (1982) compilations. This does not mean that sedimentation necessarily occurred during any given interval (be it long or short) in the unit under analysis. This method is strictly probabilistic so that, as with all techniques based on averages, one is never certain of any precise event, only the overall net result. A simplified depiction of this is illustrated in Figure 3.1. In this case, deposition occurred after a 900-year gap in most of the area shown. If this 'average' pattern were to be regularly maintained, the section would eventually be 100 per cent complete at the 1000-year level of resolution. But it would be only 10 per cent complete at the

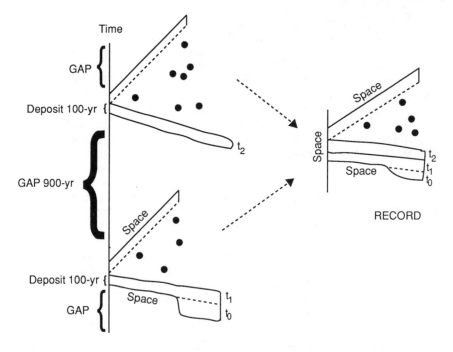

Figure 3.1   Illustration of how time–space translates into space–space dimensions of the fossil record. At time t_0, deposition occurs only locally. At time t_1, deposition of sediment and fossils (shown by large black dots) occurs over a broader area. After a 900-year hiatus, another brief episode of broad deposition occurs. Based on the Sadler–Schindel method of estimation, this example would be 100 per cent complete at the 1000-year level of resolution but only 10 per cent complete at the 100-year level, as discussed in text.

100-year level of resolution, with deposition in only one in ten 100-year intervals.

The stratigraphic completeness techniques of Sadler and Schindel have subsequently been applied by a number of workers to a variety of problems. Dingus and Sadler (1982) analysed the effects of stratigraphic completeness on evolutionary rates. McKinney and Schoch (1985) explored the use of palaeomagnetic dating of rock boundaries in stratigraphic completeness estimates. Dingus (1984) observed the effects of stratigraphic completeness on extinction rates across the Cretaceous–Tertiary boundary. McShea and Raup (1986) suggested that the stratigraphic completeness of a section of known time-span may be estimated by comparing that section's thickness against the global average thickness of sediment accumulated during that time-span. This is a very attractive tool for estimating completeness, but it has been severely criticized due to its reliance on (what is at least now) an imprecise and biased data set (Tipper, 1987). Most recently, Allmon (1989) has discussed the implications of stratigraphic (and palaeontological) completeness on phylogenetic studies. However, a number of problems have appeared in estimating stratigraphic completeness. These are discussed by Allmon (1989), but I outline them here:

1. Compaction of sediment during lithification is highly variable and very difficult to estimate at this time. Most workers simply ignore compaction when measuring the thickness of the rock unit for completeness estimates. This leads to underestimation of accumulation rate and, thus, of completeness.

2. Age determination of time boundaries usually relies on biostratigraphic zones to at least some extent. Yet zone identification does not mean that the entire zone is present because of sampling error (see below). This leads to underestimation of completeness because the local zones are treated as if they run the full length of known biozone duration; time intervals are estimated to be longer than they really are.

3. Very low sedimentation rates are not measured over short time-spans because of a lower limit of measurement precision. This causes anomalously high median short-term accumulation rates in the published rate compilations of Sadler and Schindel, again producing underestimates of completeness.

4. If the above three problems were all that occurred, estimates of stratigraphic completeness would have considerable utility as approximations, especially if we were to bear in mind that they were most probably underestimates. However, Anders *et al.* (1987) have discussed two major flaws which question the whole procedure of estimating stratigraphic completeness. The first flaw is that the lower limit of measurement precision not only causes underestimation of completeness, but can lead to a change in the relationship between sediment accumulation rate and temporal resolution. This change is very misleading since it can be caused by scale-related changes in variance alone. The problem may be avoided by stating completeness estimates in terms of rates at fixed thicknesses rather than at levels of temporal resolution. The second flaw noted by Anders *et al.* is potentially more invalidating. They point out that there is so much variation in sedimentation rate, even within the 'same' environment, that thicker sections may not be more complete than thinner sections spanning the same time interval.

Allmon (1989) has argued that these problems may not invalidate stratigraphic completeness estimates at coarser levels of resolution. The bias caused by measurement limitations applies mainly to short-term rates. I would add that it is also possible (but unproven) that coarser levels of resolution do not suffer as greatly from rate variation bias. In any case, these problems clearly require serious consideration by anyone attempting to employ the Sadler–Schindel approach to estimating stratigraphic completeness. If it is true that only coarser levels of resolution provide valid completeness estimates, the method is still very useful because many palaeontological phenomena occur at such time scales. For example, Allmon (1989) notes that most speciation and species durations are likely to involve time-frames over 10 000 years.

Causes of completeness

Many factors obviously affect how complete a given rock unit is and there is no space here for a full discussion of them. However, two key characteristics of the depositional process are so prominent that a brief mention seems necessary: first, the Markov chain model; and second, the hierarchical nature of deposition. Dacey and Lerman (1983) reviewed the concept of sediment growth and ageing, and showed that the Markov chain model can exactly describe the observed distribution of sediment masses with age. While a number of Markovian models (with differing assumptions) apply, and the good fit of any model does not prove that it is the only explanation for a process, Dacey and Lerman presented a good argument that much of the sedimentation process can be understood by focusing on the contingency of each state on the preceding state. This leads to a high serial correlation and it no doubt largely occurs because deposited sediments and fossils are often formed from the eroded and reworked remains of pre-existing particles. The main implication is that a period of high deposition (completeness) is likely to be followed and preceded by other periods of relatively high deposition. That is, deposition will be non-randomly distributed in time.

Less commonsensical (and, therefore, more intriguing) is the apparently hierarchical nature of this Markovian process of deposition. Plotnick (1986; 1988) has discussed evidence that stratigraphic hiatuses are fractal (show nested self-similarity), based upon four main points. First, there is a negative log-linear relationship between measured sedimentation rates and the intervals over which they are measured. Second, as noted above, this is probably because longer measured intervals have more and longer hiatuses. Third, the frequency distribution of hiatuses is hierarchical in that long hiatuses are much rarer than the more numerous intervening short ones. (Though not mentioned by Plotnick, this probably can be related to the exponential, geometric and other, similar models of stochastic deposition discussed by Dacey and Lerman. This also illustrates the Markovian nature of deposition in that periods of deposition tend to be non-randomly distributed.) Fourth, over a range of levels of temporal resolution, some depositional environments are more complete than others.

What causes this self-similarity across many scales? Plotnick (1986; 1988) attributed it to different processes acting at different time-scales, with higher-level processes exerting control over lower-level ones. An example would be the control of short-time-scale avulsions in a delta by longer-time-scale delta switching. In their excellent review of fractals in natural phenomena, West and Shlesinger (1990) noted that the abundance of self-similarity in nature (lung tissue, cardiac pulses and many others) occurs because so many phenomena are caused by a large ensemble of mechanisms with no prejudice about scale. That is, fractality occurs when many subtasks ('factors') must be realized for the grand task (overall 'process') to be achieved. In contrast, phenomena with fewer subtasks (contributing factors) show additive or mul-tiplicative (proportional) variations across large scales of time or space. This implies that deposition (and consequent hiatuses) is indeed a complex process, with a large number of factors that influence it across many scales of time and space. (This may sound obvious, but in fact many processes are not fractal and can be characterized by just one or a few primary interacting causes; see West and Shlesinger, 1990.)

HORIZONTAL COMPLETENESS: SAMPLING EFFECTS ON TEMPORAL COMPLETENESS

In defining total completeness, Allmon (1989) noted the oft-neglected aspect of spatial (as well as temporal) range. Thus, to take a simple palaeontological example, two groups may have the same temporal range, but vary greatly in their spatial range during that time. Organisms with high dispersal abilities (such as marine invertebrates with planktotrophic larvae) will tend to have broader geographic ranges (Jablonski and Lutz, 1983). This topic has generated much interest recently for its macroevolutionary implications (Jablonski, 1987), but the concern here is the preservational effects of geographic range. However, the two topics are not unrelated. As Russell and Lindberg (1988) have argued, groups with restricted geographic ranges will also have more restricted observed temporal ranges because of sampling effects. It is interesting to note that, because of our viewpoint as Earth historians, we tend to see temporal range as the 'primary' focus of completeness. Horizontal range is often seen as a secondary (and nettlesome) species trait that is mainly of interest as a source of sampling error in estimating temporal range. However, it is not only 'time' that is represented in the fossil record, but also 'space'. In other words, geographic completeness is of equal importance in its own right. Aside from our time-based viewpoint, another reason for its neglect is no doubt the extreme difficulty of measuring spatial completeness in the record.

The interaction of spatial and temporal range

Figure 3.2 attempts to illustrate the deceptively subtle relationship between spatial and temporal range (and completeness). First, there is a spatial component to many abiotic environmental processes, such as varying rates of sedimentation across the geographic range of a species. Where depositional rates diminish to zero or erosion later occurs, there will obviously be local temporal incompleteness. Second, there is a spatial component to the biological characteristics of the group itself. The most obvious, just noted, is overall geographic range. This affects temporal completeness in that geographically restricted groups tend to be recorded as having more restricted temporal ranges, purely through sampling bias; they are less likely to be discovered in the record (Russell and Lindberg, 1988). However, another group trait with a spatial component that is often overlooked is abundance. Even if two groups have the same geographic range, there will be a sampling bias if one is much rarer throughout its geographic range. The rarer species will tend to be seen as having shorter temporal range, even if it did not (Koch and Morgan, 1988). This is a major source of bias because rare (but often

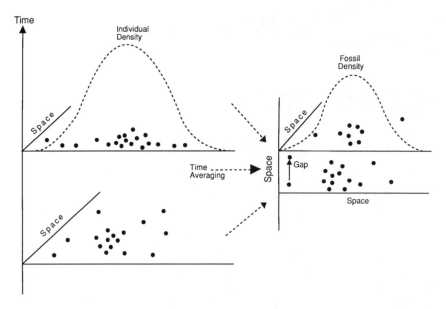

Figure 3.2   Translation of time–space relationship to the modern spatial record. Black dots represent fossils of a hypothetical species. The bell curve diagrammatically illustrates the ecological principle that, at any given time plane, abundance is highest near the centre of the geographic range of a species at that time (Brown, 1984). Therefore, sediment near the centre of the range will be richer in fossils. This has important implications for the fossil record with respect to sampling error, spatial and temporal range estimates, and biostratigraphic gaps. For instance, biostratigraphic gap size will tend to increase with increasing distance from the centre of a fossil species' geographic range.

widespread) species compose a sizeable portion of nearly all communities, including fossil ones (Buzas *et al.*, 1982). Finally, it is possible to combine both geographic range and abundance. As shown in Figure 3.2, individual abundance tends to be greatest near the centre of the geographic range of a species and progressively decreases to the margin, as optimal environmental conditions grade into less favourable ones (Brown, 1984). This seems to apply to many groups, so that even 'rare' species grow less rare at the centre of their range while 'common' species grow even more common.

In summary, geographic range and population abundance (within that range) serve as two key biotic traits that strongly affect our ability to determine the true spatial and temporal extent (total completeness) of that group. As spatial range and/or abundance decreases, our likelihood of accurately estimating total completeness of that group decreases as well (Koch, 1987; Koch and Morgan, 1988). Even under the most favourable conditions for estimating total completeness of a highly abundant group that is heavily sampled and very restricted in its true geographic range, it is highly unlikely that its full spatial extent through time will ever be known to us. The difficulty of tracking such spatial and temporal changes lies not only in the many abiotic and biotic preservational vagaries, but also in the very complexity of the true pattern. This can be visualised by noting that species ranges are not static through geological time (as is implied in the simplified example of Figure 3.2). Rather, geographic ranges expand, contract and undergo lateral shifts through migration. Just how much of such range shifts, contractions and so on the record actually captures is a crucial question that has seen little research, though it is likely to be accessible.

Biostratigraphic gap analysis

Another aspect of horizontal and vertical completeness that is illustrated in Figure 3.2 is the biostratigraphic gap. Such gaps are not at all distinct from the above discussion. Rather they represent a particular view of abundance through time and space, that of missing fossils along a temporal vector. While the gap shown in Figure 3.2 illustrates that seen in a local section, the term 'biostratigraphic gap' can refer to temporal gaps in fossils at a number of scales. Thus, Paul (1982) discussed gaps in the range of groups across long time-spans. Such coarse scales are, of course, ultimately constructed from local distributions such as that of Figure 3.2. Shaw's (1964) method of composite sections is the most direct quantitative expression of this.

Gaps can be used to estimate temporal completeness. For instance, Paul (1982) discussed how groups found above and below, but not within, an interval can quantitatively reveal the extent of preservational omissions. Another approach is to use gap distributions in local sections to derive confidence intervals that estimate the true temporal extent of a group. Because the observed vertical range almost always underestimates the true vertical range (as implied above, we are very unlikely to observe the first and last true occurrences of a group), the distribution of gaps provides a reasonable estimate of how far the true temporal range may extend beyond the beginning and end of the observed temporal range. The underlying assumptions are that the fossils have a random or known non-random distribution and that sampling has been uniform. The earlier research on biostratigraphic gap analysis by Paul (1982) and McKinney (1986a) have been greatly improved by more recent work by Springer and Lilje (1988), Strauss and Sadler (1989), and Marshall (1990).

Marshall (1990) specifically discussed how confidence intervals can estimate completeness; after making estimates of the true temporal range, one compares these with later fossil finds (more sampling) to see if the original estimates were roughly correct. Notice, however, that once again we confront a variation in the meaning of completeness. Marshall (1990) used completeness to mean the proportion of true temporal range known of a group, with no direct reference to spatial range. Yet the importance of spatial range to gaps (and therefore temporal completeness in this sense) is clear, as shown in Figure 3.2; confidence intervals may vary considerably depending on where they are estimated in a group's geographic range due to varying local abundance. Similarly, such spatial changes may greatly complicate the creation of composite sections based on biostratigraphic placement.

The overall importance of fossil gap analysis was discussed by Marshall (1990); it is a way of assessing the degree of uncertainty in estimating times of origination and extinction in many geological, palaeobiological and evolutionary problems. In particular, these include problems of correlation, speciation and extinction rates, and rates of morphological and molecular evolution. Two of the most prominent expressions of this involve the current interest in times and causes of past extinctions. Signor and Lipps (1982) pointed out that, because known biostratigraphic ranges tend to underestimate true temporal ranges, even 'instantaneous' mass extinctions may appear to be gradual ('smeared') because the recorded time of termination will vary among the groups extinguished, depending on their geographic range, abundance, size, shape, sampling intensity, and many other variables that affect both their observed and preserved completeness (discussed below). A second important example is the 'Lazarus effect', wherein taxa appear to go extinct at a certain time, but re-emerge later in the record (Jablonski, 1986). Such gaps can have a number of causes, ranging from the vagaries of preservation to range restriction (including refugia).

Observational versus preservational completeness

The distinction between observational and preservational completeness serves to summarize many of the ideas just discussed about sampling and completeness of the fossil record. Preservational completeness refers to the proportion of the referent of interest (for example, individuals, ontogenies, taxa, communities) that has been preserved, or, to cite Allmon's definition above, that is now represented by fossils. Observational completeness is a subset of preservational completeness, being defined as that proportion of the preserved referent that has actually been sampled. Usually this observed subset is a tiny fraction of the preserved, which is, in turn, only a tiny fraction of all of the referent that existed in time and space. However, as discussed shortly, this is not nearly as palaeontologically pessimistic as it sounds, because there is a huge amount of redundancy in the information of the fossil record, so that relatively tiny fractions of data can provide a great deal of information about the whole picture.

If sampling of the past was completely random, then the omission of information inherent in the sampling process would be much easier to interpret than when sampling is biased. Unfortunately, palaeontological information is highly biased and each type of completeness has its own unique biases. Biases manifested across many temporal and spatial scales are a main preoccupation of palaeontologists, for obvious reasons, and the literature is vast. My goal here is simply to draw a distinction between preservational and observational biases, and to cite a few of the many factors that cause them. Biotic factors which bias preservational completeness include durability of body parts and environment inhabited. A less well-known factor is body size, as shown by Badgley (1986) for mammals. A well-known abiotic factor in preservational completeness is the age of fossil deposition, as manifested in the 'Pull of the Recent', wherein more recent deposits tend to be more complete (Raup, 1979).

In contrast to the natural biotic and abiotic factors that cause preservational biases, some factors that cause observational biases are often less clear-cut, being due to 'accidents' of discovery (by definition non-deterministic) and subjective human motives (that favour one group being sampled and studied more than others). A good example of subjective human biases is that of temporal preference; Cainozoic fossils seem to have been studied more intensively than earlier fossils (Sheehan, 1977). Another human bias is taxonomic; gastropods and bivalves have been more intensively studied than other major groups (for example, echinoids or bryozoans; Valentine, 1989a). However, not all observational biases are due to 'accident' or human bias. Traits of the biota can serve to bias some in favour of discovery. We have already noted how greater abundance and geographic range can bias a group towards discovery. In addition, larger body size and longer shapes cause fossils to be preferentially sampled. That is, a random cross-section or core will more likely intersect larger and longer fossils, according to the laws of stereology (McKinney, 1986a; 1986b).

Given that preservational completeness is all we have of what existed in the past, how well does observational completeness approximate to that? The answer to this depends on the referents used in the completeness definition. To take extremes, not all individuals that ever existed have been preserved and not all individuals preserved have been sampled. However, looking at coarser referents such as the species level (taxonomic completeness defined above), it appears that, in at least some geographic, temporal and taxonomic areas, observational completeness approaches preservational completeness. For example, Allmon (1989) plotted the number of turritelline gastropod species described since the early 1800s for the New World Cainozoic, and early Cainozoic Gulf and Atlantic Coastal Plain. In both cases, the curve is logistic such that the number of new species approaches an asymptote over the last few decades. A similar pattern occurs in Cainozoic echinoids (Heller and McKinney, research in progress). For echinoids at least, this is not an artefact of decreased sampling intensity (fewer workers) over the last few decades. In addition to myself, a number of other workers have done extensive collecting in the Gulf and Atlantic Coastal Plain with no significant new taxonomic finds (Carter and McKinney, in press). On a larger scale, Valentine's (1989b) estimate that 12 per cent of durably skeletonized invertebrates have been described is greatly at variance with his estimate that 85 per cent of such invertebrates are initially captured by the record. It seems likely that both preservational and observational completeness play a role in this disparity. There are many taxa that are poorly described and many geographic areas (for example, parts of Asia) where the fossils are relatively unstudied. However, preservational loss must also play a role because sediment ageing causes destruction (erosion, dissolution) of those fossils that were initially captured in the record.

INCOMPLETENESS AS INFORMATION LOSS

What is information?

It is useful to view completeness of the rock and fossil record in terms of information theory. Intuitively, we all agree that loss of completeness represents a loss of 'information'; some proportion of a referent is not recorded (preservational completeness) or not sampled (observational completeness). But what is 'information'? Like energy, the term is impossible to define precisely, so operational meanings are used. Thus, information is 'the capacity to store and transmit meaning or knowledge' (Gatlin, 1972). Information represents departure from randomness and can, therefore, be described using the many tools of probability (for example, the Shannon–Wiener index). A basic unit of measurement, the bit (binary digit), is defined as the amount of information transmittable by a single binary (on/off) signal. Therefore, the first step in measuring the information content of any phenomenon is to translate it into binary code.

Information measured: the extreme incompleteness of the record

Given a general theory of information, we may ask if it is possible objectively to measure the information contained in past events in Earth history; and to compare it to the information transmitted from those events into the rock record (preservational completeness), and thence into the scientific literature (observational completeness). Consider a square·metre of the modern shallow marine benthos; how much information would be required to describe every single event that occurs in the area for a given period of time? An obvious approach would be to film the area and measure how much information is contained in the pixels (on/off units) of the resulting film. This is the method used in sending pictures back from planetary exploration, whereby the information is converted to pictures by computers. Areas of sea floor with a great deal of biological (burrowing, ecological interactions among a myriad other events) and geological activity would contain many bits in the resulting film (time series). In contrast, relatively uninhabited areas, with little activity, would be described by relatively few changes in the pixels through time.

  The idea of a film of past biotic events is no doubt alien to most palaeontologists, but a film record, like a fossil record, is a message from the past. It is alien to us only because it is a far more complete package of information about the past than we will ever encounter in the rock record. Earth historians are usually 'conditioned' to think of completeness in terms of sediment, dead organisms and the few other tangible remains of the past left for us. However, when measured objectively, in bits, these fragments are obviously an extremely tiny fraction of the bits that would be contained in a film running for thousands of years. Furthermore, even this approach underestimates the incompleteness of the record, because the use of a film to estimate the bits contained in a time series measures only that information found in the visible spectrum of light at the human scale of observation. To record changes at the microscopic scale, and in unseen (non-photic) physical parameters such as temperature, salinity, and chemical fluctuations, a host of other recording devices would be needed to operate over the entire period of time. Clearly, most of this information is also absent from the rock and fossil record, when objectively measured in bits.

On a technical point, it is often said that the rock record is a book from which the pages of the past are read. This is not quite true in the larger sense of information theory. A book is an example of a static information storage device. Instead, the rock record is what is known as 'temporal information storage'; it is the actual time series (albeit incomplete in time and space) of the information flow itself.

The brighter side of incompleteness: information filters and redundancy

The huge loss of information in the record, when measured objectively, gives no obvious cause for celebration to Earth historians. However, from an informational theory point of view, there are some beneficial aspects to this seemingly bad situation. First, information loss acts as a necessary noise filter, allowing large-scale patterns to be identified much more easily. Second, there is a huge amount of redundancy in the record so that much of the information is not truly lost.

A low-pass filter is illustrated in Figure 3.3. This is a device used in information transmission equipment that eliminates high-frequency noise. In a homologous (not merely analogous) way, the depositional process serves the same purpose; most of the information that is lost in the record is trivial from the perspective of very long-term processes. The second-to-second events that are of interest for short periods of study (for example, the ontogenetic perspective) become high-frequency 'noise' for those interested in day-to-day or seasonal events (for example, the ecological view), and even more so for the geological spans of evolutionary processes. This noise filtering is not merely a matter of convenience. The amount of information that can be analysed ('processed') by both the human mind and its aids (computers) is far less than even the limited amount available to us.

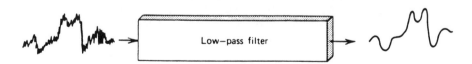

Figure 3.3   A low-pass filter, such as occurs when the fossil is formed, eliminates high-frequency information ('noise'). With less jargon: short-term events and processes are irretrievably lost, but this often allows long-term events and processes to be more clearly observed and analysed.

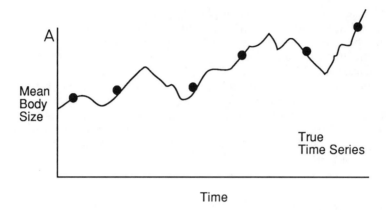

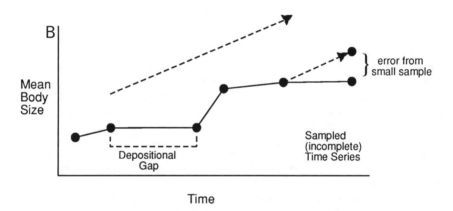

Figure 3.4   An example of the fossil record as a low-pass filter. (A) Mean body
size of a population or species increases over a long time-span, with many
short-term fluctuations. Black dots represent those times when significant numbers of individuals are preserved for later sampling. (B) Time series generated
by an investigator, based on the preserved samples. The general trend of
change is captured, although error can occur from small sample size (few
individuals).

An example of the low-pass filter effect is shown in Figure 3.4. The true
time series of mean body size in a population is only partly captured by the
time series of intermittent samples in the fossil record. Yet, the basic long-
term patterns of interest (directionality and long-term rate of change) can be
inferred, with mainly loss of background flux. Nevertheless, low-pass filtering
can also be a source of error. As Figure 3.4 shows, small samples can lead to
sampling error from loss of too much information. In addition, a basic

necessity of valid inference is that the high-frequency noise be filtered without bias (that is, using random samples). This is one of the major problems with the fossil record as a low-pass filter. It is not the omission of information in itself, which is often just the loss of noise; rather, the greatest danger to interpretation is the differential loss of information via one or more of a myriad preservational and observational biases, some of which were noted above. Of course, this can cause us to infer incorrectly that patterns due to systematic biases in the recording process (for example, 'Pull of the Recent') are due to patterns in the original events themselves (for example, increasing global diversity).

Redundancy is the repetition of information. Because so much of biology involves inheritance, there is much informational redundancy at many levels. That at the genetic level (identical genes in many organisms) is, of course, not directly accessible in the fossil record, but its manifestations at the individual (morphological) and supra-individual (species, ecosystem) levels are readily available. To cite an example discussed above, ontogenetic completeness in the record is very high because, while individual variation exists in most populations, there is a far larger component of consistency. This is what allows us to reconstruct longitudinal ontogenies (of one individual) from the cross-sectional remains of many dead individuals at different ages and size. Thus, of all the millions, and perhaps billions or trillions, of individuals that existed in any given species, we retain a relatively vast amount of information about the ontogenetic events of that species simply from biometric analysis of a few individuals of differing age and size.

Furthermore, this redundancy is also expressed in the inherited consistencies at taxonomic and ecological levels. This allows us to reconstruct much of the structure (and events) at these levels with relatively few bits of information. For example, rarefaction is based on the extrapolation of known patterns in taxonomy and ecology to make inferences where the information is incomplete. Thus, Raup (1979) used known patterns in taxonomic ratios (for example, genera per family) to infer how many lower taxa died out based on information on the number of higher taxa that died. Similarly, Koch (1987; Koch and Morgan, 1988) has used known patterns of species occurrence frequency (the log-series) to infer a great deal of information about temporal and geographic patterns of the past. Many similar attempts at palaeocommunity reconstruction based on modern analogy ('uniformitarianism') could be cited. The point is that none of this could be possible without the extremely high amount of redundant information that exists in the history of life.

To conclude, much of this last section is intuitively known to most palaeontologists, without the injected viewpoint of information theory. However, I suggest that there is an important theoretical gain in viewing such problems from a broad perspective, such as seeing completeness in the record as one of many possible examples of information loss. More importantly, this perspective opens the door to practical benefits. By trying to define and measure information more objectively, we can better estimate how much information exists, and has been lost, and generate testable hypotheses about the record itself. The discussion above is only the barest sketch of how we might begin to go about it.

ACKNOWLEDEGMENTS

This paper was prepared during my support by the Petroleum Research Fund of the American Chemical Society (Grant AC5-PRF22635-AC8).

NOTE: The following recent article reviews the theoretical and mathematical foundations of stratigraphic completeness estimates: Sadler, P. and Strauss, D., 1990, Estimation of completeness of stratigraphical sections using empirical data and theoretical models, *Journal of the Geological Society*, **147**(4): 471–485.

The following recent article makes a major contribution to biostratigraphic gap analysis and the estimation of confidence intervals to local 'true' ranges: Springer, M., 1990, The effect of random range truncations on patterns of evolution in the fossil record, *Paleobiology*, **16**(4): 512–520.

REFERENCES

Allmon, W.D., 1989, Paleontological completeness of the record of lower Tertiary mollusks, U.S. Gulf and Atlantic Coastal Plains: implications for phylogenetic studies, *Historical Biology*, **3** (2): 141–58.

Anders, M.H., Krueger, S. and Sadler, P., 1987, A new look at sedimentation rates and the completeness of the stratigraphic record, *Journal of Geology*, **95** (1): 1–14.

Badgley, C., 1986, Taphonomy of mammalian fossil remains from Siwalik rocks of Pakistan, *Paleobiology*, **12** (2): 119–42.

Bretsky, P.W., 1979, Recognition of ancestor–descendant relationships in invertebrate paleontology. In J. Cracraft and N. Eldredge (eds), *Phylogenetic analysis and paleontology*, Columbia University Press, New York: 113–64.

Brown, J.H., 1984, On the relationship between abundance and distribution of species, *American Naturalist*, **124** (3): 255–79.

Buzas, M., Koch, C., Culver, S. and Sohl, N.F., 1982, On the distribution of species occurrence, *Paleobiology*, **8** (2): 142–50.

Carter, B.D. and McKinney, M.L., in press, The biogeography of barriers and boundaries, *Paleobiology*.

Dacey, M.F. and Lerman, A., 1983, Sediment growth and aging as Markov chains, *Journal of Geology*, **91** (4): 573–90.

Damuth, J., 1982, Analysis of the preservation of community structure in assemblages of fossil mammals, *Paleobiology*, **4** (8): 343–6.

Dingus, L., 1984, Effects of stratigraphic completeness on interpretations of extinction rates across the Cretaceous–Tertiary boundary, *Paleobiology*, **10** (4): 420–38.

Dingus, L. and Sadler, P.M., 1982, The effects of stratigraphic completeness on estimates of evolutionary rates, *Systematic Zoology*, **31** (3): 400–12.

Gatlin, L.L., 1972, *Information theory and the living system*, Columbia University Press, New York.

Hoffman, A., 1989, *Arguments on evolution*, Oxford University Press, Oxford.

Jablonski, D., 1986, Causes and consequences of mass extinctions: a comparative approach. In D.K. Elliott (ed.), *Dynamics of extinction*, Wiley, New York: 183–229.

Jablonski, D., 1987, Heritability at the species level: analysis of geographic ranges of Cretaceous mollusks, *Science*, **238** (3): 360–3.

Jablonski, D. and Lutz, R., 1983, Larval ecology of marine benthic invertebrates: paleobiological implications, *Biological Reviews*, **58** (1): 21–89.

Koch, C.F., 1987, Prediction of sample size effects on the measured temporal and geographic distribution patterns of species, *Paleobiology*, **13** (1): 100–7.

Koch, C.F. and Morgan, J.P., 1988, On the expected distribution of species' ranges, *Paleobiology*, **14** (2): 126–38.

Levinton, J., 1988, *Genetics, paleontology, and macroevolution*, Cambridge University Press, Cambridge.

Marshall, C.R., 1990, Confidence intervals on stratigraphic ranges, *Paleobiology*, **16** (1): 1–10.

McKinney, M.L., 1984, The Cenozoic stratigraphic record of peninsular Florida: how complete is it?, *Florida Scientist*, **47** (1): 35–43.

McKinney, M.L., 1986a, Biostratigraphic gap analysis, *Geology*, **14** (1): 36–8.

McKinney, M.L., 1986b, How biostratigraphic gaps form, *Journal of Geology*, **94** (5): 875–84.

McKinney, M.L. and McNamara, K.J., in press, *Heterochrony: the evolution of ontogeny*, Plenum, New York.

McKinney, M.L. and Schoch, R., 1985, Titanothere allometry, heterochrony, and biomechanics: revising an evolutionary classic, *Evolution*, **39** (6): 1352–63.

McShea, D.W. and Raup, D.M., 1986, Completeness of the geological record, *Journal of Geology*, **94** (4): 569–74.

Paul, C.R.C., 1982, The adequacy of the fossil record. In K. Joysey and A. Friday (eds), *Problems of phylogenetic reconstruction*, Academic Press, London: 75–117.

Paul, C.R.C., 1985, The adequacy of the fossil record reconsidered, *Special Papers in Palaeontology*, **33**: 7–15.

Plotnick, R., 1986, A fractal model for the distribution of stratigraphic hiatuses, *Journal of Geology*, **94** (5): 885–90.

Plotnick, R., 1988, A fractal model for the distribution of stratigraphic hiatuses: a reply, *Journal of Geology*, **96** (1): 102–3.

Raup, D.M., 1979, Biases in the fossil record of species and genera, *Carnegie Museum of Natural History Bulletin*, **13**: 85–91.

Russell, M. and Lindberg, D., 1988, Real and random patterns associated with molluscan spatial and temporal distributions, *Paleobiology*, **14** (4): 322–30.

Sadler, P.M., 1981, Sediment accumulation rates and the completeness of stratigraphic sections, *Journal of Geology*, **89** (3): 569–84.

Schindel, D.E., 1982, Resolution analysis: a new approach to the gaps in the fossil record, *Paleobiology*, **6** (4): 340–53.

Shaw, A.B., 1964, *Time in stratigraphy*, McGraw-Hill, New York.

Sheehan, P.M., 1977, Species diversity in the Phanerozoic: a reflection of labor by systematists?, *Paleobiology*, **3** (3): 325–8.

Signor, P.W., III, 1985, Real and apparent trends in species richness through time. In J.W. Valentine (ed.), *Phanerozoic diversity patterns*, Princeton University Press, Princeton, New Jersey: 129–50.

Signor, P.W., III, 1990, Book review of *Mass extinctions, Bioscience*, **40** (7): 535–6.

Signor, P.W., III and Lipps, J., 1982, Sampling bias, gradual extinction patterns and catastrophes in the fossil record. In L.T. Silver and P. Schultz (eds), *Geological implications of impacts of large asteroids and comets on the Earth, Geological Society of America Special Paper*, **190**: 291–6.

Springer, M. and Lilje, A., 1988, Biostratigraphy and gap analysis: the expected sequence of biostratigraphic events, *Journal of Geology*, **96** (2): 228–36.

Strauss, D. and Sadler, P.M., 1989, Stochastic models for the completeness of stratigraphic sections, *Mathematical Geology*, **21** (1): 37–59.

Tipper, J., 1987, Estimating stratigraphic completeness, *Journal of Geology*, **95** (5): 710–15.

Valentine, J.W., 1989a, How good was the fossil record? Clues from the Californian Pleistocene, *Paleobiology*, **15** (2): 83–94.

Valentine, J.W., 1989b, Phanerozoic marine faunas and the stability of the Earth system, *Palaeogeography, Palaeoclimatology, Palaeoecology*, **75** (1): 137–55.

West, B.J. and Shlesinger, M., 1990, The noise in natural phenomena, *American Scientist*, **78** (1): 40–5.

Chapter 4

THE DIAGENESIS OF FOSSILS

Maurice E. Tucker

INTRODUCTION

The skeletons, shells and tests of organisms vary considerably in mineralogy and structure, but when the organisms die the bioclasts generated are liable to be altered to a greater or lesser extent during diagenesis. Diagenesis encompasses all the processes affecting a sediment and its contained fossils from deposition until metamorphism takes over at elevated temperatures and pressures. The diagenesis of fossils depends primarily on the original mineralogy and on the diagenetic environment, especially the pore-fluid chemistry. In some very special circumstances, the skeletons are even altered while the animals or plants are living. This chapter briefly discusses the original skeletal mineralogy and chemistry of the common animal and plant groups, and then outlines the main diagenetic changes that take place to alter the composition and structure of the fossils. The importance of the diagenetic environment is then illustrated, showing how this controls the type of preservation of the bioclasts.

THE MINERALOGY OF SKELETAL ELEMENTS

Most skeletal elements of organisms are originally made of calcium carbonate, calcium phosphate or silica, but during diagenesis alteration of the original mineralogy is common. The $CaCO_3$ mostly forms either calcite or aragonite, and some shells contain both of these minerals. Two types of calcite are recognized: low-magnesium calcite with 1–4 mole% $MgCO_3$ and high-magnesium calcite with 11–19 mole% $MgCO_3$. The carbonate minerals vaterite, monohydrocalcite and dolomite are very rare in skeletons. Carbonate-secreting organisms dominate the invertebrates and include the molluscs (bivalves, gastropods and cephalopods), the cnidarians (principally the corals), the brachiopods, the echinoderms, the bryozoans and the fora-minifera. These are all important limestone-formers and, of course, very common as fossils in other sediment types, too. Red (Rhodophyta), green (Chlorophyta) and yellow-green (Coccolithophoridae) algae and cyanobac-teria commonly have calcified skeletons and are also significant contributors to the formation of limestone. The mineralogy of the skeletons of all these groups is shown in Table 4.1. The skeletons of many of the organism groups noted above are present in abundance in shallow-water, low-latitude carbon-ate sediments, such as are accumulating in the Florida–Bahamas–Caribbean region, Trucial Coast–Red Sea, and off eastern and western Australia. In these areas, organic productivity is most prolific at depths less than 10 m. The planktonic foraminifera and Coccolithophoridae dominate the pelagic oozes which cover much of the ocean floor down to depths of several kilometres.

Opaline silica forms the tests of the zooplankton Radiolaria and the phytoplankton Diatomacea, which constitute the siliceous oozes of the deeper, open ocean floors and small marginal basins (such as the Gulf of California and Japan Sea). Both these microscopic fossils also occur in shallow-marine sediments, and diatoms are present in many lakes. Some sponges have siliceous skeletons.

Calcium phosphate forms the bones of most vertebrate animals. It is chiefly a complex carbonate hydroxyl fluorapatite, approximating to the formula $Ca_{10}(PO_4,CO_3)_6F_{2-3}$. Phosphate also occurs in some invertebrate skeletal elements, notably the small shelly fossils of Precambrian–Cambrian boundary strata and the shells of inarticulate brachiopods.

THE CHEMISTRY OF SKELETAL ELEMENTS

Skeletal parts of the various organisms have distinctive trace-element com-positions and stable isotope signatures. During diagenesis, these are com-monly altered to a greater or lesser extent. With carbonate skeletons, the elements usually measured are magnesium, strontium, sodium, iron and manganese, and the stable isotopes of oxygen ($\delta^{18}O$) and carbon ($\delta^{13}C$) are frequently analysed. Oxygen isotopes of siliceous and phosphatic skeletons can also be determined, and for the latter, rare earths, various metals and uranium series elements can provide useful environmental and diagenetic information.

Carbonate skeletons

The chemical composition of carbonate skeletons depends on the type of organism (the 'vital' effect) and on environmental conditions (water chemis-try and temperature). Where skeletons are composed of high-magnesium calcite, such as in the echinoderms and red (coralline) algae, then water

Table 4.1   *The original mineralogy of the common fossils.*
x = common mineralogy, (x) = less common

Mineralogy Organism	aragonite	low-Mg calcite	high-Mg calcite	aragonite + calcite	silica	phosphate
Mollusca:						
Bivalves	x	x		x		(x)
Gastropods	x			x		
Pteropods	x					
Cephalopods	x		(x)			
Brachiopods		x	(x)			x
Corals:						
Scleractinian	x					
Rugose + tabulate		x	x			
Sponges	x	x	x		x	
Bryozoans	x		x	x		
Echinoderms			x			
Ostracods		x	x			
Foraminifera:						
benthic	(x)		x			
pelagic		x				
Algae:						
Coccolithophoridae		x				
Rhodophyta	x		x			
Chlorophyta	x					
Charophyta		x				
Annelida	x	x	x		x	x
Arthropoda	x	x	x		(x)	x
Diatomacea					x	
Radiolaria					x	
Vertebrata	x	x	x			x

temperature is a strong control. Higher-latitude and deeper-water skeletons of these organisms have lower Mg^{2+} contents. Some skeletons in the tropical shallow-marine environment are high-magnesium calcite whereas the same or closely related organism in deep water or in a freshwater lake has a low-magnesium calcite skeleton. This is the case with certain molluscs and calcareous algae, and is a reflection of water temperature in the first case and water chemistry in the second. Aragonite skeletons have much lower magnesium contents (around 1000 ppm) compared with calcitic ones.

Strontium is an element relatively easily incorporated into the aragonite lattice compared to calcite, reflecting the higher partition coefficient of Sr^{2+} for aragonite. Skeletal aragonite has Sr^{2+} in the range 1000–10 000 ppm. Molluscs have lower Sr^{2+} (2000–4000 ppm), whereas corals and calcareous green algae have much higher values (7000–10 000 ppm). This is a vital effect. Abiogenic aragonite, as in ooids and marine cements, precipitated directly out of seawater, has 8000–10 000 ppm. Biogenic calcite has around 500–1500 ppm strontium. Sodium contents of skeletal grains are the order of several thousand parts per million and the value does appear to increase with increasing salinity. However, as well as replacing Ca^{2+} ions, like Sr^{2+}, sodium also occurs in the fluids of inclusions and in interlattice positions. Thus care has to be exerted when interpreting Na^+ contents of fossils in terms of salinity.

Iron and manganese contents of shallow marine biogenic grains are very low, normally just a few tens of parts per million, since these elements are present in low concentrations in seawater. However, they are commonly picked up by fossils during diagenesis, when the composition of pore-fluids changes (see below).

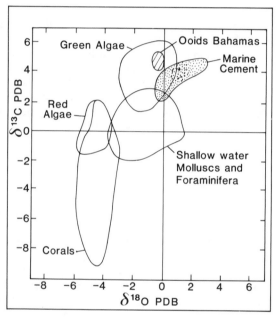

Figure 4.1 Stable isotope signatures of Recent carbonate skeletons. Also shown are the plots of the abiogenic ooids and marine cements. From Tucker and Wright (1990).

The oxygen isotopic composition of skeletal elements depends largely on the seawater $\delta^{18}O$ and on the temperature, but some organisms exert a strong vital effect. Molluscs, brachiopods and planktonic foraminifera mostly precipitate shells which are in isotopic equilibrium with ambient waters. For modern tropical cases this is generally in the range –4 to +2 per mil (see Figure 4.1), which is similar to the abiogenic precipitates of ooids and marine cements. Marine organisms exerting a vital effect include the echinoderms, corals and red algae, and generally their values are lighter (^{16}O-enriched) than expected (Figure 4.1 herein; Anderson and Arthur, 1983; Tucker and Wright, 1990).

The carbon isotopic composition of carbonate skeletons again depends on the isotopic composition of seawater and vital effects, but there is little fractionation with temperature. Biogenic grains have $\delta^{13}C$ values varying from near equilibrium (0 to +2) to quite depleted in ^{13}C (see Figure 4.1). Corals, for example, have values down to –8. The deviations from equilibrium values are mainly due to the mixing and isotopic exchange between seawater CO_2 and respiratory CO_2 (usually depleted in ^{18}O and ^{13}C) at or near the site of skeletal precipitation.

TEXTURAL, MINERALOGICAL AND CHEMICAL CHANGES OF SKELETAL ELEMENTS DURING DIAGENESIS

During diagenesis, the skeletal parts of organisms are altered and replaced to different degrees, and their initial chemistries are changed. The most obvious changes occur in the carbonate bioclasts, with aragonite shells being converted to calcite, high-magnesium calcite shells going to low-magnesium calcite, and low-magnesium calcite, high-magnesium calcite and aragonite shells being subject to wholesale dissolution, silicification, dolomitization and pyritization. Siliceous and phosphatic skeletal grains generally show more subtle changes, but they may also be converted to dolomite, calcite, silica and pyrite.

Diagenesis of aragonite fossils

Aragonite is a metastable form of $CaCO_3$ and it is rarely preserved in the geological record. Where aragonitic fossils do occur, they are usually in impermeable lithologies, such as mudstones and marls. The Kimmeridge Clay and Lower Liassic shales of western Europe, for example, contain ammonites still with aragonitic shells. Shell beds in shales in the Purbeck (Jurassic/Cretaceous, southern England) have aragonite bivalves (El-Shahat and West, 1983), and the San Cassiano patch reefs/reef blocks (Triassic, northern Italy)

Figure 4.2 Photomicrographs of bivalves preserved through complete aragonite dissolution and then later precipitation of calcite in the void. A: The upper shell is bivalve, originally of aragonite, now composed of calcite spar so that no relics of the shell structure are preserved. The shell has a conspicuous micrite envelope (arrowed). Also present, lower centre, is a brachiopod shell with perfect preservation of structure (a crossed lamellar fabric with prominent puncti), since the original mineralogy was, and still is, low-magnesium calcite. This shell is encrusted by a bryozoan on one side and fragments of bryozoans also occur in the field of view. Permian, Oman.
B: Bivalve fragments with micrite envelopes where some fracture has occurred (arrow), indicating dissolution of aragonite near the surface, some compaction to break the micrite envelope, and then calcite precipitation within the shell void. Jurassic, Yorkshire.

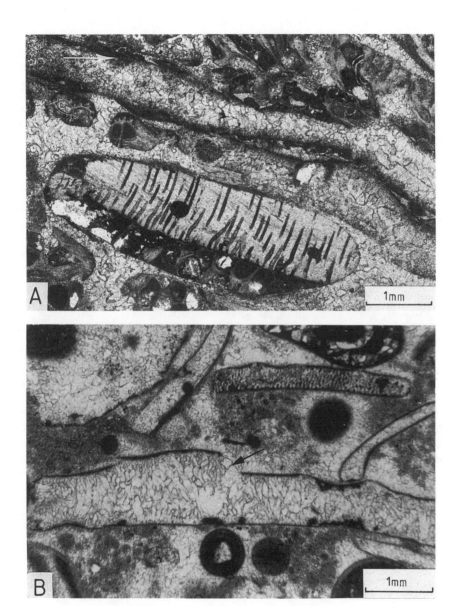

contain aragonite corals, molluscs and sponges as a result of early cementation and enclosure in low-permeability marls (Scherer, 1977; Bruni and Wenk, 1985). Still-aragonite bivalves occur within pyritic-sideritic concretions in Upper Cretaceous shales of North Dakota (Carpenter *et al.*, 1988). Aragonite fossils also occur in asphalt-impregnated limestones and sandstones, where very early oil entry prevented later meteoric or burial waters from penetrating the formation. A good example is the famous Buckhorn quarry of Oklahoma, containing perfectly preserved Upper Carboniferous aragonite cephalopods and benthic molluscs (see Brand, 1989a). Mineralogically unaltered aragonitic shells also preserve the original skeletal structure, as shown by scanning electron microscope (SEM) studies, and the geochemistry should also be pristine. In the case of the Buckhorn fossils, the strontium and stable isotope signatures are original values.

In the majority of cases, skeletal aragonite is converted to calcite and this takes place in one of two ways: either wholesale dissolution and later calcite fill of the mould, or calcitization. In the first case, complete aragonite dissolution takes place and then calcite is precipitated into the void. There may be a significant time-gap between the two events, and, of course, there are many instances where the fossils are simply left as biomoulds. The calcite fill is mostly a clear, equant spar, showing a drusy fabric of an increase in crystal size towards the cavity centre (Figure 4.2). Obviously, there is no preservation of the original skeletal structure or geochemistry in this type of preservation.

After aragonite dissolution, the original shape of the biomoulds is maintained by a partially to well-cemented host sediment, by a cement coating around the shell, or by a micrite envelope (Figure 4.2). The last is a very common feature of shells in the shallow-marine environment and it is produced by endolithic microbial organisms, notably the cyanobacteria. The outer part of the shells is altered to a micritic carbonate and there is usually some microbial encrusting as well. This sea-floor alteration process may produce completely micritized biogenic grains, where all traces of the original shell structure are destroyed. If dissolution of shell aragonite takes place early in a sediment's history, and there is a significant time-gap before calcite precipitation, then the micrite envelopes may be broken through the subsequent effects of overburden pressure (Figure 4.2B).

With some aragonite fossils, there is no wholesale dissolution, but replacement of the aragonite by calcite takes place across a thin film, with dissolution of aragonite on one side and precipitation of calcite on the other. This is the process of calcitization. In this way, some vestiges of the original skeletal structure can be preserved through the retention of organic matter or minute crystal relics of aragonite in the replacing calcite crystals (Figure 4.3). These inclusions may preserve the growth patterns of the skeleton. The calcite itself varies from small to large neomorphic crystals of irregular shape; they may have undulose extinction and are commonly pseudopleochroic from colourless to pale brown (Hudson, 1962). The calcitization of Jurassic *Neomiodon* shells has been described in detail by Sandberg and Hudson (1983), and abundant aragonite relics were seen with the SEM. Organic matter around the original crystallites probably accounts for their preservation. In the diagenesis of coral skeletons, an aragonitic chalky zone has been observed between the aragonite and replacement calcite (James, 1974; Pingitore, 1976). This may be a characteristic feature of diagenesis in the meteoric phreatic zone. The transformation process of aragonite to calcite in a Pleistocene marine gastropod (*Strombus*) has been shown to take place across a film less than 100 Å thick, without any intervening chalky zone or visible pore space (Wardlaw *et al.*, 1978).

Calcitized aragonitic skeletons preserve to varying degrees the original skeletal chemistry. The high strontium content of many formerly aragonitic shells is reduced, but several thousand parts per million may still be present. Sodium is also reduced and the oxygen and carbon isotopic values generally become negative, the amount of ^{18}O and ^{13}C depletion depending on the diagenetic environment (see below). Magnesium, iron and manganese can be enriched in the diagenetic calcite compared to the original aragonite, again depending on the diagenetic environment, particularly the redox potential for Fe^{2+} and Mn^{2+}. Figure 4.4 summarizes the main chemical changes taking place during calcitization, based on alteration of Pennsylvanian bivalves from Oklahoma (Brand, 1989a).

Diagenesis of calcitic fossils

Calcite is much more stable than aragonite in diagenetic environments and so calcite fossils are generally well preserved in the geological record, unless silicified or dolomitized. However, high-magnesium calcite (HMC) is less stable than low-magnesium calcite (LMC), so that dissolution of HMC skeletal elements, as well as silicification and dolomitization (see below), are more common than of LMC fossils. Small vugs may develop within HMC fossils, such as echinoderms (for example, Dorobek, 1987). During diagenesis, HMC fossils lose their Mg^{2+} to give diagenetic low-magnesium calcite (dLMC). This process has been referred to as 'incongruent dissolution' (Land, 1967; Bathurst, 1975), as it appears that in many cases the $MgCO_3$ is lost into solution without any disruption of the calcite lattice structure, at least at the microscopic level. At the electron microscope scale, alteration of the skeletal structure can be observed (see, for example, Sandberg, 1975, on bryozoans; Towe and Hemleben, 1976, on miliolid foraminifera; Manze and Richter, 1979, on echinoids). The process appears to operate through nanno-scale $CaCO_3$ dissolution and reprecipitation. The solubility of magnesian calcites depends on the Mg^{2+} content and on the microstructure of the particle (Walter, 1985). Calcite with 12 mole% $MgCO_3$ is of similar stability to aragonite, and with more Mg^{2+} it is more soluble.

During the transformation of HMC to dLMC small crystals of dolomite are locally produced. These so-called 'microdolomites' are common in echinoderm grains and have been described from Carboniferous crinoids by Leutloff and Meyers (1984). They are best seen through SEM examination of lightly etched polished surfaces, but they can also be observed through their cathodoluminescence. The formation of these crystals reflects the degree of openness of the diagenetic system. If there is a high water/rock ratio, then

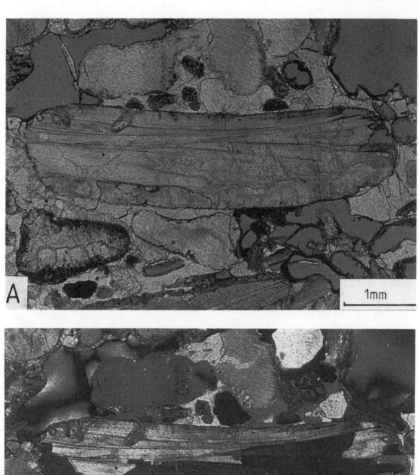

Figure 4.3 Photomicrographs of a calcitized bivalve shell. The internal structure of the originally aragonitic bivalve is partly preserved (see A), although the
shell is now composed of coarse calcite crystals, as shown by the view under
crossed polars (B). A nice micrite envelope occurs around the mollusc fragment
lower left. Pleistocene, Florida.

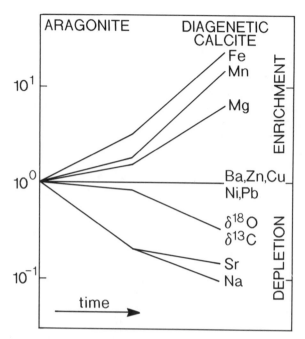

Figure 4.4 Schematic illustration of the changes in trace elements and isotopes in aragonite skeletons as they are converted to calcite during diagenesis. After Brand (1989a) based on Upper Carboniferous molluscs.

they will be rare or absent. In a more closed system (low water/rock ratio), microdolomites should be abundant.

Even though magnesian calcite skeletal grains lose Mg^{2+} during diagenesis, they may contain sufficient to indicate an original HMC mineralogy. If the stabilization of HMC to dLMC takes place in a reducing environment and there is iron available, then the resulting calcite will be ferroan. Ferroan calcite fossils are common and usually they were HMC originally. Richter and Fuchtbauer (1978) used this approach to suggest that rugose corals of the Palaeozoic had skeletons of HMC originally, contrasting with the Mesozoic–Cainozoic aragonitic scleractinian corals.

Skeletal elements originally composed of LMC, such as brachiopods, some bivalves and planktonic foraminifera, are usually perfectly preserved on the micro- and nannoscale of observation. However, there may still be some recrystallization of the constituent crystallites to form larger crystals (an aggrading neomorphism). The best technique for checking this is cathodoluminescence; there may be bright luminescent patches and zones resulting from subtle variations in manganese (the activator of luminescence) and iron (the quencher), picked up during the recrystallization (see, for example, Popp *et al.*, 1986, who illustrated this for Devonian brachiopods).

The trace-element composition of HMC skeletons is usually altered from original values. Strontium and sodium are depleted and iron and manganese are enriched relative to the living organisms. The stable isotope signatures also are likely to be altered. The chemistry of LMC shells, however, is generally unaltered from marine values or they show only slight depletions or enrichments. Al-Aasm and Veizer (1982), for example, studying the trace-element contents of Ordovician brachiopods, found that there was less than 20 per cent alteration of original values in spite of some 450 million years of contact with meteoric waters. By way of contrast, the sediment itself was much more highly altered. Brachiopods are regarded as the best shallow-water fossils to give seawater carbon and oxygen isotope values since they are altered less than most fossils. In detail, there are variations between the different brachiopod genera, largely reflecting the shell microstructure and susceptibility to recrystallization (see Popp *et al.*, 1986; Veizer *et al.*, 1986; Brand, 1989b). In pelagic deposits, the foraminifera are used for isotope stratigraphy.

Dolomitization of carbonate skeletons

Limestones are commonly dolomitized during diagenesis although there is still much debate over the mechanisms responsible. Carbonate fossils are dolomitized in several different ways, depending on their original mineralogy and texture. Dolomite crystals can faithfully pseudomorph the crystalline components of a carbonate fossil by precipitation in optical continuity with the precursor crystallites. One of the main factors for pseudomorphic replacement is the number of nucleation sites of the replacing dolomite crystals (Sibley, 1982). For most carbonate fossils, composed of millions of micro-crystals, there are numerous nucleation sites and so pseudomorphic replacement, at least at the microscopic scale, is common. HMC and aragonite fossils do tend to be dolomitized preferentially relative to LMC fossils, and HMC skeletons are commonly dolomitized with good fabric retention. Perfectly preserved but dolomitized coralline (red) algae, foraminifera such as *Amphistegina* and echinoid grains are common in Plio-Pleistocene dolomites of the Caribbean and Bahamas (for example, Land and Epstein, 1970; Sibley, 1980). In many Palaeozoic dolomites, crinoid ossicles, originally HMC, are single dolomite crystals. Most echinoderm skeletal elements are large, commonly single calcite crystals and so few nucleation sites are required here for pseudomorphic replacement. It appears that the local concentrations of $MgCO_3$ in magnesian calcite skeletons are instrumental in their dolomitization (see, for example, Blake *et al.*, 1982).

Aragonite skeletons such as those of corals, molluscs and green algae (for example, *Halimeda*) in Plio-Pleistocene dolomites are either replaced by fabric-destructive, anhedral dolomite, or more commonly their moulds are partly to completely filled by dolomite cement (an equant drusy spar), indicating that aragonite dissolution took place just before dolomite precipitation (see, for example, Land, 1973, on the Hopegate Formation of Jamaica). LMC fossils tend to resist dolomitization, unless it is all-pervasive. In the geological record, for example, there are many dolomitic limestones where the brachiopods are still calcite in a dolomite matrix. Where LMC fossils are dolomitized then it is usually a fabric-destructive, anhedral dolomite mosaic.

The timing of dolomitization is an additional factor in fossil preservation since if aragonite and HMC skeletons have already been stabilized to dLMC, before the dolomitization, then they will be dolomitized destructively. Late burial dolomitization is typically fabric-destructive for this reason. The dolomitization of skeletal grains has been studied experimentally by Bullen and Sibley (1984), who confirmed some of the general observations noted above. They found that echinoids, coralline algae and foraminifera are dolomitized with fabric retention, although the dolomite crystals are somewhat coarser than the original HMC crystallites, whereas bivalves, gastropods and corals (aragonite) were dolomitized with fabric destruction, and less quickly. Interestingly, mimic replacement of the HMC grains was still observed when they were converted to LMC before dolomitization, suggesting, in this case, a crystal size control rather than a mineralogical control.

Silicification of carbonate skeletons

Carbonate skeletons are quite commonly replaced by silica and mostly the replacing quartz is a microcrystalline mosaic of equant crystals giving a pinpoint extinction. However, fibrous quartz is a common replacement and two types can be distinguished here: the normal, length-fast chalcedonic quartz; and length-slow chalcedony, also called quartzine. The latter is commonly associated with replaced evaporites but there are now many descriptions of quartzine occurring in fossils, where it usually occurs as microspherules (see, for example, Jacka, 1974; Loope and Watkins, 1989). Intraskeletal cavities of silicified fossils are commonly filled with coarse, equant, drusy megaquartz crystals. In many limestones containing silicified fossils there is a range of susceptibilites of the various biogenic grains to silicification. Brachiopods, calcitic bivalves such as *Ostrea*, crinoids and corals are commonly silicified, whereas bryozoans, originally aragonitic molluscs and foraminifera are more rarely affected. Even within one fossil group, certain genera may be silicified preferentially (the spiriferid brachiopods, for example, appear to be more susceptible to silicification than other brachiopods).

The source of silica for fossil replacement probably comes largely from biogenic opaline silica; in many shallow-water limestones this is sponge material, and in deeper-water carbonates the siliceous plankton as well as sponges. The driving forces behind the silicification will be the pH and temperature, with the skeletal microstructure controlling which fossils are replaced.

Replacement of carbonate skeletons by pyrite, hematite, siderite and berthierine

Pyrite replacement of carbonate fossils mostly takes place in organic-rich sediments, especially mudrocks, and there is much variation in the preservation of original skeletal structures. The pyrite occurs in the form of micro-crystalline mosaics with some retention of the host internal structure, as fabric-destroying spheroidal aggregates and framboids, and as euhedral crystals, commonly nucleating along cracks in the original fossil, but also not normally preserving any structural details. Pyritization of fossils can be selective, with aragonitic fossils, such as ammonites and bivalves, being more susceptible than calcitic fossils (see the review of Brett and Baird, 1986).

The pyritization of fossils is generally an early diagenetic process taking place as a result of the bacterial reduction of sulphate in seawater in the presence of organic matter. It takes place in reducing environments, especially those set up during the shallow burial of organic-rich sediments. Many fine-grained tidal-flat and estuarine sediments have a black, reducing, organic-rich zone a few to a few tens of centimetres below the surface where pyrite is being precipitated. In very restricted environments of poor circulation (like the Black Sea), anoxic conditions may be established on the sea floor and pyrite precipitated directly there. In some instances, where an organism has an external skeleton (for example, a shell or a worm tube) pyritization may take place even while the organism is still living (for example, Clark and Lutz, 1980). This happens in tidal-flat sediments where burrowing bivalves may be living within the zone of pyrite precipitation. Some sulphate does occur within the organic membranes of skeletons and these may act as nucleation points for pyrite replacement. The iron is mostly supplied from clay minerals and other terrigenous grains. The processes of fossil pyritization have been described by Hudson (1982) and Fisher (1986) from the Jurassic of England, and Dick and Brett (1986) from the Devonian of New York state.

Pyritization does not normally occur in non-marine environments, because of the low availability of sulphate in fresh waters. Siderite is the more common iron mineral in these situations and it does replace carbonate shells. Coal measure sequences commonly have non-marine bivalve beds where shells are replaced by siderite.

Oxidation of pyrite does occur in situations where there is a change in the redox potential to positive Eh. Ferric oxide/hydroxide may then form (goethite), eventually giving hematite through an ageing effect. This can take place soon after pyritization through erosion of the fossils on to the sea floor, or much later in a sediment's history, when uplift brings the pyritized fossils into contact with oxygenated meteoric water. Loope and Watkins (1989) attributed oxidization of pyritic fossils in Upper Carboniferous strata of Nebraska and Utah to a regional regression which caused the groundwater table to fall and oxygenated meteoric vadose waters to enter the formation.

In well-oxygenated areas of slow sedimentation, carbonate skeletons may be replaced by hematite on the sea floor or just below. It is likely that hydrated iron oxides are the precursor to the hematite. Such fossil replacements are not uncommon in pelagic limestones, where they form in association with hardgrounds and sea-floor dissolution of aragonite. Hematized fossils (ammonoids and thin-shelled bivalves) occur within the Jurassic ammonitico rosso of the Mediterranean area and in the Devonian griotte of southern France (Tucker, 1973).

Berthierine-replaced fossils are common in Mesozoic ironstones and chamosite-replaced fossils in Palaeozoic ones. The conditions of formation of berthierine $(Fe^{2+}_4 Al_2) (Si_2Al_2) O_{10}(OH)_8$, are not understood, but it is thought to precipitate in the non-sulphidic, post-oxic diagenetic environment, where there is sufficient organic matter for the consumption of all dissolved oxygen by aerobic bacteria, but not enough to bring about sulphate reduction (which would cause pyritization). The Frodingham Ironstone and Raasay Ironstone of the British Jurassic contain bivalves replaced by berthierine. The Silurian Clinton Ironstone of the Appalachians contains fossils replaced by chamosite, which is simply the burial diagenetic/low-grade metamorphic equivalent of berthierine.

Reworking of berthieroidal skeletal grains on to the sea floor results in their oxidation to ferric hydroxides. These will then age to hematite during burial. The Carboniferous Rhiwbina Ironstone of South Wales and the Clinton have hematized fossil beds formed in this way (for further information, see Tucker, 1991).

Diagenesis of siliceous fossils

The siliceous skeletons are composed of opaline silica and this is an isotropic amorphous material (referred to as opal-A), containing up to 10 per cent water. Opaline silica is metastable so that it decreases in abundance back through time and is not present in Palaeozoic cherts. In the normal course of events, the diagenesis of biogenic silica is a 'maturation' of opal-A to quartz. The first stage is the formation of a crystalline variety of opal, which is an interlayering of cristobalite and tridymite (called opal-CT, also disordered cristobalite or lussatite). This is identified by X-ray diffraction. The disordered nature probably results from the small crystal size and incorporation of cations into the crystal lattice. The opal-CT replaces radiolarian, diatom and sponge opal-A and also precipitates as bladed crystals on the skeletons. It also forms minute spherical structures called 'lepispheres'. In the next stage, the opal-CT is converted to quartz, which is mostly a microcrystalline equant mosaic. This recrystallization generally obliterates the structure of the original siliceous skeletons.

The driving forces behind the opal-A to quartz transformation are the solubility differences and chemical conditions (Williams *et al.*, 1985). Biogenic silica has a solubility of 120–140 ppm, cristobalite of 25–30 ppm and quartz of 6–10 ppm, in the pH ranges of marine sediment pore water. Once the metastable opal-A dissolves, the solution is saturated with respect to opal-CT and quartz. The precipitation of opal-CT in preference to quartz is probably due to the more internally structured nature of quartz.

In view of the metastable nature of siliceous skeletons, it is not uncommon to find that they have been replaced by other minerals. This is particularly the case where the sediment as a whole is not siliceous. Sponges, radiolarians and diatoms in limestones are commonly replaced by calcite; radiolarians and diatoms in phosphatic and organic-rich sediments are commonly phosphatized. Calcite replacement of siliceous fossils is especially common. The solubility of quartz increases with increasing pH and above a pH of 10 rises dramatically. Alkaline pore-fluids do occur in organic-poor carbonate sediments during shallow burial, and this promotes opal-A dissolution.

Diagenesis of phosphatic fossils

Phosphatic skeletal elements such as bones are relatively stable in most diagenetic environments, but there are changes in their chemistry, and the mineral structure becomes more stoichiometric. The phosphate and fluorine content of bones generally increases with age and the microporosity and larger pores of the bones are usually filled with a francolite cement. Silicification of bones and phosphatic shells is not uncommon.

DIAGENETIC ENVIRONMENTS AND FOSSIL TAPHONOMY

Three major diagenetic environments are distinguished: the marine, the near-surface meteoric, and burial environments (Figure 4.5). Marine diagenesis takes place on the sea floor and just below, and on tidal flats and beaches. In the open marine environment, the processes operating depend very much on water depth, energy level and latitude, but seawater carbonate saturation state is important. Along coastlines, the climate is significant. Meteoric diagenesis can affect skeletons soon after they have been formed, if a storm throws shells up into the supratidal zone, for example, or there is a slight sea-level fall or shoreline regression allowing rainwater to fall on the sediments and their skeletal grains. In addition, fossils may be subjected to meteoric diagenesis when the sediments containing them are uplifted to the Earth's surface many millions of years after deposition. The burial diagenetic environment is from below the zone affected by surface processes, that is, tens to hundreds of metres subsurface, down to several thousands of metres or more, where metamorphic hydration reactions take over.

Skeletal diagenesis in the marine environment

On the shallow sea floor, skeletal grains are commonly altered by microbial micritization, and larger bioclasts may be bored by such organisms as clionid sponges, bivalves such as *Lithophaga* and polychaete worms. In the tropics, microbores are filled by micritic cement, usually HMC. However, in higher latitudes the microbores are usually empty of any precipitates, so that micrite envelopes and micritized grains do not develop to the same extent. Cementation of bioclasts generally occurs in higher-energy tropical environments, such as shelf margin sand shoals and reefs, where much seawater is pumped through the sediments. At the present time, the cements of low-latitude shallow-marine sediments are aragonite and HMC. However, in Jurassic–Cretaceous and mid-Palaeozoic seas, LMC was a common marine cement, especially as a fibrous calcite. Many fossils in grainstones and in reef-rocks are encrusted in such fibrous calcite (bryozoans in Carboniferous mud-

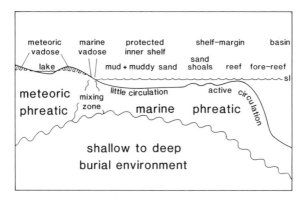

Figure 4.5 Principal diagenetic environments, schematically drawn for a carbonate shelf but applicable to siliciclastic settings, too. From Tucker and Wright (1990).

mounds are a good example). Much ancient echinoderm debris was cemented on the sea floor by syntaxial calcite overgrowths, probably of both LMC and HMC. This type of cement does not occur in modern shallow-marine sediments.

In higher latitudes, modern shallow seas are undersaturated with respect to CaCO$_3$, so that dissolution of skeletal grains does take place on the shallow sea floor (for example, Alexandersson, 1978).

When LMC was the dominant marine cement, the shallow seas were saturated with respect to calcite, but undersaturated with respect to aragonite. Aragonite shells in areas of slow sedimentation were then liable to dissolution. Examples of this have been described by Palmer *et al.* (1988) from Jurassic and Ordovician hardgrounds.

In modern deeper marine environments, seawater becomes undersaturated with respect to aragonite after some few hundreds of metres and with respect to calcite below several thousands of metres in tropical regions. Below the carbonate compensation depth (CCD), where the rate of carbonate dissolution equals the rate of carbonate supply, no calcareous fossils are preserved. Thus, in the pelagic realm, aragonite microfossils such as the pteropods are lost at relatively shallow depths, whereas calcitic micro- and nannofossils (foraminifera and coccoliths) survive to greater depths. Dix and Mullins (1988) have documented changes in the mineralogy of periplatform ooze from the slopes of the Bahamas as the result of such dissolution. In the geological record, pelagic limestones such as the ammonitico rosso and griotte commonly show the effects of sea-floor dissolution of aragonite ammonite shells, as noted earlier. The ammonites are preserved as internal moulds ('steinkerns'). There are also limestones where all evidence for the former presence of ammonite shells has gone, and only the calcitic aptychi are preserved. These sediments were deposited close to the CCD.

Once buried in marine pore waters, then the diagenesis of skeletal material is very much determined by the organic matter content of the sediments, and the Eh-pH of the microenvironment. If there is little organic matter present and pore fluids are oxic, then the skeletal material is likely to be preserved with its original mineralogy and chemistry into the burial environment. This is the normal situation for fossils occurring in open-marine limestones. In sediments with a moderate organic content berthierine and siderite replacement are possible (see above). With high organic contents, much sulphate reduction is likely in marine pore waters, leading to anoxic conditions and negative Eh. Dissolution of calcareous skeletons and pyritization of shells may occur.

Skeletal diagenesis in the meteoric environment

In the near-surface meteoric environment, the vadose zone above the water-table is distinguished from the phreatic zone below. In general, pore water moves more quickly through the vadose zone than through the phreatic. In the upper part of the vadose zone, pore fluids are usually undersaturated with respect to $CaCO_3$, but as fluids move down, dissolving calcareous material, they become saturated and calcite cement may be precipitated. Thus, in the upper part, fossils are commonly dissolved and here primary mineralogy is a control, with the aragonite material going first, the HMC bioclasts losing their magnesium, and the LMC shells being the least soluble. The aragonite skeletons will dissolve completely if the pore fluids are very undersaturated, but if the fluids are only just below saturation point, then calcitization will occur.

In the phreatic meteoric environment, the slow-moving fluids tend to result in little skeletal alteration. Calcitization of aragonite bioclasts is more common here, and chalky alteration is typical. Close to the water-table itself, where there may be substantial fluid flow, extensive dissolution does take place. Climate is an important factor in meteoric diagenesis, controlling fluid flow rates and degree of evaporation.

Many uplifted porous rocks lose their fossils in the meteoric environment. Fossils in sandstones are particularly susceptible to dissolution, especially if the sediment itself is cemented by calcite and this is removed in the early stages of uplift. Fossils in impervious mudrocks and limestones are generally unaffected by near-surface processes. Skeletal material which has already been replaced by reduced iron minerals, however, is likely to be further altered in the normally oxidizing surface meteoric environment to hydrated iron oxides (goethite-limonite).

Skeletal diagenesis in the burial environment

During burial, diagenetic processes affecting fossils are mainly those of stabilization, replacement and compaction; wholesale dissolution of fossils does occur, but normally the voids are filled by other minerals rather than being left to give moulds. Aragonite and HMC skeletal grains not converted to LMC near the surface are stabilized to dLMC during burial. Calcitization of aragonite shells is more common in this situation, where pore fluids are normally saturated with respect to $CaCO_3$, rather than complete dissolution with later calcite precipitation. Evidence for burial calcitization is seen where skeletal grains have been fractured, and the fracture surfaces follow the original shell structure, showing that the grains were still aragonite when the compaction was taking place. Post-compaction calcitization of bivalves has been described by Kendall (1975) from the Jurassic of Saskatchewan. In the burial environment, pore fluids may contain iron, in which case ferroan calcite will be the replacement mineral of aragonite and HMC shells. Replacement of calcareous skeletal material by silica, dolomite and ferroan dolomite may take place in the burial environment. The last two minerals are precipitated more easily at depth where the higher temperatures help over-come various kinetic hindrances of fluids nearer the surface (see review in Tucker and Wright, 1990).

The maturation of siliceous skeletal material from opaline silica to quartz takes place during burial. Where sediments are composed entirely of siliceous microfossils, the silica stabilization is associated with a decrease in porosity.

A common feature of the burial diagenesis of fossils is their compaction. Initially, the skeletal material is more closely packed in the sediment as the overburden pressure increases, but then the skeletons may break, especially if they are relatively thin. The nature of the host sediment is obviously sig-nificant here, especially the amount of clay present. In mudstones, muddy sandstones and muddy limestones, fracture of shells is more widespread than in clean limestones, where usually there has been a degree of early cementa-tion. Experimental studies by Shinn and Robbin (1983) showed that in muddy sediments, shells were only broken when they came into contact with each other, in spite of a sediment thickness reduction of more than 50 per cent. Where early dissolution of aragonite has taken place, then micrite envelopes and cement crusts around the former skeletons will be fractured and broken during burial compaction if the voids were not filled with calcite spar earlier.

Fossils in mudrocks especially can be preserved from the effects of compac-tion by being encased in early diagenetic nodules. There are numerous examples of such full-bodied preservation, such as crabs in Eocene London Clay septaria, ammonites in Lower Liassic calcareous nodules and sponges in chalk flints. It has been suggested that the decomposing organisms set up local chemical environments conducive to nodule formation, or that the fossils acted as nuclei for mineral precipitation. Such fossiliferous nodules have been described by Carpenter *et al.* (1988) from the Upper Cretaceous of North Dakota.

SUMMARY

This chapter has attempted to provide a review of how fossil skeletons are affected during diagenesis. The skeletons of organisms are made of a range of substances, but these may be altered to varying degrees during diagenesis. The changes may begin while the organism is still living, but mostly they take place once the animal or plant has died, and the skeleton has become available for fossilization. The mineralogy of the skeleton is a major factor in determining its diagenetic path, but the nature of the diagenetic environment, the pore-fluid composition and flow rate are also important. Studies of fossil preservation provide explanations for the nature of the fossils themselves, but in addition much useful information is obtained on the depositional environment and the subsequent history of the sediment.

REFERENCES

Al-Aasm, I.S. and Veizer, J., 1982, Chemical stabilization of low-Mg calcite: an example of brachiopods, *Journal of Sedimentary Petrology*, **52** (4): 1101–10.

Alexandersson, T., 1978, Destructive diagenesis of carbonate sediments in the eastern Skagerrak, North Sea, *Geology*, **6** (6): 324–7.

Anderson, T.F. and Arthur, M.A., 1983, Stable isotopes of oxygen and carbon and their application to sedimentologic and paleoenvironmental problems. In M.A. Arthur, T.F. Anderson, I.R. Kaplan, J. Veizer and L.S. Land, *Stable Isotopes in Sedimentary Geology, SEPM Short Course*, **10**: 1.1–1.151.

Bathurst, R.G.C., 1975, *Carbonate Sediments and Their Diagenesis*, Elsevier, Amsterdam.

Blake, D.F., Peacor, D.R. and Wilkinson, B.H., 1982, The sequence and mechanism of low-temperature dolomite formation: calcian dolomites in a Pennsylvanian echinoderm, *Journal of Sedimentary Petrology*, **52** (1): 59–70.

Brand, U., 1989a, Aragonite–calcite transformation based on Pennsylvanian molluscs, *Geological Society of America Bulletin*, **101** (3): 377–90.

Brand, U., 1989b, Biogeochemistry of Late Paleozoic North American brachiopods and secular variation of seawater composition, *Biogeochemistry*, **7** (2): 159–93.

Brett, C.E. and Baird, G.C., 1986, Comparative taphonomy: a key to paleoenvironmental interpretation based on fossil preservation, *Palaios*, **1** (3): 207–27.

Bruni, S.F. and Wenk, H.R., 1985, Replacement of aragonite by calcite in sediments from the San Cassiano Formation (Italy), *Journal of Sedimentary Petrology*, **55** (2): 159–70.

Bullen, S.B. and Sibley, D.F., 1984, Dolomite selectivity and mimic replacement, *Geology*, **12** (11): 655–8.

Carpenter, S.C., Erickson, J.M., Lohmann, K.C. and Owen, M.R., 1988, Diagenesis of fossiliferous concretions from the Upper Cretaceous Fox Hills Formation, North Dakota, *Journal of Sedimentary Petrology*, **58** (4): 706–23.

Clark, G.R. and Lutz, R.A., 1980, Pyritization in the shells of living bivalves, *Geology*, **8** (6): 268–71.

Dick, V.B. and Brett, C.E., 1986, Petrology, taphonomy and sedimentary environments of pyritic fossil beds from the Hamilton Group (Middle Devonian) of western New York, *Bulletin of the New York State Museum*, **457**: 102–27.

Dix, G.R. and Mullins, H.T., 1988, Rapid burial diagenesis of deep-water carbonates: Exuma Sound, Bahamas, *Geology*, **16** (8): 680–3.

Dorobek, S.L., 1987, Petrography, geochemistry and origin of burial diagenetic facies, Siluro-Devonian Helderberg Group (carbonate rocks), central Appalachians, *American Association of Petroleum Geologists Bulletin*, **71** (5): 492–514.

El-Shahat, A. and West, I., 1983, Early and late lithification of aragonitic bivalve beds in the Purbeck Formation (Upper Jurassic–Lower Cretaceous) of southern England, *Sedimentary Geology*, **35** (1): 15–41.

Fisher, I.S.J., 1986, Pyrite replacement of mollusc shells from the Lower Oxford Clay (Jurassic) of England, *Sedimentology*, **33** (4): 575–85.

Hudson, J.D., 1962, Pseudo-pleochroic calcite in recrystallized shell limestones, *Geological Magazine*, **99** (6): 492–500.

Hudson, J.D., 1982, Pyrite in ammonite-bearing shales from the Jurassic of England and Germany, *Sedimentology*, **29** (5): 639–67.

Jacka, A.D., 1974, Replacement of fossils by length-slow chalcedony and associated dolomitization, *Journal of Sedimentary Petrology*, **44** (2): 421–7.

James, N.P., 1974, Diagenesis of scleractinian corals in the subaerial vadose environment, *Journal of Paleontology*, **48** (4): 785–99.

Kendall, A.C., 1975, Post-compactional calcitization of molluscan aragonite in a Jurassic limestone from Saskatchewan, Canada, *Journal of Sedimentary Petrology*, **45** (2): 399–404.

Land, L.S., 1967, Diagenesis of skeletal carbonates, *Journal of Sedimentary Petrology*, **37** (3): 914–30.

Land, L.S., 1973, Holocene meteoric dolomitization of Pleistocene limestones, north Jamaica, *Sedimentology*, **20** (3): 411–24.

Land, L.S. and Epstein, S., 1970, Late Pleistocene diagenesis and dolomitization, north Jamaica, *Sedimentology*, **14** (3/4): 187–200.

Leutloff, A.H. and Meyers, W.J., 1984, Regional distribution of microdolomite inclusions in Mississippian echinoderms from southwest New Mexico, *Journal of Sedimentary Petrology*, **54** (2): 432–46.

Loope, D.B. and Watkins, D.K., 1989, Pennsylvanian fossils replaced by red chert: early oxidation of pyritic precursors, *Journal of Sedimentary Petrology*, **59** (3): 375–86.

Manze, U. and Richter, D.K., 1979, Die Veränderung des $^{13}C/^{12}C$-Verhältnisses in Seeigelcoronen die der Unwandlung von Mg-calcit in Calcit unter meteorisch-vadosen Bedingungen, *Neues Jahrbuch für Geologie und Paläontologie Abhandlungen*, **158** (3): 334–45.

Palmer, T.J., Hudson, J.D. and Wilson, M.A., 1988, Palaeoecological evidence for early aragonite dissolution in ancient seas, *Nature*, **335** (6193): 809–10.

Pingitore, N.E., Jr, 1976, Vadose and phreatic diagenesis: processes, products and their recognition in corals, *Journal of Sedimentary Petrology*, **46** (4): 985–1006.

Popp, B.N., Anderson, T.F. and Sandberg, P.S., 1986, Textural, elemental and isotopic variations among constituents in Middle Devonian limestones, *Journal of Sedimentary Petrology*, **56** (5): 715–27.

Richter, D.K. and Fuchtbauer, H., 1978, Ferroan calcite replacement indicates former magnesian calcite skeletons, *Sedimentology*, **25** (6): 843–60.

Sandberg, P.A., 1975, Bryozoan diagenesis: bearing on the nature of the original skeleton of rugose corals, *Journal of Paleontology*, **49** (4): 587–604.

Sandberg, P.A. and Hudson, J.D., 1983, Aragonite relic preservation in Jurassic calcite-replaced bivalves, *Sedimentology*, **30** (6): 879–92.

Scherer, M., 1977, Preservation, alteration and multiple cementation of aragonite skeletons from the Cassian Beds (U. Triassic, southern Alps). Petrographic and geochemical evidence, *Neues Jahrbuch für Geologie und Paläontologie Abhandlungen*, **154** (2): 213–62.

Shinn, E.A. and Robbin, D.M., 1983, Mechanical and chemical compaction in fine-grained shallow-water limestones, *Journal of Sedimentary Petrology*, **53** (1): 595–618.

Sibley, D.F., 1980, Climatic control of dolomitization, Sero Domi Formation (Pliocene), Bonaire, N.A. In D.H. Zenger, J.B. Dunham and R.L. Ethington (eds), *Concepts and Models of Dolomitization, SEPM Special Publication*, **28**: 247–58.

Sibley, D.F., 1982, The origin of common dolomite fabrics, *Journal of Sedimentary Petrology*, **52** (4): 1087–1100.

Towe, K.M. and Hemleben, K.M., 1976, Diagenesis of magnesian calcite: evidence from miliolacean foraminifera, *Geology*, **4** (6): 337–9.

Tucker, M.E., 1973, Ferromanganese nodules from the Devonian of the Montagne Noire (S. France) and West Germany, *Geologische Rundschau*, **62** (1): 137–53.

Tucker, M.E., 1991, *Sedimentary Petrology: An Introduction to the Origin of Sedimentary Rocks*, Blackwell Scientific Publications, Oxford.

Tucker, M.E. and Wright, V.P., 1990, *Carbonate sedimentology*, Blackwell Scientific Publications, Oxford.

Veizer, J., Fritz, P. and Jones, B., 1986, Geochemistry of brachiopods: oxygen and carbon isotopic records of Paleozoic oceans, *Geochimica et Cosmochimica Acta*, **50** (8): 1679–96.

Walter, L.M., 1985, Relative reactivity of skeletal carbonates during dissolution: implications for diagenesis. In N. Schneidermann and P.M. Harris (eds), *Carbonate Cements, SEPM Special Publication*, **36**: 3–16.

Wardlaw, N., Oldershaw, A. and Stout, M., 1978, Transformation of aragonite to calcite in a marine gastropod, *Canadian Journal of Earth Science*, **15** (11): 1861–6.

Williams, L.A., Parks, G.A. and Crerar, D.A., 1985, Silica diagenesis: I. Solubility controls, *Journal of Sedimentary Petrology*, **55** (3): 301–11.

Chapter 5

COLOUR PATTERN PRESERVATION IN THE FOSSIL RECORD: TAPHONOMY AND DIAGENETIC SIGNIFICANCE

Neville T.J. Hollingworth and Michael J. Barker

INTRODUCTION

Although there are several short papers in the literature describing preserved colour patterns in various groups from the fossil record, there has been little research on the taphonomic and diagenetic processes leading to their preservation. The geochemistry of fossilized pigments or their derivatives is also poorly known, this being mainly due to the rarity of material. As most descriptions have been based upon single specimens, this has precluded geochemical analysis.

Hoare (1978) summarized most of the published occurrences of fossils with preserved colour patterns and it is evident that molluscs are by far the most abundant group represented. It is worth noting that colour patterns have also been described from brachiopods (Boucot and Johnson, 1968), trilobites (Garretson 1953) and insects (Cockerell, 1906; 1916; 1922; 1927). Some crinoids (Bather, 1892; Blumer, 1965) have also been recorded with both well-preserved colour patterns and pervasive pigmentation.

Apart from the Mollusca, colour pattern preservation in other groups is an exception rather than the rule and it is beyond the scope of this chapter to describe every individual occurrence. Within the Mollusca, gastropods are the most common group encountered with preserved colour patterns and it is clear that the controlling factors affecting the preservation potential of colour patterns are: the chemistry and stability of colour pigment; the taphonomic processes leading to colour pattern preservation; the original shell mineralogy; and any post-burial diagenetic effects.

CHEMISTRY AND STABILITY OF COLOUR PATTERNS

Colour patterns on shells result from the deposition of different types of pigment within the outer shell and periostracum. The pigments, or biochromes, in modern gastropod shells are predominantly either a melanin or a tetrapyrole such as the cyclic tetrapyrole, uroporphyrin-1 (Comfort, 1951). Melanins are generally insoluable in organic and acid-based solvents (Comfort, 1951). They therefore have a higher preservation potential than tetrapyroles, which, although thermodynamically stable, are usually soluble in most liquid media (Fox, 1976). The majority of archaeogastropods and some bivalves deposit melanins (see Figure 5.1A) within the outer shell layer (Comfort, 1951), although the neritids and trochids are known to deposit porphyrins.

Colour in modern organisms is produced by a large number of organic pigments. These are mostly associated with soft tissue and, with the exception of any melanin produced, their preservation potential is extremely low. In many pigments, the colouration is incidental to some obvious physiological function in the organism, such as chlorophylls in photosynthesis and haemoglobins in aerobic metabolism. Other functional pigments are involved in warning or camouflage where the pigments are produced in epidermal chromatophores (as melanins, ommochromes, etc.) and iridocytes (as guanine, pterin, etc.). Apparently non-functional pigmentations are frequently attributed to the degradation of hormones or the storage of waste products of digestion. In some cases, a link with diet can be established, as in the accummulation of carotenoids in many birds, crustacea and in the sea fan *Eugorgia ampla*. In other cases, no such link has been established, as in the production of dibromindigotin or Tyrian purple in the hypobranchial gland of many prosobranch gastropods (especially the Muricidae). These occurrences have all been comprehensively reviewed by Fox (1976), Fox and Vevers (1960) and Kennedy (1979).

From the palaeontological perspective, those pigments which are intimately associated with hard parts are the most likely to be preserved, and then only if the organic chemistry of the pigment and the subsequent diagenetic history of both the pigment and the hard parts are favourable. Biochromes which are most commonly incorporated in hard tissues belong to the following groups: melanins; tetrapyroles; ommochromes; sclerotin; pterins; quinnones; and napthazarin pigments.

Melanins

Melanin is produced by oxidation and subsequent polymerization of the amino acid tyrosine. It is highly insoluble and frequently attached to a protein. Depending upon the concentration, it produces colours ranging from black-brown-red to yellow, although other black pigments of unknown composition and structure have been erroneously called melanin, including sclerotins and ommochromes (see below). Melanin is widely distributed in the

Figure 5.1   (A) *Pleurotomaria hirasei* (Recent), Sea of Japan. Note colour pattern of light and dark bands. Dark bands are probably due to melanins ($\times$ 1.2). (B) *Mourlonia antrina* (Kazanian), Tunstall Hills, Sunderland, Tyne and Wear.

Colour pattern preserved as a series of light and dark bands especially on the early whorls (× 1.5). (C) *Beecheria hastatus* (Dinantian), Treak Cliff, Castleton, Derbyshire. Colour pattern preserved as light brown radial bands (× 1). (D) *Pseudamusium* sp. (Dinantian), Treak Cliff, Castleton, Derbyshire. Colour pattern preserved as distinctive radial bands (× 1).

animal kingdom both in hard and soft tissue. The best-known occurrence of melanin is in cephalopod ink (both fossil and Recent). The dark melanin pigments of insects are involved in photoprotection against ultraviolet light (Fox and Vevers, 1960). Epidermal chromatophores which produce melanin (melanophores) and melanocytes are widely distributed and function as thermoregulators, camouflage, etc. in several animal groups. According to Comfort (1950; 1951), melanin is also widely distributed among the Mollusca, not only in soft tissues such as photoreceptors and mantle integument, but also in the general colour of the shells, as in the pulmonate gastropod *Lymnaea* or the intertidal prosobranch gastropod *Littorina*, where melanin is associated with the protein of the periostracum. Melanins are also widely distributed in all the Echinodermata except for the Asteroidea (Kennedy, 1979).

Tetrapyroles

Two groups of tetrapyroles are found as biochromes in hard tissues: the cyclic class, or the porphyrins; and the linear/open chain class, or the bilochromes (bilins or bile pigments). Overwhelmingly, the shell colouration of molluscs has been attributed to the secretion of the porphyrin, uroporphyrin-1 (Fox, 1966). The shell of the extant brachiopod *Terebratella rubicunda* also displays the characteristic red fluorescence of porphyrins under ultraviolet light. Among the echinoderms, the starfishes *Uraster (=Asterias) rubens*, *Luidia ciliaris* and *Astropecten irregularis* have been reported to contain porphyrins in their integuments, but the only reported occurrence of porphyrins in the tests of echinoderms is in the echinoid *Arbacia lixuda* (Kennedy and Vevers, 1972).

   According to Kennedy (1979), the Mollusca are the only phylum in which the linear tetrapyroles, the biliproteins, have been detected and the presence of bile pigments in the shells of *Haliotis*, *Turbo* and *Trochus* is well known (Tixier, 1945; Rüdiger, 1970). However, these molluscs are voracious feeders of red algae (Rhodophyceae) and their pigment is probably derived from their food preferences, namely the red-coloured phycoerythrobilin of the rhodophytic algae. The fact that the shell colours of *Haliotis* and *Turbo* can be varied by feeding them with different algal diets confirms this (Underwood, 1979). A clear link between shell colour and diet cannot, however, be established in other gastropod species.

Ommochromes

Ommochromes are only found in invertebrates and, like most biochromes, they are conjoined to proteins. The skin chromatophores of *Sepia officinalis* and other cephalopods secrete ommochromes, but they are more usually associated with eyes and other light receptors. They are found in the epidermal tissues of the Crustacea and Insecta. For example, the red-brown wing pigmentation of the Nymphalidae (tortoiseshell butterflies) is due to the ommochrome, ommatin (Fox and Vevers, 1960).

Sclerotin

This is a quinnone-tanned protein responsible for the dark colouration of some insect and crustacean cuticles. It has been reported from the byssus and periostracum of *Mytilus*, and the ligament of *Anodonta* (Fox and Vevers, 1960).

Pterins

When present as integumentary pigments, pterins are found in poikilothermic vertebrates (especially fishes and amphibians) and, among the invertebrates, they are particularly prominent in the Lepidoptera and Hymenoptera.

Quinnones and naphthazarins

These are the echinochromes (polyhydroxynaphthaquinnone) and spini-chromes (substituted naphthazarins) associated with the tests of echinoderms. Their distributions have been reviewed by Anderson *et al.* (1969) and Thompson (1971). Some species of Hawaiian echinoid have been shown to contain as many as 30 pigments, half as spinichromes (Moore *et al.*, 1965). Recent crinoids contain a different group of quinnone pigments, the anthra-quinnones (see below).

TAPHONOMIC AND DIAGENETIC PROCESSES LEADING TO THE PRESERVATION OF COLOUR PATTERNS

Colour patterns in the majority of molluscs are deposited in the outer shell layers or in the periostracum, seemingly as a byproduct of metabolic processes. Post-mortem transportation commonly results in abrasion and the loss of the pigmented layers. In many vagile gastropods the early (apical) whorls are commonly abraded during life and the pigmented outermost shell layers are usually worn away. Pigments are also degraded through bacterial action and boring by endolithic algae. Prolonged exposure of the shells to sunlight results in loss of pigment through breakdown by ultraviolet light. Rapid burial is therefore an important factor controlling the preservation of colour patterns and this is reflected in the distribution of molluscs with preserved patterns within different lithofacies.

As already mentioned, the melanin and the porphyrin tetrapyrole pigments exhibit great stability and both have been reported from fossil occurrences. Naturally, since fossils which do preserve some of their original colour pattern/pigments are rare, there is a reluctance to undertake any analysis of specimens which would result in their destruction. Consequently, only shell porphyrins which fluoresce red under ultraviolet light have been identified with certainty in fossils and, since these are the dominant shell biochromes in the molluscs, there is a considerable, albeit scattered, fossil record for molluscan shell porphyrins. Normally, their distribution even among the Mollusca is both sporadic and unpredictable. However, among the archaeogastropod superfamily Neritacea, colour preservation is reasonably common and they are the only fossil group in which colour preservation can confidently be expected. The pigment is produced by the activity of secretory epithelial cells either in the mantle edge gland or the periostracal cells at the extreme mantle edge. Ermentrout *et al.* (1986) have modelled shell pattern generation upon a neural model in which the secretion of pigments is by cells under control of the nervous system. In some shells, most of the colour pigment resides in the conchiolin of the periostracum where it is conjoined to the protein; in others, the pigment is more penetratively emplaced in the intercrystalline organic matrix of the outer shell. Its subsequent stability is then dependent upon the stability and diagenetic history of the shell depositional site. In the Neritacea, the outer shell layer is calcite (stable) and the inner layers are composed of crossed-lamellar aragonite. Consequently, provided the passage of groundwater (with attendent changes in Eh and pH) is minimized, either by early cementation of the sediment voids or the presence of clays, and the geothermal history is favourable, then the pigment is usually preserved in neritid gastropods.

Neritid gastropods are a superfamily which were originally marine, and, while they have maintained a marine presence to the present day (usually high tidal/intertidal, as in *Nerita peloronata*), their main evolutionary radiation has been into brackish/freshwater (family Neritidae, such as *Neritina virginea* (brackish) and *Theodoxus fluviatilis* (freshwater)) and terrestrial habitats (the Helicinidae, the tree snails). This habitat distribution is also reflected in the fossil record. Various colour-banded 'species' of *Neritoma* and *Neridomus* have been recorded from the marine Bathonian White Limestone Formation of England and its French equivalent (Cox and Arkell, 1948–50; Barker, 1977; Fischer, 1969). They are not common, but in the brackish water faunas of the Middle Bathonian Sharps Hill Formation, colour-banded specimens of *Neridomus cooksonii* are abundant (Barker, 1969). Similarly, the brackish and freshwater environments of the Upper Eocene/Lower Oligocene Solent Group of the Isle of Wight contain abundant colour-banded neritids: *Theodoxus concavus* (Figures 5.2D–5.2F) from the brackish-water Headon Hill Formation and *T. planulatus* (Figures 5.2B and 5.2C) from the freshwater Bembridge Limestone Formation (Keen, 1977; Insole and Daley, 1985). The morphometrics and colour variation of these two species have been studied by Hill (1986).

Figure 5.2   (A) *Naticopsis* sp. (Bathonian), Taynton Limestone, Snowshill Hill, Gloucestershire. Colour pattern preserved as red/brown stripes. (B, C) *Theodoxus planulatus* (Palaeogene), Bembridge Limestone Formation, Isle of

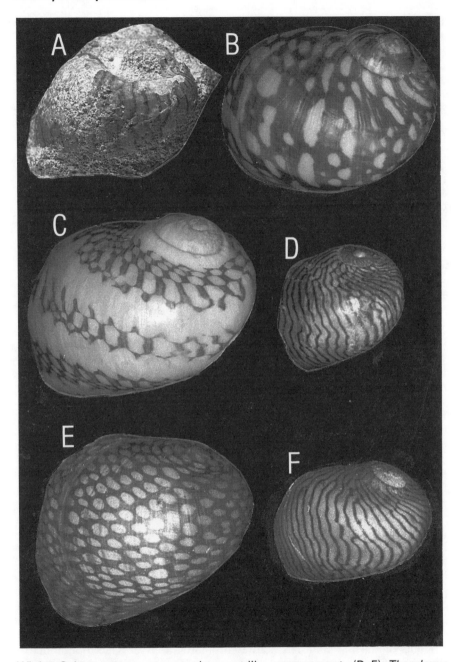

Wight. Colour pattern preserved as net-like arrangement. (D–F) *Theodoxus concavus* (Palaeogene), Headon Hill Formation, Isle of Wight. Colour pattern preserved as sinuous and zigzag stripes, as well as in a complex net-like arrangement. All figures × 15.

Table 5.1   *Colour-banded species examined using a Link system 860 Series II energy-dispersive X-ray analyser combined with a Jeol 35C scanning electron microscope*

Age	Genus/species	Shell mineralogy
Recent	*Pleurotomaria australis*	aragonite outer shell
	Neritina (Vitta) sp.	calcite outer shell
	Theodoxus (Pictoneritina) communis	calcite outer shell
Upper Eocene/	*Theodoxus concavus*	calcite outer shell
Lower Oligocene	*Theodoxus planulatus*	calcite outer shell
Middle Bathonian	*Neridomus cooksonii*	calcite outer shell
Upper Permian	*Mourlonia antrina*	aragonite outer shell neomorphosed to calcite
Lower Carboniferous	*Euconospira conica*	aragonite outer shell neomorphosed to calcite

Table 5.1 lists those taxa of colour-banded fossil gastropod that we have investigated geochemically. In all specimens, atomic spectra (for atomic weights greater than 12) were obtained from both the pigmented and non-pigmented regions of the shells. The spectra of the pigmented and non-pigmented shells, both fossil and Recent, were remarkably similar and no significant differences were recorded (see Figure 5.3). Plas (1969), Křiž and Lukeš (1974), and Kennedy (1979) have suggested that the original pigments in fossil molluscs and brachiopods have been replaced by inorganic compounds—probably ferric oxides. This is clearly not the case, since the atomic spectra in our fossil samples do not record the presence of unusual amounts of iron in the pigmented shells. Furthermore, the fact that these pigments still fluoresce weakly under ultraviolet light strongly suggests that they are still porphyrins, even if somewhat altered.

Since the seminal work of Treibs (1934; 1936), organic geochemists have ascribed great significance to the occurrence of porphyrins (petroporphyrins) in petroleum- and organic-rich sediments. These compounds account for only a very small proportion of the total organic carbon, but their significance lies in their postulated derivation from chlorophyll into desoxophylloerythroetioporphyrin (the Treibs hypothesis). The study of the petroporphyrins and other fossil porphyrin pigments in sediments has been problematic due to the inherent difficulties of isolating these stable, intractable compounds of high molecular weight. Any isolating procedure usually involves relatively large amounts of starting materials before chromatography or spectrophotometry (Baker, 1969). However, as noted by Fox and Vevers (1960), the amount of porphyrin present in shells is often in such small quantities that only the red fluorescence can be seen and there is insufficient material to extract for geochemical analysis. In fossil shells, the aromatic nucleus of the porphyrin

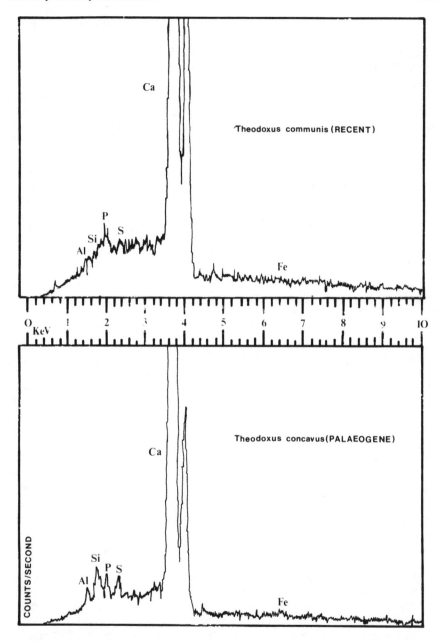

Figure 5.3 Energy dispersive X-ray spectra of pigmented shell in *Theodoxus concavus* (Upper Eocene/Lower Oligocene) and *Theodoxus (Pictoneritina) communis* (Recent). Note similarity of spectra and absence of Fe or other related peaks, suggesting that no inorganic replacement of shell pigment or chelation has taken place.

pigment ensures geological survival under favourable circumstances, but the attached peripheral groups will alter depending upon the environment to which the material has been exposed. Certainly these processes will render extraction more difficult. This may be alleviated by advances in mass spectrometry in which solid samples are directly introduced into the ionizing beam; together with the more abundant fossil material now available, this should yield positive results. Petroporphyrins from sediments are almost invariably found as metallochelates with either vanadium or nickel, and such chelation is thought to stabilize the porphyrins in sediments (Baker, 1969). The atomic spectra of fossil pigments that we have sampled showed no vanadium or nickel peaks. There is some evidence to suggest that the uroporphyrin pigment in archaeogastropods is a chlorophyll derivative (Underwood and Creese, 1976); in other cases the porphyrin may be derived from haems (such as the gut haemochromogens; Kennedy, 1979). The absence of chelation in fossil shell porphyrins presumably reflects the absence of suitable chelating ions in the 'sealed systems' in which the pigments have been preserved.

PIGMENT GEOCHEMISTRY AND COLOUR PATTERN DISTRIBUTION IN THE FOSSIL RECORD

Křiž and Lukeš (1974) suggested that the colour patterns preserved on the gastropod *Platyceras* and the brachiopod *Merista* from the Silurian and Devonian of Czechoslovakia are melanins as the fossils were collected from decalcified micritic limestones which have been subject to intense diagenetic alteration. However, these pigments could equally be altered porphyrins, since proven porphyrin shell pigments from at least the Jurassic onwards show marked stability. Evidence to suggest that the pigments in some, if not all, pleurotomariids are melanins is based upon the fact that recent pleurotomariids, such as *Pleurotomaria hirasei* (Figure 5.1A), do not display red fluorescence under ultraviolet light. Comfort (1951) also noted the absence of porphyrins from the shells of Recent '*Pleurotomaria*'. Cretaceous specimens of *Pleurotomaria fausta* from Hokkaido, Japan, which retain their colour patterns despite complete replacement of the shell microstructure to analcime have been recorded by Kanie *et al.* (1980).

Porphyrins characteristically display bright red fluorescence under longwavelength ultraviolet light, and Nuttall (1969), Wilson (1975), Dockery (1980), Swann and Kelley (1985), and Kelley and Swann (1988) have described a number of Tertiary bivalves and gastropods that are strongly fluorescent. It is probable, however, that the fossil pigments are diagenetic alteration products of the original pigments.

The Palaeozoic bivalves, brachiopods and gastropods illustrated in Figures 5.1B–5.1D have been collected by the authors from Lower Carboniferous and Upper Permian algal/bryozoan build-ups where early marine cementation was ubiquitous. All of the specimens were completely circumcrusted by aragonite which has subsequently been altered to calcite during diagenesis. The precipitation of an early marine cement on the bioclasts soon after death prevented abrasion that would have been caused by post-mortem transportation, and would have effectively sealed the shells from further bacterial and chemical degradation. The decay of soft tissues and the periostracum may have acted as a locus for early cement precipitation. Despite subsequent calcitization of the shells the pigments are still preserved even though neomorphism was fabric-destructive (see Figure 5.4).

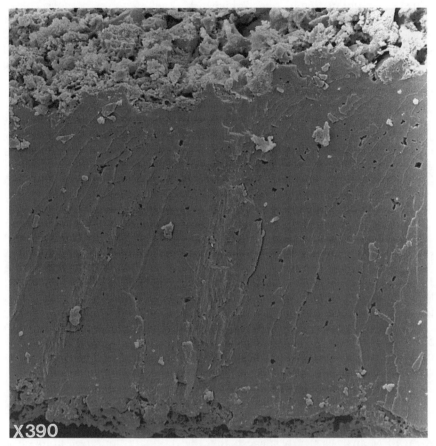

Figure 5.4 Scanning electron micrograph of *Mourlonia antrina* (Upper Permian), × 390. Shell replaced by blocky calcite spar with no relict shell microstructure but colour pattern preserved (see Figure 5.1B).

Early marine cementation (Marshall and Davies, 1981) is also a common feature of many Triassic carbonate build-ups. Neritids with beautifully preserved colour patterns described from the St Cassian Formation in the Alps by Kitti (1899) and Zardini (1978) may have been circumcrusted, and the pigments locked in place by early marine cements.

Early marine cements are generally uncommon in Mesozoic and Tertiary sediments. Fossil colour patterns in molluscan taxa are usually found in fine-grained sediments such as mudstones or shales, especially if the aragonite is still preserved. Several specimens of the ammonite *Asteroceras obtusum* described from the Sinemurian Black Ven Marls of the Dorset coast by Lang and Spath (1926) still preserve colour patterns, because the aragonite shells have remained unaltered and the host sediment has not been subject to extensive post-burial, pervasive diagenetic processes.

It is worth noting that in 'open framework', high-energy carbonates, colour patterns are only found in fossil molluscs where the shells are composed of more stable calcite or possess a bimineralic structure with an outer calcitic shell layer. Such an example is found in the specimen of *Naticopsis* illustrated in Figure 5.2A, which was found in Middle Jurassic shallow-water oolitic grainstones (Taynton Limestone). Although the shell is completely neo-morphosed, the outer calcite shell layer was stable and, combined with rapid burial, effectively sealed and preserved the pigment.

Most of the published descriptions of fossil colour patterns are from Tertiary sediments (see earlier references). They all describe colour patterns in biotas which retain their original shell mineralogy and microstructure and which have not been subject to multi-phase diagenetic events. For example, the specimens of the neritid *Theodoxus concavus* (Figure 5.2D–5.2F) from the Palaeogene Headon Hill Formation were collected from unconsolidated lignitic sandy clays and retain an inner aragonite and calcite outer shell layer (Figure 5.5).

The only published analysis of colour pigment geochemistry in other fossil groups was by Blumer (1965) who has described the preservation of crystal-lized pigments (called 'fringelites') and aromatic hydrocarbons from the Jurassic crinoid *Apiocrinites*. He suggested that both the hydrocarbons and the fossil pigments were derived from the original anthraquinnone pigment of the crinoid by reduction.

Red pigmentation is frequently found in the fossil rhodophytic alga *Solenopora* (Harland and Torrens, 1982) and is probably a derivative of an original biloprotein pigment such as phycoerythrobilin, which is found in extant representatives.

SUMMARY

X-ray/SEM analysis clearly suggests that the residual colouration in fossils is not a mineralogic replacement of the original pigment, but a residual organic compound. Where fluorescence is present, porphyrins are clearly indicated, the degree of fluorescence being proportionate to the amount of degradation and alteration that these cyclic tetrapyroles have undergone during dia-genesis. In the absence of fluorescence, melanins are probably the most common residual pigment, but other biochromes cannot yet be discounted until satisfactory geochemical procedures for the analysis of these intractable pigment residues can be established. The preservation of shell pigmentation is dependent upon the original pigment composition (tetrapyroles and melanins being the most stable), the mineralogy and site of emplacement within the

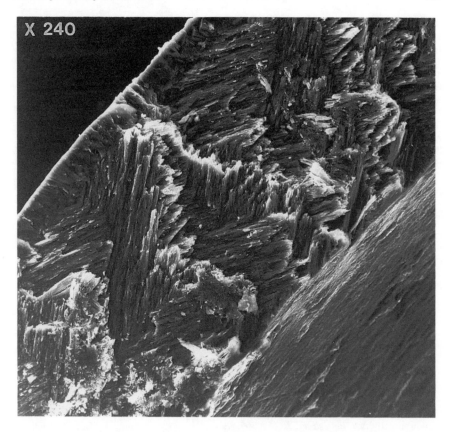

X 240

Figure 5.5 Scanning electron micrograph of *Theodoxus concavus* (Upper Eocene/Lower Oligocene), × 240. Note inner crossed lamellar layers (aragonite) and outer prismatic layer (calcite). See Figures 5.2D–5.2F for colour patterns.

shell, the rapidity of burial and subsequent diagenetic history. In exceptional cases, pigments may survive neomorphism or even complete shell replacement.

ACKNOWLEDGEMENTS

We thank Peter Bond of the SEM Unit at Portsmouth Polytechnic for his help.

REFERENCES

Anderson, M.A., Mathieson, J.W. and Thompson, R.H., 1969, Distribution of spinichrome pigments in echinoids, *Comparative Biochemistry and Physiology*, **28** (1): 333–45.

Baker, E.W., 1969, Porphyrins. In G. Eglinton and M.T.J. Murphy (eds), *Organic geochemistry: methods and results*, Springer-Verlag, Berlin: 464–97.

Barker, M.J., 1969, Snowshill Hill quarry. In H.S. Torrens (ed.), *International field symposium on the British Jurassic—Guides to field excursions from London*, University of Keele: 16–18.

Barker, M.J., 1977, *A stratigraphical, palaeoecological and biometrical study of some English Bathonian Gastropoda (especially Nerineacea)*, Ph.D. thesis, University of Keele (unpublished).

Bather, F.A., 1892, British fossil crinoids, *Annals and Magazine of Natural History*, 9 (6): 202–26.

Blumer, M., 1965, Organic pigments: their long term fate, *Science*, 149 (3685): 722–6.

Boucot, A.J. and Johnson, J.G., 1968, Evidence of color banding in a Lower Devonian rhynchonellid brachiopod, *Journal of Paleontology*, 42 (1): 1208–9.

Cockerell, T.D.A., 1906, Fossil saw-flies from Florissant, Colorado, *Bulletin of the American Museum of Natural History*, 22 (2): 499–501.

Cockerell, T.D.A., 1916, British fossil insects, *United States National Museum Proceedings*, 49 (2119): 469–99.

Cockerell, T.D.A., 1922, Some Eocene insects from Colorado and Wyoming, *United States National Museum Proceedings*, 59 (2358): 29–39.

Cockerell, T.D.A., 1927, Tertiary insects from Kudia River Maritime Province, Siberia, *United States National Museum Proceedings*, 64 (2503): 13.

Comfort, A., 1950, Biochemistry of molluscan shell pigments, *Proceedings of the Malacological Society of London*, 28 (2): 79–85.

Comfort, A., 1951, The pigmentation of molluscan shells, *Biological Reviews of the Cambridge Philosophical Society*, 26 (3): 285–301.

Cox, L.R. and Arkell, W.J., 1948–50, A survey of the Mollusca of the British Great Oolite Series—primarily a nomenclatorial revision of the monographs by Morris and Lycett (1851–55), Lycett (1863) and Blake (1905–07), *Monograph of the Palaeontographical Society*, London.

Dockery, D., 1980, Color patterns of some Eocene molluscs, *Mississippi Geology*, 1 (1): 3–7.

Ermentrout, B., Campbell, J. and Oster, G., 1986, A model for shell patterns based on neural activity, *The Veliger*, 28 (4): 369–88.

Fischer, J.-C., 1969, Géologie, paléontologie et paléoécologie du Bathonien au sud-ouest du Massif Ardennais, *Mémoires du Musée National d'Histoire Naturelle, Nouvelle Série, Série C, Sciences de la Terre*, 20.

Fox, D.L., 1966, Pigmentation of molluscs. In K.M. Wibur and C.M. Yonge (eds), *Physiology of Mollusca II*, Academic Press, New York: 249–74.

Fox, D.L., 1976, *Animal biochromes and structural colours* (2nd edn), University of California Press, Berkeley.

Fox, H.M. and Vevers, G., 1960, *The nature of animal colours*, Macmillan, New York.

Garreston, M.W., 1953, Color in trilobites of Trenton age, *Science*, 117 (3027): 17.

Harland, T.L. and Torrens, H.S., 1982, A redescription of the Bathonian red alga *Solenopora jurassica* from Gloucestershire, with remarks on its preservation, *Palaeontology*, 25 (4): 905–12.

Hill, M.L., 1986, *The morphometrics, palaeobiology and taxonomy of some fossil neritid gastropods*, M. Phil. thesis, Portsmouth Polytechnic (unpublished).

Hoare, R.D., 1978, Annotated bibliography on preservation of color patterns on invertebrate fossils, *The Compass of Sigma Gamma Epsilon*, 55: 39–63.

Insole, A.N. and Daley, B., 1985, A revision of the lithostratigraphical nomenclature of the Late Eocene and early Oligocene strata of the Hampshire Basin, southern England, *Tertiary Research*, 7 (3): 67–100.

Kanie, Y., Takahashi, T. and Mizuno, Y., 1980, Color patterns of Cretaceous pleurotomariid gastropods from Hokkaido, *Science Report of the Yokosuka City Museum*, 27 (290): 37–42.

Keen, M.C., 1977, Ostracod assemblages and the depositional environments of the Headon, Osborne and Bembridge Beds (Upper Eocene) of the Hampshire Basin, *Palaeontology*, **20** (2): 405–45.

Kelley, P.H. and Swann, C.T., 1988, Functional significance of preserved color patterns of mollusks from the Gosport Sands (Eocene) of Alabama, *Journal of Paleontology*, **62** (1): 83–7.

Kennedy, G.Y., 1979, Pigments of marine invertebrates, *Advances in Marine Biology*, **16**: 309–81.

Kennedy, G.Y. and Vevers, H.G., 1972, Tetrapyrrole pigments in the test of the echinoid *Arbacia lixula*, *Journal of Zoology*, **168** (4): 527–33.

Kittl, E., 1899, Die Gastropoden der Esinokalke, nebst einer Revision der Gastropoden der Marmolatakalke, *Annalen des K.K. Naturhistorischen Hofmuseums*, **14** (1–2): 1–237.

Křiž, J. and Lukeš, P., 1974, Color patterns on Silurian *Platyceras* and Devonian *Merista* from the Barrandian area, Bohemia, Czechoslovakia, *Journal of Paleontology*, **48** (1): 41–8.

Lang, W.D. and Spath, L.F., 1926, The black marl of Black Ven and Stonebarrow, in the Lias of the Dorset coast, *Quarterly Journal of the Geological Society of London*, **82** (2): 144–87.

Marshall, J.F. and Davies, P.J., 1981, Submarine lithification on windward reef slopes, Capricorn-Bunker Group, southern Great Barrier Reef, *Journal of Sedimentary Petrology*, **49** (3): 953–60.

Moore, R.E., Singh, H., Chang, C.J.W. and Scheuer, P.J., 1965, Isolation of eleven new spinichromes from echinoids of the genus *Echinothrix*, *Journal of the American Chemical Society*, **87** (17): 4023–4.

Nuttall, C.P., 1969, Colouration. In R.C. Moore (ed.), *Treatise on Invertebrate Paleontology, Part N, Mollusca* **6** (1), Geological Society of America and University of Kansas Press, New York and Lawrence: N70–2.

Plas, L.P., 1969, Some considerations of fluorescence in fossil gastropods and pelecypods, *Term paper, Paleontology*, **217**, University of California, Berkeley.

Rüdiger, W., 1970, Gallenfarbstoffe bei wirbellosen Tieren, *Naturwissenschaften*, **57** (7): 331–7.

Swann, C.T. and Kelley, P.H., 1985, Residual color patterns in mollusks from the Gosport Sand (Eocene), Alabama, *Mississippi Geology*, **5** (3): 1–9.

Thompson, R.H., 1971, *Naturally occurring quinnones*, Academic Press, London.

Tixier, R., 1945, Contribution à l'étude de quelques pigments pyrroliques naturels des coquilles de mollusques, de l'oeuf de l'émeu et du squelette du corail bleu, *Heliopora coerulea*, *Annales de l'Institut Océanographique de Monaco*, **22**: 343–97.

Treibs, A., 1934, Chlorophyll-und Häminderivate in bituminösen Gesteinen, Erdölen, Erdwachsen und Asphalten, *Annalen der Chemie*, **510**: 42.

Treibs, A., 1936, Chlorophyll-and Häminderivate in organischen Mineralstoffen, *Angewandte Chemie*, **49**: 682.

Underwood, A.J., 1979, The ecology of intertidal gastropods, *Advances in Marine Biology*, **16**: 111–210.

Underwood, A.J. and Creese, R.G., 1976, Observations on the biology of the trochid gastropod *Austrocochlea constricta* (Lamark) (Prosobranchia) II. The effects of available food on shell-banding pattern, *Journal of Experimental Marine Biology and Ecology*, **23**: 229–40.

Wilson, J.L., 1975, *Carbonate facies in geologic history*, Springer-Verlag, New York.

Zardini, R., 1978, *Fossili Cassiani (Trias Medio-Superiore). Atalante dei gasteropodi della Formazione di S. Cassiano raccolti nella regione dolomitica attorno a Cortina d'Ampezzo*, Ediziono Ghedina, Cortina d'Ampezzo.

Chapter 6

THE TAPHONOMY OF SOFT-BODIED ANIMALS

Peter A. Allison and Derek E.G. Briggs

INTRODUCTION

The rarity of soft-bodied and lightly skeletoned organisms in the geological record is the most obvious of the preservational biases which result from the process of fossilization. Such organisms only occur as fossils in exceptional sedimentological conditions where decay and biogenic disturbance are inhibited (Seilacher, 1970; Seilacher et al., 1985; Allison, 1988a).

In modern marine communities macroorganisms composed solely of soft-parts can make up as much as 60 per cent of the species and individuals (Jones, 1968). Several studies have attempted to assess the effect of soft-part decay on community preservation potential (Lawrence, 1968; Dörjes, 1972; Driscoll and Swanson, 1973; Schopf, 1978; Table 6.1 herein). Schopf's (1978) study was based on an intertidal fauna at Friday Harbour in Washington State, USA. He combined data from three methods of analysis (morphology, presence of a fossil record of the genus, and presence/absence of remains in sediment samples) in a consideration of mud, sand and rock faunas. The different approaches yielded broadly similar results and Schopf (1978) concluded that 60 per cent of the fauna would be lost during the fossilization process. Surprisingly, the study also showed that similar percentages of organisms would be preserved from each of the different substrates, although Schopf acknowledged that very few rocky intertidal environments are preserved in the rock record.

Mineralized skeletons increase the likelihood of preservation. Thus, a temporal variation in preservation potential of organisms can be expected across the Precambrian–Cambrian boundary to coincide with the evolution of

Table 6.1 *Percentages of species and individuals composed wholly of soft-parts in selected marine environments*

Author	Jones (1968)	Dörjes (1972)	Driscoll and Swanson (1973)
Location	S. California, 0–22 km offshore, 0–180 m depth.	Georgia shelf	Living epifauna colonizing bivalves. Buzzards Bay, Massachusetts.
	Number of species	*Number of species and*	
		% of each group that is preservable	
Polychaetes	523	76 (56%)	57 (4%)
Crustaceans	419	67 (15%)	4 (25%)
Molluscs	408	90 (3.5%)	8 (12%)
Echinoderms	64		1 (0%)
Others	59		36 (80%)
Total	1473	268	106
Likely % of preservable fauna	‹ 33% of total species	21%	39%

hard skeletons (Lowenstam and Weiner, 1989). Temporal variation in preservation potential may lead to a distortion in the assessment of Phanerozoic metazoan diversity (Plotnick, 1986). Sepkoski (1981) estimated that as much as 20 per cent of marine metazoan clades originated from only three of the principal localities for soft-bodied biotas (Burgess Shale, Mazon Creek and Hunsrück Slate). The Middle Cambrian Burgess Shale biota (Conway Morris, 1986), for example, is dominated by non-mineralized organisms.

Decay of soft tissues may also lead to erroneous taxonomic and phylogenetic analysis. Such a bias is frequently encountered when dealing with, for example, fossil plants where leaf, stem, fruit, seed, and root may all be separated from one another (see Greenwood, this volume). A similar bias affects animals. Soft-bodied organisms with isolated skeletal elements are particularly susceptible, since decay liberates preservable spicules, plates, ossicles, etc., to the sediment. For this reason the origins of conodonts were obscure for over a century after their initial discovery. Their affinities were only resolved by the discovery of exceptionally preserved fossils from the Carboniferous of Scotland which included details of soft-part anatomy (Briggs *et al.*, 1983).

Although biotas which yield soft-bodied organisms are rare in comparison with those yielding skeletal remains, they are more common than is normally realized. For example, over 40 major Phanerozoic soft-bodied biotas have been recognized (Conway Morris, 1986) and many more contain a few non-mineralized taxa. Such biotas are deemed to have formed under exceptional circumstances which have conserved soft-parts.

CONDITIONS NECESSARY FOR THE PRESERVATION OF SOFT-PARTS

The conditions most often cited as being responsible for this type of preservation include: minimal transport to limit hydrodynamic damage; anoxia to preclude scavengers and inhibit decay; rapid burial as a means of inhibiting scavenging and promoting anoxia; rapid mineralization as a means of 'freezing' decay-induced information loss; deposition in a bacterially sterile environment; and encapsulation within a microbial mat to promote heightened sediment stability.

Transport

The occurrence of soft-bodied organisms within a biota has been used to infer minimal transport for the Middle Cambrian Burgess Shale of western Canada (Whittington, 1971, p. 1196; Conway Morris 1979, p. 156) and for the Devonian Hunsrück Slate of the Bundenbach district of Germany (Stürmer and Bergström, 1973, p. 106). Such an assumption constrains the spatial relationship between the environment where the organism lived and that in which it was entombed. In addition, it defines the confidence limits for using palaeoenvironmental data derived from the site of deposition in palaeoecological reconstruction. Experimental work (Allison, 1986), however, showed that lightly skeletoned and soft-bodied organisms, when freshly killed, are capable of withstanding lengthy periods of transport in a rotating tumbling barrel without appreciably fragmenting. Conversely, carcasses which have decayed disarticulate even in the absence of agitation. Similar experiments on regular echinoid carcasses have confirmed that disarticulation and fragmentation are dependent upon degree of decay prior to tumbling (Kidwell and Baumiller, 1990). These observations show that the degree of fragmentation and disarticulation of soft-bodied and lightly skeletoned organisms is a function of both decay and transport. Preservation, therefore, cannot be used to infer distance of transport.

Anoxia

Sediments rich in organic matter are often finely laminated and are generally assumed to have been deposited under anaerobic conditions. This association is often used as evidence to support the notion that anoxia is a decay inhibitor (Curtis, 1980). Early taphonomic investigations appeared to confirm this view. Hecht (1933) conducted experiments near Wilhelmshaven on the North Sea. He observed fishes decaying in 50 l aquaria filled with seawater which were either open (aerobic decay) or enriched with hydrogen sulphide (anaerobic decay). Unfortunately he used different species, young gadidae (codfish) in his aerobic experiment, small *Galeus vulgaris* (shark) in his anaerobic experiment, hence only general comparisons can be made between the two. However, the fishes in his aerobic experiments were completely destroyed within four weeks, although decay products had poisoned the tank in half that time. In the presence of inoculated hydrogen sulphide, on the other hand, the shark carcasses apparently decayed at a slow pace and appeared in good condition for over a year, at which point they were disturbed and fell apart. Hecht's experiments emphasized the poor preservation potential of soft-parts and, he believed, demonstrated the differential in decay rate between aerobic and anaerobic systems. However, neither the temperature nor the size of the animals used in the experiment was documented, both key factors in influencing decay rate. In addition, Hecht did not take into account morphological differences between the species, and it is possible that the tough skin of the shark may have helped the carcasses in the anaerobic control to remain intact.

Zangerl and Richardson (1963) conducted similar experiments in Louisiana, enclosing fishes in fine-mesh cages which were deposited underwater upon the sediment surface. They noted extremely rapid decomposition in temperatures of approximately 24–37°C even though the carcasses were covered with black 'obviously reducing' mud. However, since these results were in conflict with those of Hecht (1933), Zangerl and Richardson (1963, p. 168) assumed that some oxygen must have been present and suggested that a certain amount of aerobic decomposition had taken place.

In the absence of oxygen, microbes in sediment utilize a series of alternative oxidants for metabolism. In an idealized sediment profile (Figure 6.1) these oxidants are consumed in a specific order by different bacterial communities (Redfield, 1958; Berner, 1981; Coleman, 1985). Anaerobic processes first appear in most sediments in anoxic microenvironments within otherwise aerobic or dysaerobic conditions. The initial oxidant to be utilized is manganese and this occurs very close to the anoxic–oxic boundary in otherwise dysaerobic sediments. Once this has been used up, nitrate, iron and then sulphates are successively utilized until pore-water depletion occurs (Figure 6.1). Eventually carbon dioxide is consumed in the manufacture of methane (the so-called carbonate-reduction zone). An additional form of microbial respiration is that of fermentation in which neither oxidation nor reduction occurs. This process probably takes place throughout the sediment profile, but is the dominant degradative process beyond the zone of carbonate reduction. Each of these reactions has a different free-energy yield which probably governs their ordering in a sediment profile. Those reactions providing the greatest yield of free energy occur closer to the sediment–water interface and those providing less occur lower (Berner, 1981). The notion that anaerobic processes are less energy-efficient has generally been accepted as the reason for the accumulation of organic carbon in anaerobic environments. This concept has been challenged by Henrichs and Reeburgh (1987) and Cannfield (1989) who compared a series of published rates of degradation from anaerobic and aerobic settings. These authors concluded that there was little intrinsic difference in overall rate between the two degradative pathways. Thus, organic-rich sediments are anoxic because of the carbon they contain; they do not contain the carbon because they are anoxic. Cannfield (1989) suggested that a rate diffential of a factor of only 2 or 3 is sufficient to facilitate the accumulation of carbon within anaerobic sediments.

Rate of degradation of disseminated organic carbon falls with depth below the sediment–water interface (Westrich and Berner, 1984). This, however, is

AEROBIC DECAY

$(CH_2O)_{106}(NH_3)_{16}H_3PO_4 + 106\ O_2$ → $106\ CO_2 + 16\ NH_3 + H_3PO_4 + 106\ H_2O$

ANAEROBIC DECAY

Manganese Reduction

$(CH_2O)_{106}(NH_3)_{16}H_3PO_4 + 212\ MnO_2 + 332\ CO_2 + 120\ H_2O$ → $438\ HCO_3^- + 16\ NH_4^+ + HPO_4^{2-} + 212\ Mn^{2+}$

Nitrate Reduction

$(CH_2O)_{106}(NH_3)_{16}H_3PO_4 + 84{\cdot}8NO_3^-$ → $7{\cdot}2CO_2 + 98{\cdot}8HCO_3^- + 16NH_4^+ + 42{\cdot}4N_2 + HPO_4^{2-} + 49H_2O$

Iron Reduction

$(CH_2O)_{106}(NH_3)_{16}H_3PO_4 + 424Fe(OH)_3 + 756CO_2$ → $862\ HCO_3^- + 16\ NH_4^+ + HPO_4^{2-} + 424\ Fe^{2+} + 304\ H_2O$

Sulphate Reduction

$(CH_2O)_{106}(NH_3)_{16}H_3PO_4 + 53\ SO_4^{2-}$ → $39\ CO_2 + 67\ HCO_3^- + 16\ NH_4^+ + HPO_4^{2-} + 53\ HS^- + 39\ H_2O$

Methanogenesis (Carbonate Reduction)

$(CH_2O)_{106}(NH_3)_{16}H_3PO_4 + 14\ H_2O$ → $39\ CO_2 + 14\ HCO_3^- + 53\ CH_4 + 16\ NH_4^+ + HPO_4^{2-}$

Fermentation

$12\ (CH_2O)_{106}(NH_3)_{16}H_3PO_4$ → $106\ CH_3CH_2COOH + 102H_2 + 106\ CH_3COOH + 192\ NH_3 + 212\ CH_3CH_2OH + 318\ CO_2 + 12H_3PO_4$

Figure 6.1   Idealized decay profile in sediment.

not due to depositional controls on rate of degradation, but to the chemical nature of the carbon. Carbon near the sediment–water interface has undergone relatively little degradation and is less refractory than degraded residues a few tens of centimetres below. Introduction of fresh carbon at depth leads to an increase in rate (Westrich and Berner, 1984; Parkes *et al.*, 1990). This observation shows that bacterial degradation can be food-limited (the G-model of Westrich and Berner, 1984).

The evidence that anoxia is an intrinsic inhibitor of decay is equivocal. However, even if anoxia does promote a reduction in decay rate, modelling suggests that the overall differential is well below an order of magnitude (Cannfield, 1989). If this is the case it is clear that anoxia alone is insufficient to promote soft-part preservation. This has been confirmed by laboratory experiments which have documented rate of decay of freshly-killed invertebrate carcasses in both aerobic and anaerobic environments (Plotnick, 1986;

Allison, 1988a; 1990; Kidwell and Baumiller, 1990). In each case the experiments showed that anaerobic decay was very rapid. It is probable that most carcasses decompose anaerobically regardless of the state of pore-water oxygenation (Allison, 1988a). The oxygen demand for aerobic microbial respiration is greater than 600 cm^3 of gas (at STP) for every gram (dry weight) of soft tissue. Increased mass/surface area ratios for larger carcasses clearly limit diffusion and promote anoxia. The critical size for this phenomenon will be dependent upon temperature, since this controls rate. For example, oxygen demand for aerobic microbes in a cool environment will be low and it may be possible to aerate a large carcass. However, if temperature is raised, rate of degradation and oxygen demand will increase. Should demand exceed diffusive supply, then anaerobic degradative processes will ensue. Further experimentation is required if the critical size is ever to be defined.

Although anoxia does not function as a long-term inhibitor of microbial degradation it may impede scavenging and bioturbation, particularly if the anoxic–oxic boundary extends above the sediment–water interface. Such a condition does not guarantee soft-part preservation, however, but it does promote the preservation of articulated hardparts.

Rapid burial

The importance of rapid burial as a means of promoting exceptional preservation is well known (Seilacher, 1970; Seilacher *et al.*, 1985). Catastrophic burial inhibits the diffusion of oxygen and therefore promotes carcass anoxia although, as previously outlined, this is insignificant for long-term preservation potential. More importantly, the pile of sediment on top of the carcass protects it from scavenging and disturbance by burrowers (Plotnick, 1986). For catastrophic burial to function as an effective barrier, the depth of sediment must be deeper than the zone of bioturbatory disturbance which will develop when the substrate is recolonized. This depth varies according to facies and depositional system and has changed through geological time as different organisms evolved. The overall diversity of trace fossils increased throughout the Phanerozoic (Seilacher, 1977), as did the intensity and depth of bioturbation. There was, for example, a general increase in bioturbation between the Lower and Upper Palaeozoic (Sepkoski, 1982; Larson and Rhoads, 1983). Detailed analysis has suggested a gradual development in intensity of bioturbation during the Palaeozoic, with a more marked increase during the Mesozoic and Cainozoic (Thayer, 1983). As a general rule this may be true, although some deep burrowing trace-makers were active during the early Palaeozoic (Sheehan and Schiefelbein, 1984; Miller and Byers, 1984).

Rapid mineralization

Allison (1988a; 1988b) emphasized that while scavengers are eliminated by rapid burial, low oxygen levels do not seriously inhibit decay; indeed, most decay is anaerobic. Fundamental to soft-tissue preservation, therefore, is prevention of the information loss which decay causes. In the case of the most readily decayed soft-tissues (muscle, gut, etc.), this normally only occurs through the replication of the soft-part morphology by early diagenetic minerals. Mineral formation may be promoted by soft-part decay, the minerals most often associated with soft-part preservation being pyrite, carbonate and phosphate. The major environmental controls on the formation of these minerals are different and form the basis for a classification of soft-part preservation based on mineral paragenesis (Allison, 1988b).

Pyrite forms as a result of the activity of sulphate-reducing bacteria, most commonly in fine-grained marine sediments where there is an adequate supply of sulphate (see, for example, Berner, 1970; 1984). Much of the pyrite found associated with fossils coats or replaces shells. The formation of early authigenic pyrite can be limited by the supply of carbon (the bacterial substrate), iron or sulphate (Berner and Raiswell, 1984) and the precise combination of conditions required rarely allows pyritization early enough to lead to the preservation of soft tissues. Where such exceptional preservation does occur it appears to be favoured by isolated concentrations of organic matter in fine-grained sediments where sufficient ions can diffuse toward a carcass to allow enough pyrite to form quickly enough to replicate soft tissues (Allison, 1988b).

Soft-part preservation is most frequently associated with carbonate mineralization. The decomposition of organic carbon under anaerobic conditions yields bicarbonate ions which can combine with calcium or iron (Berner, 1968; Raiswell, 1976). Calcium carbonate is more likely to form in marine conditions, but in freshwater conditions iron may reach a sufficient concentration, precipitating siderite (see, for example, Berner, 1981; but see also Pye *et al.*, 1990). The preservation of early diagenetic carbonates is favoured by rapid burial of large quantities of organic matter which are necessary to provide the high concentration of bicarbonate ions required (Allison, 1988b).

The most spectacular soft-part preservations are those in phosphate. In order for authigenic phosphate to form, however, concentration of the phosphate ion must exceed that of carbonate (Gulbrandsen, 1969). Various models have been put forward to explain such occurrences. The decomposition of biogenic remains is the most likely source for phosphate; both soft tissues (Lucas and Prévôt, 1984) and vertebrate skeletal remains (Van Cappelan and Berner, 1988) will yield soluble phosphate. Even so, some concentration mechanism is still required to exceed the bicarbonate which may be released by decaying soft tissues. This can occur if phosphates become adsorbed onto ferric hydroxides in aerobic sediment. Their release at the anoxic–oxic boundary occurs as a result of microbial iron reduction and may promote precipitation (Benmore *et al.*, 1983). The concentrations required imply high organic input (for example, as a result of seasonal blooms) and very slow rates of sedimentation. Areas of oceanic upwelling may be important sites of phosphate formation. Inorganic sources of phosphate (such as volcanic activity) have also been postulated in some cases.

The most volatile of soft-tissues are only preserved as a result of replacement or replication by authigenic minerals. The mineral which forms (pyrite, carbonate, phosphate) depends mainly on sedimentological and geochemical factors. The types of tissue and morphological details which survive depend on the relative rates of decay versus mineralization (Allison, 1988b). Certain unmineralized tissues, however, are less susceptible to decay than others, and these refractories can survive for geologically significant periods of time (Tegelaar *et al.*, 1989). They include resistant macromolecular structures in cell walls of some algae, and cuticular membranes and periderms of higher plants (for example, suberan, cutan, algaenan, lignins). Resistant macromolecules may also endow the tissues of some animals with a relatively high preservation potential (such as chitin in arthropod exoskeletons). This explains the high preponderance of plant and insect biotas preserved in the fossil record in the form of altered organic cuticles, but without authigenic mineralization. In some cases, however, even more volatile tissues may survive as highly altered organic carbon as in the case of the Burgess Shale. Here the organic tissues clearly provided a template for the clay minerals (now complex alumino-silicates) which cover them. It has even been suggested that the clay minerals may in some way have inhibited decay (Butterfield, 1990).

Sterile environments

Truly sterile sedimentary environments are rare and primarily terrestrial, and include amber, permafrost, tarsands, peat, dry caves and salt deposits. A sterile depositional system is formed when environmental conditions exceed the ecological tolerances of degradative bacteria. This may occur when carcasses are deposited within naturally antiseptic systems such as within peat, tree resin (which becomes amber) or salt. Peat, for example, contains fulvic and humic acids which inhibit microbes and may permit soft-part preservation. The most striking example is that of the Iron Age bog people of northern Europe whose skin, hair and internal body organs have been preserved (Brothwell, 1987). Preservation in amber is largely restricted to small organisms such as insects and spiders, although vertebrates are known (Poinar, 1988). Amber biotas are insignificant before the Cretaceous. Fossilization within amber is sometimes restricted to the outer integument of the organism and internal structures are usually absent; sterilization may also preserve the soft-tissues of the organism. Mammals preserved in salt at a Pleistocene site at Starunia in Galicia, Poland, include a woolly rhinoceros which is three-dimensional and preserves hair and skin (Niezabitowski, 1911). The carcasses have been impregnated with a mixture of petroleum and salt.

Microbial degradation may also be slowed or stopped by the imposition of environmental extremes such as dessication or freezing. Naturally dried and mummified human cadavers entombed over 5000 years ago have been recovered from the Nile Valley of Egypt (Brothwell, 1987). Frozen mammoths and bison in permafrost (Kurtén, 1986; Guthrie, 1990) include external soft-parts such as skin, hair and some muscle. The inner body cavity and organs, however, have often decomposed. Conditions of water and temperature stress may change as climatic belts migrate, and decay may be reactivated. These preservation conditions are not, therefore, long-term decay inhibitors, but are temporally ephemeral.

Microbial mats

The current velocity required to erode sediment bound by microbial mats is between two and five times greater than that for sediment alone (Scoffin, 1968). Seilacher *et al.* (1985) suggested that the growth of microbial films on sediment surfaces and on top of carcasses heightened sediment stability and thereby prevented hydrodynamic disarticulation and fragmentation. Exceptional preservation of medusae and worms within shales of the Triassic Grès-à-Voltzia is presumed to be due to the formation of microbial 'veils' (Gall, 1990). These coverings are thought to have formed shortly after death, and protected the cadavers from hydrodynamic and biogenic damage.

Microbial films not only inhibit sediment movement, but can also pseudomorph soft-parts. Soft-part outlines of ichthyosaurs from the Oxford Clay have been shown to be composed of small rod and oval-shaped microbes (Martill, 1987). The soft-parts themselves are absent and the preserved microbes may be a remnant of the original degrader community. A similar preservation type has been recorded from the Eocene Messel Shale of Germany where soft-part outlines of frogs and bats are preserved by sideritized films of coccoid and rod-shaped bacteria (Wuttke, 1983).

KONSERVAT-LAGERSTÄTTEN

Occurrences of soft-bodied fossils have long attracted considerable attention due to the much greater amount of palaeobiological information they preserve compared to the shelly fossil record. They are normally categorized as *Lagerstätten* using the German terminology adopted by Seilacher *et al.* (1985). *Fossil-Lagerstätten* are divided into two types: *Konservat-* or conservation *Lagerstätten*; and *Konzentrat-* or concentration *Lagerstätten*. The preservation of *Konservat-Lagerstätten* ranges through a spectrum from those which preserve the mineralized skeletons of echinoderms or vertebrates, but in a complete articulated state (see Donovan, this volume; Martill, this volume), to those which retain evidence of the soft tissues of organisms. It is only the latter which concern us here.

Exceptionally preserved fossil deposits can be grouped in various ways. Seilacher *et al.* (1985) emphasized what can be termed broadly sedimentological or environmental criteria. They recognized three major influences: stagnation (oxygen levels); obrution (the effect of rapid burial); and bacterial sealing through the formation of films. Most soft-bodied occurrences can be plotted on a triangular diagram with one of these factors at each corner. Outside this triangle the so-called conservation traps are plotted.

Soft-bodied fossils may be preserved either as altered original organic material, or by replication in authigenic minerals. In either case the effects of decay have been nullified at an early stage. Non-mineralized (including soft-bodied and lightly sclerotized) tissues, however, afford a range of resistance to decay which varies not only with their composition but also with the environment. Thus, *Konservat-Lagerstätten* can be categorized not just on the basis of sedimentary environments, or mineral paragenesis, but also in terms of taphonomic thresholds. The thresholds outlined here are very broad and grade into one another. They are described simply to illustrate the approach, which can be refined to give a much greater number of preservation levels. The nature of thresholds is such that they form nested sets. Biotas displaying the highest level of preservation will also include specimens preserving less detail.

Taphonomic thresholds

Three-dimensional volatiles
Collapse of a carcass due to decay occurs rapidly. Although a fossil arthropod cuticle, for example, may appear unimpaired morphologically, it will have lost any structural strength or rigidity within a few weeks of death unless it was mineralized in life. Collapse of less refractory tissues occurs even earlier and likewise before they lose their morphological integrity. Thus, the preservation of volatiles in three-dimensions indicates very rapid decay inhibition and authigenic mineralization.

Three-dimensional preservation of soft-parts is most frequently found in association with phosphate. One of the best-known examples is provided by the Santana Formation (Lower Cretaceous) of the Chapada do Araripe, northeast Brazil (Martill, 1988). Here a diversity of fishes occur in phosphate within calcium carbonate nodules. Even gill filaments and striated muscle tissue fibres are preserved (Martill, 1990; Martill and Harper, 1990) and comparisons with the effects of decay in experiments on Recent fish suggest that phosphatization must have taken place within a few hours. Other less spectacular examples of three-dimensional preservation in phosphate include muscles of concavicarids (enigmatic bivalved arthropods) from the Upper Devonian of Western Australia (Briggs and Rolfe, 1983) and squid (Figure 6.2A) from the Oxford Clay of England (Allison, 1988c). It has yet to be explained why phosphate should be the most rapidly formed authigenic mineral and therefore that yielding some of the most spectacular soft-part preservations. Soft-part outlines, including tentacles of teuthoids and octopus, are preserved in three-dimensional relief by replacement with sparry calcite, limonite and pyrite in calcareous nodules from the Callovian of Voulte-sur-Rhône in France (Fischer and Riou, 1982a; 1982b; Secretan and Riou, 1983).

Two-dimensional volatiles
Organisms normally collapse, losing body fluids, at a stage when volatile soft tissues are still retained. Overburden compaction, on the other hand, is a long-term phenomenon which postdates any authigenic mineralization of soft tissues. Its effects can be distinguished from decay collapse by the distortion and flattening of the mineral infill, replacement or coating. Two-dimensional volatiles may be preserved in a variety of ways: as organic residues, as apparently occurs in the Burgess Shale and possibly in other Cambrian Konservat-Lagerstätten (Butterfield, 1990); as carbonized bacterial films such as those preserving the soft-tissue outlines in vertebrates from the Jurassic

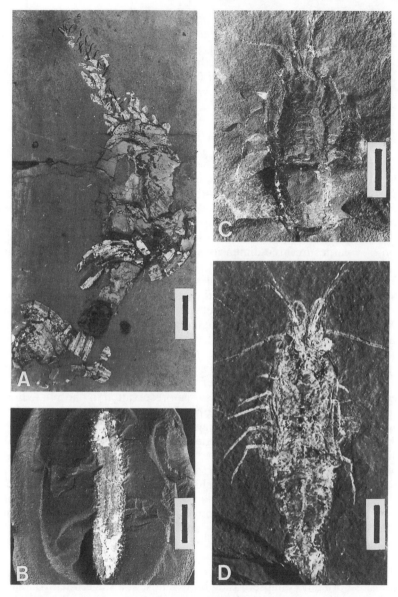

Figure 6.2   Selected examples of taphonomic thresholds for soft-part preservation. (A) Soft-parts including musculature (volatiles) of squid preserved in three dimensions from the Jurassic Oxford Clay of England (Allison, 1988c). (B) Impression of worm soft-parts (volatiles) in siderite nodule from the Carboniferous Mazon Creek biota of Illinois, USA. (C) Three-dimensional preservation of arthropod cuticle (refractory) from Mazon Creek. (D) Two-dimensional preservation of refractory shrimp cuticle from the Carboniferous Midland valley of Scotland. Scale bars on A and B are 1 cm, and on C and D are 3 mm.

Posidonienschiefer at Holzmaden (Martill, 1987); as mineral replacements such as those rare examples preserving histological details in silicified material from the Eocene of Geiseltal (Voigt, 1988); as mineral films, for example, the siderite preserving skin and blood vessels in the frogs from Eocene Lake Messel (Wuttke, 1983); as impressions, like the specimens of the jellyfish *Rhizostomites*, from the Jurassic Solnhofen Limestone, or within diagenetic concretions such as those from Mazon Creek (Figure 6.2B).

Three-dimensional refractories
This taphonomic threshold represents a stage at which volatile tissues have decayed, but refractory tissues (resistant cuticle, for example) remain three-dimensional. Where subsequent tectonic compaction has occurred this may be difficult to distinguish from the effects of decay collapse. The term 'refractory' is rather loosely applied in studies of decay and fossilization. There is a continuum between refractory and volatile tissues, and, therefore, a real risk of circularity in describing non-mineralized tissues as 'refractory' simply because they are preserved in the fossil record. In practice, however, the distinction is usually clear. Refractory tissues are normally structural in function (for example, cuticles) and are made up of resistant biomacromolecules (such as lignin, sporopollenin and cellulose in various plant cuticles, and chitin in arthropod cuticles). In the context of soft-tissue preservation the term 'refractory' is applied to the tissue (cuticle) rather than the macromolecules, which may not have the same preservation potential within other structures. Refractories also vary in their resistance to decay. They may be preserved as organic residues, such as scorpions from the Carboniferous East Kirkton Quarry near Edinburgh (Jeram, 1990a; 1990b), and arthropods which have been extracted from coals (Bartram *et al.*, 1987), and an increasing number of fine-grained continental lithologies (Shear *et al.*, 1984). Such organisms show varying degrees of collapse and compaction. Insects and other organisms preserved in amber of various ages are a special case. The animals may be decayed internally, presumably due to autolysis and endogenous bacteria; the external morphology is retained. Ironically, amber resin itself is one of the more decay-resistant products of vascular plants (Langenheim, 1990). Refractories may also be preserved as a result of authigenic mineralization. Refractories are more likely to retain some three-dimensionality if they have been mineralized. Hence, Carboniferous arthropods from the Mazon Creek of Illinois are three-dimensional due to encapsulation within siderite nodules (Figure 6.2C). Although specimens of the pyritized trilobite *Triarthrus* from Beecher's Trilobite Bed have suffered some decay collapse, the limbs are to some degree three-dimensional, the pyrite apparently having formed as an infill (Whittington and Almond, 1987).

Two-dimensional refractories
The number of fossil occurrences which fall within this threshold is orders of magnitude greater than those within the previous three. It includes the major part of the plant fossil record (in the case of leaves, for example, the distinction between three- and two-dimensional plant fossils is somewhat difficult to apply). It also incorporates most of the record of terrestrial

arthropods, such as the insects. A number of the spectacular shrimp biotas (see, for example, Figure 6.2D) also belong here (Briggs and Clarkson, 1989), including the Lower Carboniferous example at Glencartholm (Schram, 1983).

Recurrent associations

Previous authors have compared various examples of exceptional preservation (for example, Seilacher *et al.*, 1985; Allison, 1988b), but more work is required before a comprehensive genetic classification of *Konservat-Lagerstätten* can be produced. A number of recurrent associations can be identified, however, and some of the more important examples are outlined below. These deposits continue to be the focus for extensive research, not only into their taphonomy, but also into their biotic composition and potential evolutionary significance.

Ediacaran type
Ediacaran biotas range through the latest Precambrian and occur at a range of localities in Australia, Africa, North and South America, Europe and Asia (Glaessner, 1984; Fedonkin, 1990; Conway Morris, 1990). Examples formed mainly in shallow water (for example, Ediacara Hills, South Australia), but deeper-water occurrences (such as Mistaken Point, Avalon Peninsula, Newfoundland) are known. They share similar organisms, which are preserved as impressions, often on the base of a sandstone unit lying immediately above a siltstone. Although the organisms clearly lacked a mineralized skeleton, the nature of the tissue is the subject of some controversy (Seilacher, 1989). In view of the associated diverse trace fossils it is clear that conditions were not anoxic. It is difficult to see how decay might have been inhibited unless decay processes were different or the tissues were to some degree refractory, that is, some sort of tough integument. The nature of the organisms is the primary factor linking the Ediacaran type of preservation and, as few survive into the Cambrian, may explain why this type of preservation is confined to the latest Precambrian.

Burgess Shale type
Although there are unusual animals among the Burgess Shale type assemblages, they include representatives of most of the major modern groups. They are distributed through the Lower and Middle Cambrian and occur in North America, Spain, Poland, South Australia, China and Siberia (Conway Morris, 1989; 1990). The majority are preserved near the edge of open shelf environments where they could be caught up and buried by clouds of suspended sediment moving downslope into deeper water. In at least the Burgess Shale itself, near Field in British Columbia, organic material survived (Butterfield, 1990) as a template for diagenetic minerals, now mainly silicates. Why this type of preservation is largely confined to the Cambrian is not yet known. The widespread occurrence of Burgess Shale type fossils in the

Cambrian contrasts with a paucity of soft-bodied biotas in the Ordovician (Cisne, 1973) and Silurian (Mikulic *et al.*, 1985a; 1985b). There is no apparent fundamental difference in preservation potential between many Cambrian and later organisms.

Beecher's Trilobite Bed type

Pyrite commonly occurs as a component of fine-grained, organic-rich marine sediments. Its formation in sediment, as a result of the activity of sulphate-reducing bacteria under anaerobic conditions, is well understood (see, for example, Berner, 1970; 1984). It occurs to a limited degree in the Burgess Shale (Conway Morris, 1986) and examples persist through the entire fossil record. Very rarely, however, is pyrite involved in the extensive preservation of volatiles, or even refractory organic tissue. The two most significant examples are Palaeozoic: Beecher's Trilobite Bed from the Upper Ordovician of New York State (Cisne, 1973); and the Hunsrückschiefer from the Lower Devonian of Germany (Stürmer *et al.*, 1980). The former preserves mainly refractory tissues, the latter includes the soft tissues of trilobites, cephalopods and ctenophores, for example. Both *Konservat-Lagerstätten* are from fine-grained, deeper-water marine sediments. A similar style of preservation is common in graptolites, but it is not yet known why soft-part preservation in pyrite is not more widespread.

Orsten type

The Upper Cambrian Orsten of Sweden are characterized by the three-dimensional preservation of tiny arthropods in calcium phosphate within calcareous nodules. Similar remarkably preserved arthropods are known from a number of localities including the Upper Devonian of the Carnic Alps, the Lower Triassic of Spitzbergen and the Lower Cretaceous Santana Formation of Brazil (Müller, 1985). Their distribution suggests that the conditions necessary for rapid precipitation of authigenic phosphate, although rare, persisted through the Phanerozoic and that further examples are likely to yield to systematic preparation of suitable lithologies. In each case there appears to be an upper limit of about 2 mm on the size of the arthropods with preserved appendages. The exceptional preservation of larger, three-dimensional organisms such as fishes (Santana Formation: Martill, 1988; 1990) and squid (Oxford Clay: Allison, 1988c), on the other hand, must reflect very high concentrations of phosphate.

Mazon Creek type

The Mazon Creek biota in the Upper Carboniferous (Westphalian D) of northern Illinois includes the non-marine Braidwood Fauna and the more marine-influenced Essex Fauna. Its remarkable diversity of soft-bodied fossils reflects the range of habitats in this transitional setting. A similar preservation style is known from several Upper Carboniferous localities in the mid-continental United States (Baird *et al.*, 1985), but it also occurs in the British Coal Measures and at Montceau-les-Mines in France (Gall, 1983). A closely comparable biota occurs in the Triassic Grès-à-Voltzia of northeastern

France, but there only the incipient stages of nodule formation are present (Gall, 1971). Siderite nodule preservations are most prevalent in the Upper Carboniferous. Broad coastal delta plains in tropical latitudes provided a setting in which restricted water bodies, such as interdistributary bays and lagoons, could form. These were sites of episodic sedimentation, high organic input and variable salinity (due to mixing of normal marine and freshwater). Associated with them were other types of soft-bodied preservation, forming a continuum of Carboniferous Konservat-Lagerstätten (Schram, 1979; Briggs and Gall, 1990).

Solnhofen type
The Solnhofen is the best known (not least through the presence of *Archaeopteryx*) of a number of fine-grained lithographic limestones which yield soft-bodied fossils. They formed in shallow, restricted marine basins, and a semi-arid tropical climate resulted in high salinity and a stratified water body. Benthic organisms are rare. Bacterial films on the sediment surface have played a critical role (Seilacher *et al.*, 1985). Similar preservations occur in the Middle Triassic at Alcover-Montral in northeastern Spain, the Upper Jurassic at Cerin in the French Jura, and in the Cretaceous at Hakel and Sahel-Alma in the Lebanon.

Green River type
The Eocene Green River Formation extends over a wide area of Wyoming, Utah and Colorado. It represents the deposits of large, permanently stratified lakes with a biota dominated by fishes, insects and plants (Grande, 1984). The sediments are laminated, with alternating pale calcite mudstone and organic rich layers. Similar, but less extensive, lake biotas occur elsewhere in the Eocene of North America, particularly in British Columbia and Washington State (Wilson, 1980). Other varved lake deposits also yield extraordinary preserved fossils, although the lithology may vary, as it does within the Green River Formation. There is a higher proportion of organic material in the Eocene oil shale at Grube Messel near Darmstadt, Germany, for example, and volcanic ash is interbedded with the organic-rich clays in the Middle Miocene Clarkia deposit of northern Idaho (Smiley and Rember, 1985). This type of deposit is not confined to the Tertiary, but lake biotas of this age provide the means to document the radiation of insects, fishes and birds (Gray, 1988), and, being geologically young, the sedimentary sequences are more likely to be preserved in the rock record.

Baltic amber type
Amber is produced as resin by a range of plant taxa, which hardens through organic diagenesis into a highly inert substance (Langenheim, 1990). When plant and animal remains, particularly insects, are trapped in this resin they preserve perfect details including, in some cases, cell organelles. Insects in amber have also yielded traces of DNA (Pääbo *et al.*, 1989). The earliest amber is Late Carboniferous, but it does not become abundant until the Early Cretaceous. The most diverse faunas in amber are Tertiary, from the Late

Eocene/Early Oligocene of the Baltic and the Eocene to Miocene of the Dominican Republic.

SUMMARY

What, then, are the conditions necessary for the preservation of soft-tissues? Minimal transport is not a prerequisite for exceptional preservation if a carcass is freshly killed (Allison, 1986), nor is anoxia a significant decay inhibitor for most volatile tissues (Plotnick, 1986; Allison, 1988a; but see Butterfield, 1990), although it may promote soft-part mineralization (Allison, 1988b). Anoxicity will, however, preclude scavengers if it extends above the sediment–water interface, as will rapid burial, if the deposited sediment layer is sufficiently thick. The latter may cast soft-part outlines, but is more likely to promote the formation of articulated skeletal remains.

Some recurrent associations of soft-part preservation are limited in temporal extent. Both the Ediacaran and Burgess Shale types are characterized by preservational styles which are principally known from very restricted time intervals. The former occurs in what was an oxygenated environment and preservation may be due to either a unique body construction (Seilacher, 1989) or the fact that deep-burrowing organisms had not yet evolved. The Burgess Shale biota includes altered residues of original tissue which are encased in diagenetic clay minerals, the origin of which is enigmatic. Why this biota includes carbon residues when others do not remains to be determined.

Three of the preservational types are characterized by early diagenetic mineralization. The Beecher's Trilobite Bed type includes soft-parts preserved in pyrite. The geochemistry of pyrite formation is well known (see, for example, Berner, 1984) and early diagenetic pyrite is fairly common, but this type of preservation occurs rarely. Clearly the depositional conditions necessary are rarely achieved. The Orsten type is characterised by three-dimensional preservation in phosphate. The level of preservation is clear evidence of early diagenesis (Martill and Harper, 1990), but the mechanism and controls on mineral precipitation are unknown. Laboratory experiments have shown that this type of replacement can occur within days (see Allison, 1988c). The Mazon Creek type of preservation occurs within siderite nodules which are known to have formed rapidly (Baird *et al.*, 1985). Although the general mechanism for siderite precipitation has been known for some time (see, for example, Berner, 1981), a modern analogue of rapid (within weeks or months) concretion growth has only recently been described (Pye *et al.*, 1990). This type of preservation is particularly important in the Carboniferous, probably due to the abundance of paralic basins which formed as a result of a warm, humid climate in low-lying, swampy depositional systems.

The Solnhofen and Green River preservational types are most abundant in the Mesozoic and Tertiary, respectively, and the causes of preservation are reasonably well understood. Both types involved deposition within laminated sediments in a presumably stratified water column which prevented scavenging and biogenic disturbance. Although some diagenetic replacement of soft tissues is known from these localities, tissues are more commonly preserved as faint impressions. Stagnation and the growth of algal films are responsible for preservation.

The Baltic amber type is the only geologically significant sterile medium for soft-part preservation, but it is temporally restricted to post-Palaeozoic rocks.

Actualistic experimentation has demonstrated the limited role of anoxia in softpart preservation and highlighted the importance of early diagenetic mineralization. The exact controls on decay inhibition and on mineral precipitation, however, will only be determined by further experimentation and by detailed petrographic and geochemical analysis of fossils and of the sedimentary sequences which yield them.

ACKNOWLEDGEMENTS

We thank Steve Donovan for inviting us to participate in this project. This chapter is Reading University PRIS Contribution Number 110.

REFERENCES

Allison, P.A., 1986, Soft-bodied organisms in the fossil record: the role of decay in fragmentation during transport, *Geology*, **14** (12): 979–81.

Allison, P.A., 1988a, The role of anoxia in the decay and mineralization of proteinaceous macro-fossils, *Paleobiology*, **14** (2): 139–54.

Allison, P.A., 1988b, *Konservat-Lagerstätten*: cause and classification, *Paleobiology*, **14** (4): 331–44.

Allison, P.A., 1988c, Phosphatized soft-bodied squids from the Jurassic Oxford Clay, *Lethaia*, **21** (4): 403–10.

Allison, P.A., 1990, Variation in rate of decay and disarticulation of Echinodermata: implications for the application of actualistic data, *Palaois*, **5**: in press.

Baird, G.S., Sroka, S.D., Shabica, C.W. and Beard, T.L., 1985, Mazon Creek-type fossil assemblages in the U.S. midcontinent Pennsylvanian: their recurrent character and palaeoenvironmental significance, *Philosophical Transactions of the Royal Society of London*, **B311**: 87–99.

Bartram, K.M., Jeram, A.J. and Selden, P.A., 1987, Arthropod cuticles in coal, *Journal of the Geological Society of London*, **144** (3): 513–17.

Benmore, R.A., Coleman, M.L. and McArthur, J.M., 1983, Origin of sedimentary francolite from its sulphur and carbon isotope composition, *Nature*, **302** (5908): 516–18.

Berner, R.A., 1968, Calcium carbonate concretions formed by the decomposition of organic matter, *Science*, **159** (3811): 195–7.

Berner, R.A., 1970, Sedimentary pyrite formation, *American Journal of Science*, **268** (1): 1–23.

Berner, R.A., 1981, Authigenic mineral formation resulting from organic matter decomposition in modern sediments, *Fortshritte der Mineralogie*, **59** (1): 117–35.

Berner, R.A., 1984, Sedimentary pyrite formation—an update, *Geochimica et Cosmochimica Acta*, **48** (4): 605–15.

Berner, R.A. and Raiswell, R. 1984, A new method for distinguishing freshwater from marine sedimentary rocks, *Geology*, **12** (6): 365–8.

Briggs, D.E.G. and Clarkson, E.N.K., 1989, Environmental controls on the taphonomy and distribution of Carboniferous malacostracan crustaceans, *Transactions of the Royal Society of Edinburgh: Earth Sciences*, **80** (3/4): 293–301.

Briggs, D.E.G., Clarkson, E.N.K. and Aldridge, R.J., 1983, The conodont animal, *Lethaia*, **16** (1): 1–14.

Briggs, D.E.G. and Gall, J.-C., 1990, The continuum in soft-bodied biotas from transitional environments: a quantitative comparison of Triassic and Carboniferous *Konservat-Lagerstätten*, *Paleobiology*, **16** (2): 204–18.

Briggs, D.E.G. and Rolfe, W.D.I., 1983, New Concavicarida (new Order: ?Crustacea) from the Upper Devonian of Gogo, Western Australia, and the palaeoecology and affinities of the group, *Special Papers in Palaeontology*, **30**: 249–76.

Brothwell, D., 1987, *The bog man and the archaeology of people*, Harvard University Press, Cambridge, Massachusetts.

Butterfield, N.J., 1990, Organic preservation of non-mineralizing organisms and the taphonomy of the Burgess Shale, *Paleobiology*, **16** (3): 272–86.

Cannfield, D.E., 1989, Sulfate reduction and oxic respiration in marine sediments: implications for organic carbon preservation in euxinic environments, *Deep Sea Research*, **36** (1): 121–38.

Cisne, J.L., 1973, Beecher's trilobite bed revisited: ecology of an Ordovician deep water fauna, *Postilla*, **160**: 1–25.

Coleman, M.L., 1985, Geochemistry of diagenetic non-silicate minerals: kinetic considerations, *Philosophical Transactions of the Royal Society of London*, **A315**: 39–54.

Conway Morris, S., 1979, Burgess Shale. In R.W. Fairbridge and D. Jablonski (eds), *The encyclopedia of paleontology*, Dowden, Hutchinsen and Ross, Stroudsberg, Pennsylvania: 153–60.

Conway Morris, S., 1986, The community structure of the Middle Cambrian phyllopod bed (Burgess Shale), *Palaeontology*, **29** (3): 423–67.

Conway Morris, S., 1989, The persistence of Burgess Shale-type faunas: implications for the evolution of deeper-water faunas, *Transactions of the Royal Society of Edinburgh: Earth Sciences*, **80** (3/4): 271–83.

Conway Morris, S., 1990, Late Precambrian and Cambrian soft-bodied faunas, *Annual Reviews of Earth and Planetary Sciences*, **18**: 101–22.

Curtis, C.D., 1980, Diagenetic alteration in black shales, *Journal of the Geological Society of London*, **137** (2): 189–94.

Dörjes, J., 1972, Distribution and zonation of macrobenthic animals, *Senckenbergiana Maritima*, **4** (2): 183–216.

Driscoll, E.G. and Swanson, R.A., 1973, Diversity and structure of epifaunal communities on mollusc valves, Buzzard Bay, Massachussetts, *Palaeogeography, Palaeoclimatology, Palaeoecology*, **14** (3): 229–47.

Fedonkin, M.A., 1990, Precambrian metazoans. In D.E.G. Briggs and P.R. Crowther (eds), *Palaeobiology—a synthesis*, Blackwell Scientific Publications, Oxford: 17–24.

Fischer, C. and Riou, B., 1982a, Le plus ancien Octopode connu (Cephalopoda, Dibranchiata): *Proteroctopus ribeti* nov. gen., nov. sp., du Callovien de L'Ardèche (France), *Compte Rendu Academie Sciences Paris, Série II*, **295** (20 September): 277–80.

Fischer, C. and Riou, B., 1982b, Les teuthoides (Cephalopoda, Dibranchiata) du Callovien Inférieur de la Voulte-sur-Rhône (Ardèche, France), *Annales de Paléontologie*, **68** (4): 295–325.

Gall, J.-C., 1971, Faunes et paysages du Grès-à-Voltzia du Nord des Vosges. Essai paléoécologique sur le Bundsandstein supérieur, *Mémoire service Carte géologique Alsace Lorraine, Strasbourg*, **34**: 1–318.

Gall, J.-C., 1983, Interpretation paléoécologique de la faune des nodules fossiliferes du Stéphanien de Montceau-les-Mines, *Memoires Géologiques de l'Université de Dijon*, **8**: 51–4.

Gall, J.-C., 1990, Les voiles microbiens. Leurs contribution à la fossilisation des organismes au corps mou, *Lethaia*, **23** (1): 21–8.

Glaessner, M.F., 1984, *The dawn of animal life: a biohistorical study*, Cambridge University Press, Cambridge.

Grande, L., 1984, Paleontology of the Green River Formation, with a review of the fish fauna (2nd edn), *Bulletin of the Geological Survey of Wyoming*, **63**: 1–333.

Gray, J. (ed.), 1988, Aspects of freshwater paleoecology and biogeography, *Palaeogeography, Palaeoclimatology, Palaeoecology*, **62**: 1–623.

Gulbrandsen, R.A., 1969, Physical and chemical factors in the formation of marine apatite, *Economic Geology*, **64** (4): 365–82.

Guthrie, R.D., 1990, *Frozen fauna of the Mammoth Steppe, the story of Blue Babe*, University of Chicago Press, Chicago.

Hecht, F., 1933, Der Verbleib der organischen Substanz der Tiere bei meerischer Einbettung, *Senckenbergiana*, **15** (3/4): 165–249.

Henrichs, S.M. and Reeburgh, W.S., 1987, Anaerobic mineralization of marine sediment organic matter: rates and the role of anaerobic processes in the oceanic carbon economy, *Geomicrobiology Journal*, **5** (3/4): 191–235.

Jeram, A.J., 1990a, Book-lungs in a Lower Carboniferous scorpion, *Nature*, **343** (6256): 360–1.

Jeram, A.J., 1990b, When scorpions ruled the world, *New Scientist*, **126** (1721): 52–5.

Jones, G.F., 1968, The benthic macrofauna of the mainland shelf of southern California, *Allan Hancock Monographs in Marine Biology*, **4**: 1–219.

Kidwell, S.M. and Baumiller, T., 1990, Experimental disintegration of regular echinoids: roles of temperature, oxygen and decay thresholds, *Paleobiology*, **16** (3): 247–71.

Kurtén, B., 1986, *How to deep-freeze a mammoth*, Columbia University Press, New York.

Langenheim, J.H., 1990, Plant resins, *American Scientist*, **78** (1): 16–24.

Larson, D.W. and Rhoads, D.C., 1983, The evolution of infaunal communities and sedimentary fabrics. In M.J.S. Tevesz and P.L. McCall (eds), *Biotic interactions in Recent and fossil benthic communities*, Plenum, New York: 627–48.

Lawrence, D.R., 1968, Taphonomy and information losses in fossil communities, *Geological Society of America Bulletin*, **79** (10): 1315–30.

Lowenstam, H.A. and Weiner, S., 1989, *On biomineralization*, Oxford University Press, Oxford.

Lucas, J. and Prévôt, L., 1984, Synthèse de l'apatite par voie bactérienne à partir de matière organique phosphatée et de divers carbonates de calcium dans des eaux douces et marines naturelles, *Chemical Geology*, **42** (2): 101–18.

Martill, D.M., 1987, Prokaryote mats replacing soft tissues in Mesozoic marine reptiles, *Modern Geology*, **11** (4): 265–9.

Martill, D.M., 1988, Preservation of fish in the Cretaceous Santana Formation of Brazil, *Palaeontology*, **31** (1): 1–18.

Martill, D.M., 1990, Macromolecular resolution of fossilized muscle tissue from an elopomorph fish, *Nature*, **346** (6280): 171–2.

Martill, D.M. and Harper, L., 1990, An application of critical point drying to the comparison of modern and fossilized soft tissues of fishes, *Palaeontology*, **33** (2): 423–8.

Mikulic, D.G., Briggs, D.E.G. and Kluessendorf, J., 1985a, A new exceptionally preserved biota from the lower Silurian of Wisconsin, U.S.A., *Philosophical Transactions of the Royal Society of London*, **B311**: 75–85.

Mikulic, D.G., Briggs, D.E.G. and Kluessendorf, J., 1985b, A Silurian soft-bodied biota, *Science*, **228** (4700): 715–17.

Miller, M.F. and Byers, C.W., 1984, Abundant and diverse early Paleozoic infauna indicated by the stratigraphic record, *Geology*, **12** (1): 40–3.

Müller, K.J., 1985, Exceptional preservation in calcareous nodules, *Philosophical Transactions of the Royal Society of London*, **B311**: 67–73.

Niezabitowski, E.L., 1911, Die Überreste des in Starunia in einer Erdwachsgrube mit Haut und Weichteilen gefundenen *Rhinoceras antiqualis* Blum. (Tichorhinus Fisch.) (Vorläufige Mitteilung.), *Bulletine Internationale Akademie Umiej Krakow*, **1911**(B): 240–67.

Pääbo, S., Higuchi, R.G. and Wilson, A.C., 1989, Ancient DNA and the polymerase chain reaction, *Journal of Biology and Chemistry*, **264** (17): 9709–12.

Parkes, R.J., Cragg, B.A., Fry, J.C., Herbert, R.A. and Wimpenny, J.W.T., 1990, Bacterial biomass and activity in deep sediment layers from the Peru margin, *Philosophical Transactions of the Royal Society of London*, **A331**: 139–53.

Plotnick, R.E., 1986, Taphonomy of a modern shrimp: implications for the arthropod fossil record, *Palaios*, **1** (3): 286–93.

Poinar, G.O. Jr., 1988, The amber ark, *Natural History*, **97** (12): 43–7.

Pye, K., Dickson, J.A.D., Schiavon, N., Coleman, M.L. and Cox, M., 1990, Formation of siderite-Mg-calcite-iron sulphide concretions in intertidal marsh and tidal flat sediments, north Norfolk, England, *Sedimentology*, **37** (2): 325–44.

Raiswell, R., 1976, The microbiological formation of carbonate concretions in the Upper Lias of N.E. England, *Chemical Geology*, **18** (3): 227–44.

Redfield, A., 1958, The biological control of chemical factors in the environment, *American Scientist*, **48** (3): 206–26.

Schopf, T.J.M., 1978, Fossilization potential of an intertidal fauna: Friday Harbor, Washington, *Paleobiology*, **4** (3): 261–70.

Schram, F.R., 1979, The Mazon Creek biotas in the context of a Carboniferous continuum. In M.H. Nitecki (ed.), *Mazon Creek fossils*, Academic Press, New York: 159–90.

Schram, F.R., 1983, Lower Carboniferous biota of Glencartholm, Eskdale, Dumfrieshire, *Scottish Journal of Geology*, **19** (1): 1–15.

Scoffin, T.P., 1968, An underwater flume, *Journal of Sedimentary Petrology*, **38** (1): 244–6.

Secretan, S. and Riou, B., 1983, Un groupe énigmatique de crustacés. Ses représentants du Callovien de la Voulte-sur-Rhône, *Annales de Paléontologie*, **69** (2): 59–97.

Seilacher, A., 1970, Begriff und Bedeutung der *Fossil-Lagerstätten*, *Neues Jahrbuch für Geologie und Paläontologie Abhandlungen*, **1970** (1): 34–9.

Seilacher, A., 1977, Evolution of trace fossil communities. In A. Hallam (ed.), *Patterns of Evolution*, Elsevier, Amsterdam: 359–76.

Seilacher, A., 1989, Vendozoa: organismic construction in the Proterozoic biosphere, *Lethaia*, **22** (3): 229–40.

Seilacher, A., Reif, W.-E. and Westphal, F., 1985, Sedimentological, ecological and temporal patterns of *Fossil Lagerstätten*, *Philosophical Transactions of the Royal*

Society of London, **B311**: 5–23.

Sepkoski, J.J., Jr., 1981, A factor analytic description of the Phanerozoic marine fossil record, *Paleobiology*, **7** (1): 36–53.

Sepkoski, J.J., Jr., 1982, Flat-pebble conglomerates, storm deposits, and the Cambrian bottom fauna. In G. Einsele and A. Seilacher (eds), *Cyclic and event stratification*, Springer-Verlag, Berlin: 371–85.

Shear, W.A., Bonamo, P.M., Grierson, J.D., Rolfe, W.D.I., Smith, E.L. and Norton, R.A., 1984, Early land animals in North America; evidence from Devonian age arthropods from Gilboa, New York, *Science*, **224** (4648): 492–4.

Sheehan, P.M. and Schiefelbein, D.R.J., 1984, The trace fossil *Thalassinoides* from the Upper Ordovician from the eastern Great Basin: deep burrowing in the early Paleozoic, *Journal of Paleontology*, **58** (2): 440–7.

Smiley, C.J. and Rember, W.C., 1985, Physical setting of the Miocene Clarkia fossil beds, northern Idaho. In C.J. Smiley (ed.), *Late Cenozoic history of the Pacific northwest*, American Association for the Advancement of Science, Pacific Division: 11–31.

Stürmer, W. and Bergström, J., 1973, New discoveries on trilobites by X-rays, *Paläontologische Zeitschrift*, **47** (1/2): 104–41.

Stürmer, W., Schaarschmidt, F. and Mittmeyer, H.-G., 1980, Versteinertes Leben im Röntgenlicht, *Kleine Senckenberg-Reihe*, **11**: 1–80.

Tegelaar, E.W., de Leeuw, J.W., Derenne, S. and Largeau, C., 1989, A reappraisal of kerogen formation, *Geochimica et Cosmochimica Acta*, **53** (11): 3103–6.

Thayer, C.W., 1983, Sediment-mediated biological disturbance and the evolution of the marine benthos. In M.J.S. Tevesz and P.L. McCall (eds), *Biotic interactions in Recent and fossil benthic communities*, Plenum, New York: 480–625.

Van Capellen, P. and Berner, R.A., 1988, A mathematical model for the early diagenesis of phosphorus and fluorine in marine sediments: apatite precipitation, *American Journal of Science*, **288** (4): 289–333.

Voigt, E., 1988, Preservation of soft tissues in the Eocene lignite of the Geiseltal near Halle/S, *Courier Forschungsinstitut Senckenberg*, **107**: 325–43.

Westrich, J.T. and Berner, R.A., 1984, The role of sedimentary organic matter in bacterial sulfate reduction: the G model tested, *Limnology and Oceanography*, **29**: 236–49.

Whittington, H.B., 1971, The Burgess Shale: history of research and preservation of fossils, *Proceedings of the First North American Paleontological Convention*, **1**: 1176–1201.

Whittington, H.B. and Almond, J.E., 1987, Appendages and habits of the Upper Ordovician trilobite *Triarthrus eatoni*, *Philosophical Transactions of the Royal Society of London*, **B317**: 1–46.

Wilson, M.V.H., 1980, Eocene lake environments: depth and distance-from-shore variation in fish, insect, and plant assemblages, *Palaeogeography, Palaeoclimatology, Palaeoecology*, **32** (1): 21–44.

Wuttke, M., 1983, 'Weichteil-Erhaltung' durch lithifizierte Mikroorganismen bei mitteleozänen Vertebraten aus den Ölschiefern der 'Grube-Messel' bei Darmstadt, *Senckenbergiana Lethaea*, **64** (5/6): 509–27.

Zangerl, R. and Richardson, E.S., Jr., 1963, The paleoecological history of two Pennsylvanian black shales, *Fieldiana Geology Memoir*, **4**: 1–132.

Chapter 7

THE TAPHONOMY OF PLANT MACROFOSSILS

David R. Greenwood

INTRODUCTION

Taphonomy is defined as the study of the transition of organic remains from the living organism to fossil assemblages (Efremov, 1940). As such, plant taphonomy incorporates the processes of the initial abscission of plant parts, their transport (by air and/or water) to a place of eventual deposition, entrapment and eventual burial, and subsequent lithification. Within these processes a number of factors can be identified which influence both the character of the eventual assemblage and its taxonomic composition. These factors produce a taxonomic mix within the assemblage which is a subset of the taxonomic composition of the original source plant community or communities. The organographic character of the assemblage may also be biased (Collinson, 1983). Spicer (1989, p. 99) defined a plant fossil assemblage as 'an accumulation of plant parts, derived from one or several individuals, that is entombed within a volume of sediment that is laid down in essentially the same conditions'. His definition will be used here.

Plant macrofossil assemblages are fundamentally different from most animal assemblages in that they are almost entirely composed of disarticulated parts. This results in part from the continual production throughout a plant's life cycle of generally temporary modular plant organs—leaves, stems, flowers and fruits—which are shed from the plant when their usefulness is complete, through traumatic loss, or for the dispersal of propagules. Each of these parts is given a separate name until connection with other parts can be demonstrated (organ taxa; for example, *Stigmaria* and *Lepidophylloides* are the root system and leaves, respectively, of the same Carboniferous lycopod: Thomas and Spicer, 1987; Thomas, 1990). A further factor is that plant communities, in contrast to most animal communities, are composed of organisms that remain in one place throughout their whole life cycle. In common with animal remains, the behaviour of the disarticulated plant organs as sedimentary particles varies, as does their preservation potential.

Consideration of the above factors in the interpretation of plant fossil assemblages has only occurred comparatively recently. However, early work by Chaney (1924; Chaney and Sanborn, 1933) pioneered the idea of using modern leaf assemblages to interpret Tertiary localities. Over the last decade a number of seminal works have appeared on plant taphonomy (Hickey, 1980; Spicer, 1980; 1981; Ferguson, 1985; Scheihing and Pfefferkorn, 1984; Gastaldo, 1986; 1988; Burnham, 1989). Spicer (1980; 1989) and Hickey (1980) stressed the need to consider plant macrofossil assemblages within their stratigraphic and sedimentological context. In a series of experiments Spicer (1981) and Ferguson (1985) have investigated the behaviour of plant parts, mainly leaves, as sedimentary particles to determine the factors which control their dispersal. Other plant taphonomists have concentrated on the influence of the character of the standing vegetation and sedimentary environments on the resulting plant macrofossil assemblages (Drake and Burrows, 1980; Scheihing and Pfefferkorn, 1984; Gastaldo, 1986; 1988; Hill and Gibson, 1986; Taggert, 1988; Burnham, 1989).

These studies have contributed to a change in how plant assemblages are sampled and interpreted. This chapter is not intended as a review of plant taphonomy, as thorough reviews have appeared recently (Gastaldo, 1988; Spicer, 1989). Rather, this account emphasizes the influence of vegetation on plant taphonomy, although necessarily some major points are reviewed.

The nature of the plant macrofossil record

Most plant macrofossil assemblages are in fact fossilized litter (Figures 7.1, 7.2). Plant macrofossils constitute any plant organ or organ part which is large enough to be recognizable without microscopy. Thus, plant macrofossils include leaves and other foliar organs (such as stipules and raches), flowers, fruit and other reproductive organs (for example, cones, sporangia, etc.), stems and stem fragments (including woody axes), and root systems. These organs are produced in considerably different proportions, with, for example, most trees producing only a single trunk throughout their life, but thousands of leaves in any given year. Although strictly not macrofossils, the cuticle of fossil leaves (as dispersed cuticle) and its associated fungal microflora are also studied by palaeobotanists, providing both biostratigraphic and palaeoclimatic information (Kovach and Dilcher, 1984; Lange, 1976; 1978; Wolfe and Upchurch, 1987).

Some palaeobotanists have stressed the varying preservation potential of plant organs (Gastaldo, 1988; Spicer, 1989). Heavily lignified plant tissue, such as woody stems and some fruits, decays far slower than, for example, the delicate perianth and stamens of most flowers, and so has a higher preservation potential than the latter. The chemically resistant cuticle of leaves also has a very high preservation potential and may persist extremely well in many different sediment types. However, in contrast to many animals, most plant organs have a similar preservation potential. In many animals hardparts such as shells or bones have high fossilization potential, whereas the softparts (visceral mass) have a very low preservation potential and are only preserved in fossil-Lagerstätten (Seilacher, 1990). However, the relative preservation potential of related plant organs varies enormously between different taxa, instituting an important bias in the plant fossil record.

In general, most plant organs have a poor preservation potential compared to vertebrate bones or shelly fossils, and so have a very limited potential to be reworked from older sediments into much younger sediments. Thus, time-averaging—the emplacement of fossils from older stratigraphic position to a younger part of the sequence—is far less common in plant macrofossil assemblages than animal assemblages. Fossilized wood, either as unaltered logs or as petrifications, and (more so) pollen and spores, are highly resistant to decay, and may experience significant reworking (see, for example, Kemp, 1972). More rarely, blocks of sediment may be reworked, carrying portions of an older macroflora intact into younger sediment (see, for example, Hill and Macphail, 1983).

Plant fossils are preserved primarily within clastic sediments (including volcanoclastics): as compressions with much of the organic matter preserved (but flattened); as impressions with the organic matter mostly lost, leaving an imprint or stain on the rock; as fusain through the conversion of the cell wall into charcoal by burning; and as permineralizations and petrifications (Schopf, 1975; Scott, 1990). Compression fossils vary from essentially mummified remains preserving considerable anatomical detail to preserving only the cuticular envelope of leaves. The sediments may become lithified, resulting in the chemical transformation of the plant remains into petrifications with little or none of the original organic material retained. This often occurs through the secondary deposition of silica, carbonate or pyrite, preserving internal or external moulds of cells, tissues or whole organs (Schopf, 1975; Scott, 1990). Examples range from silicified transported leaf litter (Peters and Christophel, 1978; Greenwood *et al.*, 1990) to *in situ* plant communities (Knoll, 1985). Silicified plant assemblages often preserve a high degree of internal anatomical detail (see, for example, Peters and Christophel, 1978).

The cell wall of plants is primarily cellulose and generally only preserves in acidic anoxic conditions. These conditions are most commonly found at the bottom of lakes, (generally abandoned) stream channels, and in swamps or river deltas. The plant macrofossil record is therefore random and strongly biased towards plant communities representative of high rainfall environments, or environments associated with water courses and other water bodies. Important plant macrofossil assemblages are also found in volcanoclastic deposits (see, for example, Burnham and Spicer, 1986). These include many important North American Palaeogene floras (Wing, 1987). Individual plant fossil beds consist primarily of disarticulated intermixed individual organs of many different taxa from the local vegetation. Only rarely are whole plants or whole plant communities preserved in a form which approximates the original plant or vegetation.

Figure 7.1   The forest-floor litter of Complex Notophyll Vine Forest from north-east Queensland (equivalent to Paratropical Rainforest of Wolfe, 1979).

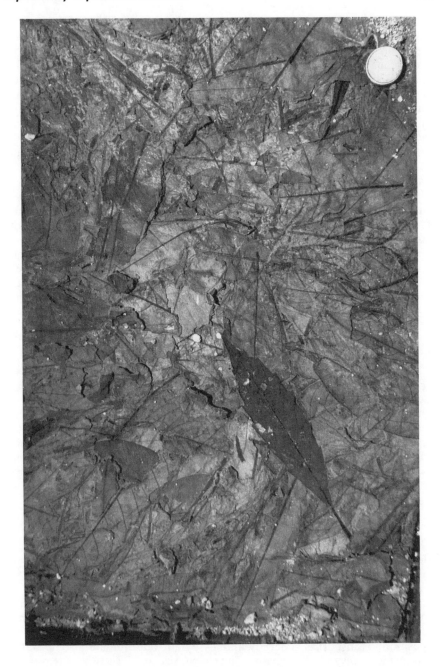

Figure 7.2 The oxidized leaf-mat from Golden Grove (Middle Eocene, South Australia).

The traditional approach in palaeobotany has been to view plant macro-fossil localities as a single 'flora' and primarily as depositories of specimens. The fossils were viewed as examples of extinct plant species and so, therefore, as evidence of morphological evolution (phylogenetic palaeobotany). The representation of particular taxa or lineages of taxa was then used as evidence for the presence of particular analogous modern plant communities and/or climates. This approach ignores the reality that most individual localities are composed of a number of separate depositional structures, each representing discrete events, separated in time and/or space. Even within apparently uniform assemblages such as seen in low-rank coals (lignites), individual lithotypes often contain separate macrofloras, reflecting differences in edaphic conditions and ecological processes such as succession.

In a review of North American Eocene macrofloras, Wing (1987, p. 751) stated that many floras 'actually consist of the summed floral lists of a number of separate quarry sites', and cautioned that such floras are not strictly comparable with floras derived from a single excavation. This type of approach is not always avoidable, however, as the diversity at a single exposure may often be very low, and outcrop of a single fossiliferous unit extensive and continuous over a wide area. It is common in these cases to see regional accounts of the resulting floral summation. It must be remembered, however, that such summed floras potentially contain several plant communi-ties. Separation of sedimentary facies is essential. Burnham (1989, p. 27) found that the size of individual North American floras varied dramatically because of similar factors, ranging from an average of as few as 64.3 speci-mens to 1667 specimens.

Ferguson (1985) and Gastaldo (1988) presented flowcharts which summar-ized the formation of a plant macrofossil deposit and emphasized the role of taphonomic biases occurring during transport and deposition. According to their model, the selective nature of the leaf-rain is further emphasized by these post-abscission biases. Analyses of leaf accumulations within modern sedimentary environments have demonstrated also that the type of sedimen-tary facies and the character of the local vegetation have a strong influence on the composition of the fossil macroflora (Taggert, 1988; Burnham, 1989). A brief review of the influences of each of these factors is presented below.

TRANSPORT AND DEPOSITION OF PLANT PARTS

Transport of plant parts

A potentially strong source of taphonomic bias is the transport of the dis-articulated organs (particularly leaves) from the source plant(s) to the site of deposition. Transport is generally initially by air, but ultimately some water transport is involved in most cases. The shape, size and structure of leaves, as well as the relative density of leaf tissue (weight per unit area), are intrinsic factors which may influence the passage of leaves through either air or water (Spicer, 1981; 1989; Ferguson, 1985). However, the character and behaviour of the source vegetation have an initial controlling effect by determining the composition of the litter-rain.

In modern Australian humid tropical forests leaves from the canopy tree species dominate the litter-fall (Brasell *et al.*, 1980; Spain, 1984; Stocker *et al.*, in press). Leaves from other forest synusiae—shrubs, terrestrial herbs, vines, parasites (for example, Loranthaceae, Viscaceae) and giant monocots (Zingiberaceae, Musaceae, etc.)—are rare components of the leaf-rain. However, where vines constitute a major component of the canopy their leaves may be a significant fraction of the leaf-fall. The volume and taxonomic composition of the leaf-rain is also variable, with a peak in absolute volumes just prior to the wet season, although leaf-fall in some species peaks at other times of the year (Spain, 1984; Rogers and Barnes, 1986; Stocker *et al.*, in press). The presence of flowers and fruits in the litter-rain is highly variable, and generally represents a minor fraction of the resultant debris.

The initial composition of the litter-rain is therefore biased, with the leaves of the canopy trees swamping the leaves of other synusiae, and non-foliage organs are a relatively minor component. Potentially, the taxonomic composition also varies through the year. Leaf-litter from the forest floor reflects this variation and the dominance by canopy trees. In temperate deciduous forests Ferguson (1985) found that the taxonomic composition of leaf litter was strongly influenced by the nearest trees. Greenwood (1987a) found similar results in Australian humid evergreen tropical forests and evergreen temperate rainforests. Most trees in deciduous forests lose their leaves in a short space of time; however, trees in evergreen forests lose their leaves over the whole year with peak periods of leaf loss for the forest and particular species (Brasell *et al.*, 1980; Spain, 1984; Rogers and Barnes, 1986). Residence time on the forest floor of litter in tropical forests is variable (Anderson and Swift, 1983); however, a significant amount of litter remains from previous months, and much longer during dry weather (Brasell *et al.*, 1980; Spain, 1984). The litter volumes in the Australian humid tropical forests can therefore be expected to contain a time-averaged sample of the litter-rain.

The forest-floor litter (see, for example, Figure 7.1) in the Australian forests is dominated by leaves from the principal canopy tree species (Greenwood, 1987a), reflecting the bias seen in the litter-rain (Brasell *et al.*, 1980; Spain, 1984; Stocker *et al.*, in press). In the low-diversity forests (temperate rainforest from New South Wales) little change was seen between samples, although the relative proportion of leaves from the non-canopy synusiae was variable, reflecting spatial variability in the distribution and abundance of these plants (Greenwood, 1987a). Variation between samples from the high-diversity forests (humid tropical lowland and upland forests from northeast Queensland) was high, reflecting high spatial variation in abundance and distribution of species in all woody synusiae. This evidence implies that the litter was derived from only the immediately adjacent trees (see, for example, Ferguson, 1985).

The distance that the components of the litter-rain may travel is controlled by the behaviour of the individual leaves, flowers and fruits as sedimentary particles (Spicer, 1981; 1989). In a series of experiments using both leaf models and actual leaves, Spicer (1980; 1981; 1989; and references therein) and Ferguson (1985) have examined leaves as sedimentary particles to determine the relative significance of taphonomic biases created during transport. The distance travelled by leaves in air was found to be largely controlled by the weight per unit area of the leaf (Spicer, 1981; 1989); however, leaf morphology and overall weight are contributing factors (Spicer, 1981; Ferguson, 1985). The general consequence of this is that lighter, usually smaller leaves, travel further than heavier (and denser), usually larger leaves. However, as Spicer (1981; 1989, p. 108) pointed out, the coriaceous 'sun leaves' of the upper canopy have a higher weight per unit area than the comparatively larger but membranous 'shade leaves' of the understory, and so can be expected to travel less far than the 'shade leaves'.

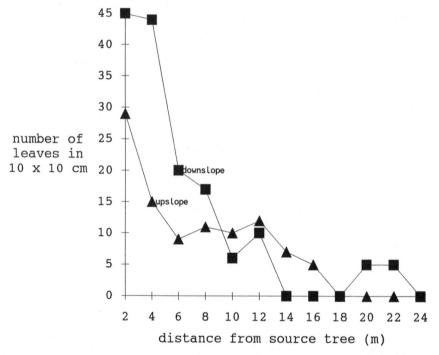

Figure 7.3   Number of leaves found in forest floor-litter at distance from the source tree (*Prumnopitys amara* — Podocarpaceae) in 10 × 10 cm quadrats along two transects—upslope and downslope—in Complex Notophyll Vine Forest (*sensu* Webb, 1959).

In unrestricted fall, fewer leaves are found with increasing distance from a source tree in a negative exponential relationship (Figure 7.3; Spicer, 1981; Ferguson, 1985; Greenwood, 1987a). In a forest situation, the structure of the trunk space and the presence of any screening foliage may also significantly influence the passage of leaves. Ferguson (1985) concluded that, in general, leaves are unlikely to travel further in a forest interior laterally than the height at which they abscissed from the tree. Measurement of the dispersal of leaves around their source trees in tropical forests (mesothermal, Notophyllous Forest *sensu* Wolfe, 1979) in northeastern Queensland demonstrated that the complex trunk-space of these forests substantially reduced the effective distance travelled by leaves (Figure 7.3; Greenwood, 1987a).

The dominance of the leaf-fall by canopy trees and the local nature of the taxonomic composition of the leaf-fall controls the physiognomic character of the resulting leaf assemblage. Correlations between the foliar physiognomy of modern vegetation (leaf size and margin type) and climate (mainly mean annual temperature (MAT): Wolfe, 1979; Upchurch, 1989; Spicer, 1990) have been used to predict palaeoclimate from leaf assemblages (see, for example, Wolfe, 1985; 1990; Wolfe and Schorn, 1989). The main criticism of this approach has been the extent to which a decodable climatic signal is preserved in the foliar physiognomy of leaf assemblages (Dolph and Dilcher, 1979; Roth and Dilcher, 1978; Burnham, 1989; Christophel and Greenwood, 1988; 1989). A key assumption, however, has been that the leaves of canopy trees will dominate fossil leaf assemblages. Based on the evidence of leaf-rain and forest-floor litter in Australian forests, this assumption would appear well founded. Greenwood (1987a; Christophel and Greenwood, 1988; 1989) and Burnham (1989) have examined whether a decodable physiognomic signal can be detected in modern leaf assemblages.

The forest-floor litter examined by Greenwood (1987a) contained leaves much smaller than expected based on canopy values for the same forests. Using taxon-based observations (leaf size index (LSI)[1]), the litter was found also to have fewer taxa with larger leaf-size classes than expected from canopy observations (that is, smaller LSI values). In general, the bias was in the order of 3:2 smaller LSI values for litter compared to canopy (Figure 7.4), with departures between litter and canopy decreasing with concomitant decreases in mean leaf size in the canopy. However, the proportion of species with entire leaf margins in the litter was generally different than that recorded for the canopy alone (Table 7.1), reflecting the influence of non-canopy synusiae. Similar biases in actual fossil assemblages would bias estimations of MAT by similar ratios.

Burnham (1989) sampled litter from a number of sedimentary subenvironments in a riverine environment within Paratropical Rainforest (*sensu* Wolfe, 1979) in Mexico. In general, Burnham found that the foliar physiognomy (LSI and leaf-margin analysis) of most of the subenvironments reflected the regional climatic signal. Burnham determined the leaf-size class of each taxon found in the litter from canopy-collected herbarium samples, and so gave no measure of any changes in LSI due to biases from the actual leaf-rain. However, in the studies of both Greenwood and Burnham it was demonstrated that a decodable climatic signal (Burnham, 1989), or 'foliar physiognomic signature' for particular forest types (Greenwood, 1987a; Christophel and Greenwood, 1988; 1989) was preserved in leaf assemblages, which, when properly constrained, can be used to reconstruct palaeoclimates. An important constraint is the influence of additional transport by water prior to deposition.

Transport of plant parts in water is controlled by the rate at which they

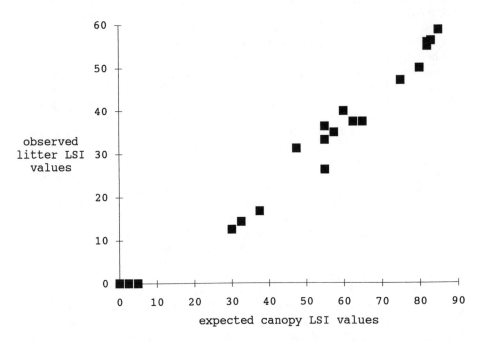

Figure 7.4   Leaf Size Index (LSI) of forest-floor litter and rainforest canopies for four Australian rainforest types. Data derived from Greenwood (1987a) and hitherto unpublished data.

Table 7.1   *Comparison of the number of species with entire leaf margins found in the canopy and litter (percentage of taxa). Key: Qld = Queensland; NSW = New South Wales*

Forest type	Litter	Canopy (Webb, 1959)
Complex Mesophyll Vine Forest	82.4–95.0	70–90
Complex Notophyll Vine Forest	60.0–81.3	70–85
Simple Notophyll Vine Forest Qld	84.2–90	40–70
Simple Notophyll Vine Forest NSW	14.3–25	40–70
Microphyll Fern Forest	33.3–40	0

become waterlogged, their hydrodynamic properties and the turbulence of water flow (Spicer, 1989). The rate of decay also contributes to the transport and preservation potential of the part, with different rates of decay demonstrated for the leaves of different species (Ferguson, 1971; 1985; Greenwood, 1987a), and also for leaves of differing physiognomy (Heath and Arnold, 1966). Ferguson (1971; 1985) found that whereas trees of *Ilex europea* were common in local forest, leaves of this species were reduced to almost unrecognizable bags of cuticle after only short periods of immersion. Similarly, litter collected in a stream surrounded by forest with a canopy dominated by *Doryphora sassafras* (Monimiaceae) and *Ceratopetalum apetalum* (Cunoniaceae) was found to be dominated by leaves of *C. apetalum*, with few or no leaves of *D. sassafras*. Litter collected from the forest floor nearby contained leaves of both species in nearly equal proportions (Greenwood, 1987a). Preferential decay in water of some species may, therefore, bias fluvial and lacustrine leaf assemblages.

However, Hill and Gibson (1986) found that leaves from a number of species from Tasmanian subalpine evergreen vegetation were essentially intact after six months of immersion, in contrast to the high level of decay detected for leaves from deciduous temperate species after two to four months (Spicer, 1981; Ferguson, 1985). This difference probably reflects differences in the chemical and anatomical nature of the two sets of leaves (Ferguson, 1985; Spicer, 1989), with the evergreen leaves typically coriaceous and rich in both lignin and tannins (and other phenolic compounds), and the deciduous leaves typically papery and containing fewer phenolics. Spicer (1989) discussed how leaves may undergo limited transport within the water column, and so the period leaves remain relatively intact free of sediment is a controlling factor. However, the initial behaviour of leaves (and other parts) as they enter a water body largely controls water transport distances.

The leaves of most temperate deciduous trees are shed dry (Spicer, 1981), whereas many evergreen species shed essentially unaltered leaves. A dry leaf will remain on the surface of still water for more than several weeks. Experiments with aquaria have shown that thin papery leaves float for much shorter periods of time than thick coriaceous leaves (Spicer, 1981; 1989; Ferguson, 1985). Duration of floating appears to be controlled by the rate at which the leaf tissue becomes waterlogged. Cuticle thickness, stomatal frequency and size (for example, hydathodes), damage to the leaf lamina and petiole, and the conditions of the water (temperature and chemistry) would appear to be the main factors controlling water uptake by the leaves, and thus floating times (Spicer, 1989).

Hill and Gibson (1986) found that the majority of leaves of *Eucalyptus coccifera* and *Orites acicularis* sank within two days. The leaves of both of these species are sclerophyllous (markedly coriaceous). The majority of other Tasmanian subalpine taxa examined by Hill and Gibson (1986) had significant numbers of leaves floating after much longer periods of time. Spicer (1981; 1989) also found variable floating times, with the thin papery leaves of the deciduous species *Alnus glutinosa* sinking within hours, while significant numbers of evergreen (coriaceous) leaves of *Rhododendron* remained floating after several days. Hill and Gibson (1986) concluded that sinking rates may be a significant factor determining the distribution and abundance of leaves in lake sediments.

Christophel and Greenwood (1988; 1989; Greenwood, 1987a) found that leaves collected in allochthonous deposits downstream were significantly smaller than leaves from an autochthonous (essentially forest-floor) leaf assemblage, approximately 100 m upstream in Simple Notophyll Vine Forest (northeast Queensland). Little difference in the relative width was observed, although the leaves in both assemblages tended towards narrow elliptic. This suggests that there was a trend towards stenophylly in the leaves contributing to the stream litter load, perhaps reflecting a bias towards input from river-margin vegetation (Greenwood, 1987a; Christophel and Greenwood, 1988; 1989). Both assemblages were more diverse than equivalent forest-floor samples, with the allochthonous sample the most diverse with 36 leaf taxa from 249 specimens. Transport within streams, therefore, may significantly alter the physiognomic character of leaf assemblages, and results in enhanced sampling of the local vegetation (see Burnham, 1989).

The variable form and size of fruits, seeds and flowers contributes to a wide variation in observed floating times by these organs (Collinson, 1983; Spicer, 1989). Collinson (1983) concluded that deposits composed primarily of seeds and fruits (that is, thick-walled, durable plant material) probably occur through the selective biodegradation of the less durable plant material (that is, leaves and flowers). Plant organs which remain floating are more likely to be transported further by streams than quickly sinking organs. Spicer (1989, p. 119) has highlighted the problems attendant with the highly transportable nature of tree logs (see, for example, Frakes and Francis, 1990) and suggested that transported logs of unknown provenance pose problems when used as palaeoclimatic indicators. Significant accumulations of plant detritus occur at the mouths of rivers, including a rich assortment of more durable plant parts, such as logs, twigs, seeds and fruits. Much of this material has probably been transported considerable distances from upstream.

Modern analogues of plant fossil assemblages

A primary distinction can be made between plant fossil deposits formed from the gradual accumulation of plant material *in situ* (autochthonous assemblages) and assemblages formed by the accumulation of transported plant materials (allochthonous assemblages). This dichotomy has important taphonomic consequences as the autochthonous assemblages reflect primarily the plants growing within the depositional site, including the potential preservation of *in situ* whole or nearly complete plants. These assemblages are therefore likely to represent only the immediate vegetation. The allochthonous assemblages contain plant material which may have been transported from a number of separate plant communities within the local depositional basin and so could represent several plant communities. Taphonomic biases caused by transportation effects are also likely to be more profound in allochthonous assemblages.

Autochthonous assemblages: peat bogs and swamps (mires)
The abundant Mesozoic and Cainozoic coal sequences represent fossilized peat. Peat results from a long-term accumulation of plant matter and usually occurs through the suppression of decay processes in the subsoil and humic layer of soils. The conditions necessary for the accumulation of peat are usually found where soil water levels are high, maintaining anoxic soil conditions. As such, they can be found from the wet tropics to the arctic tundra (Moore, 1989). The presence of coal does not imply high gross productivity by these plant communities, but rather high net accumulation (Moore, 1989). Petrological differences between coals reflect a combination of differing source sedimentary environments and biological communities as well as subsurface processes subsequent to burial (Cameron *et al.*, 1989). In particular, differences between lithotypes within a single seam, particularly in low-rank coals, can be attributed to the presence of a variety of plant communities in the original peat-forming vegetation (Luly *et al.*, 1980; Cameron *et al.*, 1989).

Peat-forming plant communities range from essentially treeless peat-bogs dominated by bryophytes (typically *Sphagnum*) and pteridophytes or swamps of herbaceous monocots (often including *Sphagnum*) to swamps containing a significant cover of woody plants, including trees (for example, the Kerangas of Borneo; Brünig, 1983). Tree-dominated swamps, or bog forests, are usually dominated by conifers at mid- to high latitudes (for example, *Taxodium*, *Picea* and *Larix* in the northern hemisphere; *Dacrydium sensu lato* and *Dacrycarpus* in the southern hemisphere), but contain significant angiosperm components in tropical areas. A mosaic of different peat-forming communities may occur within a single basin. Many of the plant species may be restricted to these communities, or occur only rarely in other plant communities.

Peat-forming communities can be collectively referred to as 'mires' and can be classified according to the primary source of water to the community (Moore, 1989). Mires where all of the water is sourced from rain are termed 'ombrogenous' (ombrotrophic) (raised or domed mires), whereas mires with additional water supplied from ground water or inflowing streams are termed 'rheogenous' (rheotrophic) (Cameron *et al.*, 1989; Moore, 1989) or 'topogenous' (Macphail and Hope, 1985). The rheogenous mires produce peats with much higher fractions of inorganic matter (mostly clastics) to organic matter than ombrogenous mires due to transported materials brought in by inflowing streams. Additional plant material, particularly pollen and leaves, may also be transported into a rheogenous mire, thus altering the floristic character of the macrofossil suite preserved.

Peat-forming environments occur within a number of sedimentary settings (Flores, 1981; Gastaldo *et al.*, 1987; Gastaldo, 1988; Moore, 1989). Peats that form in alluvial floodplains, lake margins, deltaic wetlands and mangroves generally accumulate macrodetritus solely from the *in situ* vegetation (Gastaldo, 1988). Peats may also form from solely allochthonous accumulations due to reworking of macrodetritus in channels and coastal settings (Gastaldo *et al.*, 1987), and are generally composed of highly fragmented plant material (Gastaldo, 1988).

There is some debate over the types of peat community and sedimentary setting which give rise to very large coal seams. On the island of Borneo, and elsewhere, modern and Holocene peat-forming (ombrogenous) raised mires have been studied as possible analogues for large-scale coal formation (Cameron *et al.*, 1989). Floristically, the peat-forming communities in Borneo resemble the Miocene plant communities which produced extensive economic reserves of lignite in southern Australia (Christophel and Greenwood, 1989). Smaller peat (and ultimately coal) structures that form within fluvial settings in floodplain swamps and in deltas (Flores, 1981) are likely to contain some, and may be primarily composed of, transported material (Gastaldo *et al.*, 1987; Gastaldo, 1988).

Peat macrofossil assemblages are generally dominated by seeds and fruits of the mire vegetation, their roots systems (especially the rhizomes of reeds, sedges and other monocots) and their stems or woody axes (GreatRex, 1983; Raymond, 1987; Gastaldo, 1988). Leaf material tends to be rare. Material from all of the plants present (herbaceous and woody) is incorporated and herbaceous plants may contribute as much as 25 per cent of the accumulated material (Raymond, 1987). GreatRex (1983) found that seeds and fruits were generally transported no further than 1 m from their source plant, but that seeds and fruits which were adapted to wind or water transport may be transported further. Coal, however, generally shows a high level of alteration and often very little plant material is recognizable. In low-rank coals in the Latrobe valley, southeastern Australia (see, for example, Luly *et al.*, 1980), peat-surface horizons may be seen with *in situ* tree bases and rarely also with leaf litter beds. The relative abundance of root fossils in autochthonous peat communities is correlated with the abundance of the source taxon in the original plant community (Raymond, 1987).

Allochthonous assemblages: fluvial and lacustrine deposits
Plant macrofossil assemblages are commonly formed in fluvial, deltaic and lacustrine settings. Fluvial plant fossil assemblages are typically small in size and may occur within fine-grained sediments included within coarser fluvial sediments. Several depositional facies exist within fluvial environments: channel, levee and crevasse splay, floodplain, and infilled abandoned channels (oxbow ponds). In addition, swamps are associated with some river systems, representing areas of impeded drainage beyond the levee banks which may be seasonally flooded by the river. Where streams flow into larger water bodies, such as lakes or the sea, deltas form. Significant accumulations of plant macrodetritus may occur in deltaic deposits. These will be only briefly considered here.

Channel deposits of meandering river systems can be divided into channel lag and point bar deposits, representing, respectively, deposition in the active channel and on the inside loops of meandering river channels (Collinson, 1986). In general, plant macrofossil assemblages are rare in channel lag deposits, although larger plant parts such as tree logs may accumulate (Scheihing and Pfefferkorn, 1984; Wing, 1988). However, the upper portion of point bar structures typically preserves leaves, flowers and other delicate plant structures, whereas the lower part of these structures is typically barren of plant fossils, or only contains more durable plant remains (Wing, 1988). Levees form through the preferential deposition adjacent to the existing channel course of suspended sediment by floodwaters. Crevasse splay deposits occur when the levee is breached, depositing a fan of sediment onto the adjoining floodplain. Typically both of these deposits are characterized by extensive rooting by vegetation and other characteristics of soil formation (Collinson, 1986) and so are poor sites for plant macrofossil preservation (Scheihing and Pfefferkorn, 1984; Wing, 1988; Spicer, 1989), although occasionally forest-floor litter, tree bases and whole herbaceous plants may be preserved in crevasse splay deposits, producing very rich assemblages (Scheihing and Pfefferkorn, 1984; Gastaldo, 1986; Gastaldo *et al.*, 1987).

Floodplain environments contain a number of subenvironments, including swamps, inactive channels (oxbow ponds) and alluvial dryland plant communities. The interfluvial areas may be seasonally inundated, receiving a thin, extensive layer of generally fine-grained sediment. These sediments are generally strongly rooted by the *in situ* vegetation and may also show signs of bioturbation. However, where the floodplain remains waterlogged for a substantial part of the year, plant debris buried by the sediment may produce assemblages of finely preserved, often quite delicate structures (Wing, 1984; 1988; Spicer, 1989). Where plant growth is luxuriant, thick sequences of peat may develop in this environment (Flores, 1981; Collinson, 1986).

Infills of abandoned channels are generally of little lateral extent and have a distinctly lenticular shape when viewed in section (see, for example, Potter and Dilcher, 1980). Individual 'lenses' are usually dominated by flat-bedded, fine-grained sediments, mainly clays, with coarser sediments at their base representing the old channel bedload. The clay lenses are a rich source of well-preserved plant macrofossils in Eocene sediments in both North America and southern Australia (Potter and Dilcher, 1980; Wing, 1987; Christophel, 1981; Christophel and Greenwood, 1989). Fossilization is usually as compressions or impressions (for example, Golden Grove, Christophel and Greenwood, 1987: see Figure 7.2 herein), although subsequent lithification may produce other modes of preservation such as silicification (Ambrose *et al.*, 1979; Greenwood *et al.*, 1990).

Generally, numerous clay lenses are found at individual quarry sites (see, for example, Christophel *et al.*, 1987), or in closely associated sites (Christophel and Greenwood, 1987; Potter and Dilcher, 1980), and in many cases were coeval or closely coeval, representing the meander track or braided channel zone of single river systems. The Australian Eocene clay lenses are generally finely laminated, with individual layers often defined by rich accumulations of mummified leaves and other plant macrodetritus (Christophel and Greenwood, 1987; Barrett and Christophel, 1990). The presence of very delicate plant structures, such as staminate conifer cones with *in situ* pollen (see, for example, Greenwood, 1987b), suggests that little transport of plant parts occurred in many of these assemblages.

Sequences formed on lake beds are generally more uniform in thickness and laterally continuous than fluvially deposited sediments. These lake sediments are consequently tabular in section and may be finely laminated (lamina from less than 1 mm to 1 cm; Wing, 1988). Associated sedimentary environments, such as delta fronts and swamp facies, may interrupt lacustrine sequences. The foreset and toeset beds of lacustrine deltas preserve abundant plant remains, mainly leaves (Spicer, 1980; 1981; Spicer and Wolfe, 1987). The occurrence of plant remains in the finely laminated lake sediments generally shows a strong relationship between diversity and abundance, and distance from shore; near-shore samples are generally much more diverse and with more specimens for the same amount of sediment than samples further from shore (Wilson, 1980; Spicer, 1981; Hill and Gibson, 1986; Spicer and Wolfe, 1987).

A number of studies have examined the extent to which leaf assemblages in the lakes reflected the local plant communities. Very few have considered input of plant material into fluvial environments. Birks and Birks (1980) reviewed many studies on the input of plant materials into lakes, but the majority of these concerned seeds and fruits. Several other studies have concentrated on leaves entering lakes from nearby woody vegetation (McQueen, 1969; Drake and Burrows, 1980; Spicer, 1981; Hill and Gibson, 1986; Spicer and Wolfe, 1987). The general consensus is that lake sediments are dominated by leaves from the local waterside flora and nearby vegetation. Additional transport of plant macrodetritus from upstream sources (in the case of lakes with stream infeeds) or by wind transport (storm effects and chance dispersal) may introduce elements from extralocal vegetation (McQueen, 1969; Drake and Burrows, 1980; Hill and Gibson, 1986; Spicer and Wolfe, 1987).

Carpenter and Horowitz (1988) and Burnham (1989) have examined modern fluvially deposited leaf assemblages to determine potential bias in their taxonomic membership relative to the riparian vegetation. In Carpenter and Horowitz's study, Tasmanian high-rainfall tall evergreen forest dominated by *Eucalyptus obliqua* with an understory of cool temperate rainforest containing *Nothofagus cunninghamii* and associated species was sampled. The composition and relative abundance of taxa in stream-bed and drift samples of litter were compared to the surrounding forest. In general, the litter matched the riparian forest; however, very few leaves of *E. obliqua* were found in any samples, possibly due to rapid sinking rates for leaves of this species (for example, Hill and Gibson, 1986). Leaves from the interfluvial forest were also encountered in significant numbers, indicating that the streambed litter represented more than just the riparian vegetation.

Burnham (1989) found that most fluvial subenvironments reflected the local Paratropical Rainforest flora well, but that the adjacent trees overwhelmingly dominated the leafbeds. Overproduction of leaflets by leguminous trees with compound leaves was a significant factor (Burnham, 1989). The channel deposits contained a biased sample of the local vegetation, possibly reflecting the smaller sample sizes from this subenvironment (Burnham, 1989). Litter from Simple Notophyll Vine Forest in New South Wales and northeast Queensland collected by Greenwood (1987a; Christophel and Greenwood, 1988; 1989) in stream beds indicated a small but significant influence by a specialized riparian flora, although the canopy trees of the local forest overwhelmingly dominated the deposits. Notably, in the New South Wales example, leaves of one of the canopy dominants, *Doryphora sassafras*, were scarce or rare in stream litter, but represented a significant component of nearby forest-floor litter (see discussion above).

It can be seen from the discussion above that observations of modern depositional analogues demonstrate that a consequence of the localised leaf-rain and hydrodynamic influences is that fluvial and lacustrine leaf beds generally only reflect the immediate local vegetation and thus local plant communities. In contrast, palynofloras generally reflect the regional flora and resolution of individual plant communities is often dependent on comparison with modern floristic associations.

Biofacies

Plant communities are not evenly distributed in space or time over modern landscapes and it is reasonable to assume that this has been the case for much of the Mesozoic and Cainozoic. Similarly, sedimentary environments vary spatially and temporally, and this heterogeneity is expressed by the presence of discrete facies (Reading, 1986). Within each sedimentary facies quite characteristic plant assemblages can often be recognized. These assemblages are often termed 'biofacies', but this term is used in two different ways. In an ecological sense (as above) biofacies refer to a fossil assemblage which characterizes a region or body of rock; in a stratigraphic sense, biofacies refer to a 'body of rock which is characterised by its fossil content which distinguishes it from adjacent bodies of rock' (Moore, 1949; Brenchley, 1990, p. 395). The ecological sense of the term emphasizes differences between environments, whereas the stratigraphic sense emphasizes horizontal (?biogeographic) and vertical changes in the biota (temporal changes in communities). In this discussion, 'biofacies' is used in the ecological sense.

Separate biofacies may be caused by a number of interacting factors and it is somewhat artificial to separate them. For ease of discussion, they are here separated into influences arising from the sedimentary behaviour of various plant organs, and those caused by the heterogeneity of vegetation. In fluvial-paludal sedimentary settings several sedimentary subenvironments can be recognized: within-channel, levee bank, point bar, floodplain swamp, abandoned channels, deltas and crevasse splays. Hydrodynamic sorting within and between these environments, and the individual character of the local vegetation surrounding them, will all contribute to forming separate biofacies in each of these environments (see, for example, Scheihing and Pfefferkorn, 1984; Burnham, 1989).

The fate of leaves, fruit, wood and other plant parts
The unequal dispersal characteristics and preservation potential of different organs (leaves, fruits, flowers, wood fragments) and of different forest synusiae (trees, vines, herbs) contribute to taphonomic biases acting on the organographic character of plant fossil beds (Collinson, 1983; Spicer, 1989) and the representation of structural components of the standing vegetation (Scheihing and Pfefferkorn, 1984; Gastaldo, 1988). The differential potential

for incorporation and fossilization of some taxa also ensures that most plant fossil beds preserve a taxonomically biased subset of the local vegetation.

Vegetational heterogeneity
Differential sorting and ecological separation of the standing vegetation between individual sedimentary environments causes sedimentary facies within a single sedimentary unit to contain different taxonomic subsets of the local vegetation, representing the subenvironments in the local vegetational mosaic (Scheihing and Pfefferkorn, 1984; Taggert, 1988; Burnham, 1989). At finer sedimentary scales, temporal variability in the character of the leaf-rain (phenological seasonality) may result in cyclic changes in the taxonomic and organographic character of individual horizons within a single sedimentary sequence. Similarly, lateral floristic heterogeneity in the local vegetation may be expressed in the lateral taxonomic heterogeneity of leaf beds (Taggert, 1988).

Several authors have attempted to portray the influence of different sedimentary environments on the composition of plant macrofossil assemblages (for example, Wing, 1988; Spicer, 1989). Ferguson (1985) and Gastaldo (1988) summarized the factors which influence dispersal, incorporation (biostratinomy; Gastaldo, 1988) and diagenesis. Their summaries emphasized the role of these processes and the segregation that occurs between sedimentary facies. Only limited attention has been paid to the initial influence of the vegetation itself (see, for example, Taggert, 1988; Spicer, 1989). In most instances, plant macrodetritus deposition reflects strongly the local flora, with the specialized riparian or water's-edge vegetation and 'dryland' vegetation represented to varying degrees. Figure 7.5 summarizes the interaction between vegetational heterogeneity and the presence of separate sedimentary environments in a hypothetical landscape.

In this hypothetical example, a landscape of generally moderate relief is crossed by a meandering river which has numerous cutoff meander lakes (oxbow ponds) in its lower reaches. Lenticular clay bodies are shown in section through the vertical fluvial sequence. Infeed from highlands is dammed forming a small lake and, in the interfluvial floodplain area, impeded drainage has formed a small internal basin within which lacustrine conditions fluctuate with swamp (peat-forming) conditions. Intertonguing lacustrine fine clastic and lignitic sediments are shown in sequence beneath the present lake and marginal swamp. The vegetation is simplistically divided into a 'wetland' (mostly swamp) flora associated with water's-edge and mire communities, a 'dry-lowland' flora subdivided into riparian elements and interfluvial forest, and a highland forest flora. Each of these units is indicated by different grades of shading (see legend, Figure 7.5).

The relative contribution of each of the floristic elements to leaf assemblages formed in particular sedimentary facies in this hypothetical landscape is indicated by a series of pie charts. The relative percentages given in these pie charts is partly based on actual examples, but should be viewed as approximate and used as a rough guide only. It can be seen in this example that each sedimentary facies contains a different subset of the regional flora. The overall representation of species from each floristic element is respect-

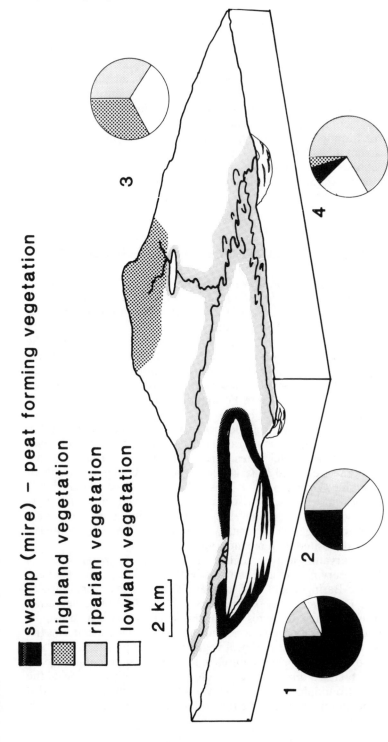

■ swamp (mire) – peat forming vegetation

▨ highland vegetation

□ riparian vegetation

□ lowland vegetation

2 km

Figure 7.5 Block diagram showing a vegetational mosaic superimposed on a fluvial sedimentary environment with separate sedimentary facies shown diagrammatically. The proportion of taxa found in leaf assemblages from each regional vegetation type (see key) in selected sedimentary facies is shown in pie charts. Proportions are approximate and given as a guide only.

ively diluted or enhanced in different facies due to selective taphonomic processes.

Essentially autochthonous deposition in the swamp-coal facies (1) produces a flora biased to the 'wetland' flora, including a significant component of herbaceous plants, whereas the lake deposits (2) may include significant elements from the lowland dryland flora due to fluvial transport and also hydrophytes (see, for example, Collinson, 1988). Similarly, the second lake (3) will probably contain significant macrofossil representation of the high-land flora, in addition to the lowland flora, due to infeed from upland fluvial sources. The abandoned channel infill deposits (4) in the main river-channel zone will essentially be dominated by the riparian and lowland flora. Successional mosaics within this area may also be expressed by differences in the macrofloras of the clay lenses.

THE PALAEOECOLOGY OF PLANT MACROFOSSIL ASSEMBLAGES

Several plant macrofossil localities have been studied in some detail taphono-mically (by, for example, Gastaldo, 1986; Burnham, 1990). However, Eocene floras from southern Australia are the most familiar to the author and so these will be used to discuss some of the points made above. The main Eocene plant macrofossil localities from Australia are Middle Eocene and occur in the southern part of the present continent; the Maslin Bay and Golden Grove floras of the North Maslin Sands (St Vincents Basin: Lange, 1970; 1982; Christophel, 1981; 1988; Christophel and Greenwood, 1987; 1989) and the Anglesea flora of the Eastern View Formation (Christophel *et al.*, 1987; Christophel and Greenwood, 1989; Figure 7.2 herein).

These Eocene macrofloras were all deposited within fluvial settings and in most instances represent either temporary lacustrine conditions within aban-doned channels, or other channel facies. The Golden Grove, Maslin Bay and Anglesea macrofloras are found in large lenticular clay bodies within cross-bedded coarse-grained sandstones (Christophel and Greenwood, 1987; Lange, 1970; Blackburn, 1981; Christophel *et al.*, 1987), indicating low-energy deposition within a larger high-energy fluvial setting, possibly braided streams, although the Anglesea deposits were interpreted by Christophel *et al.* (1987) as being deposited by a meandering river system.

The level of preservation at the Anglesea and Golden Grove localities is generally high, preserving delicate flowers with attached intact stamens (Christophel, 1984; Basinger and Christophel, 1985) and leaf domatia with oribatid mites (O'Dowd *et al.*, in press). The high standard of preservation and delicate nature of much of the fossil material suggests very little transport of the plant material prior to deposition, and thus it may be inferred that the leaf assemblage will only reflect the local vegetation. The presence of arbor-eal, as opposed to soil/litter, oribatid mites in the fossil leaf domatia suggests that leaves were abscissed directly into the Golden Grove clay lens from the canopy of the surrounding forest (O'Dowd *et al.*, in press). Bulk dis-aggregation of the sediment containing the mummified plant matter generally releases most of the original leaves, and thus avoids the problem of under-estimation of taxon abundance due to overlapping specimens seen in impression floras (Ferguson, 1985; Spicer, 1988).

Each of the Eocene macrofloras share common floristic characters, with a high incidence of leaves from the Lauraceae and Elaeocarpaceae, and low-diversity (but consistent) representation of leaves from Myrtaceae, Podocarpaceae and Proteaceae. Macrofossils of *Nothofagus* (Fagaceae) are rare or absent from (mainland) Australian Palaeogene macrofloras (Christophel, 1988; Christophel and Greenwood, 1989). In contrast, the palynofloras associated with these macrofloras are dominated by *Nothofagidites* (palyno-morph attributed to *Nothofagus*) and a highly diverse assemblage of grains assignable to the Myrtaceae, Podocarpaceae and Proteaceae. Pollen grains of the Lauraceae are absent. This discrepancy between the microfloras and macrofloras has led to differences in regional vegetational reconstructions for the Australian Palaeogene, with palynologists stressing the role of *Nothofagus* and microthermal Podocarpaceae in Palaeogene vegetation (Martin, 1978; 1981; 1982; Truswell and Harris, 1982), whereas the macrofos-sil workers have stressed the presence of mesothermal rainforests of similar floristic and physiognomic character to the modern tropical rainforests of northeast Queensland (Christophel, 1988; Christophel and Greenwood, 1987; 1988; 1989).

A single fossiliferous clay lens is known to occur at Golden Grove (Christophel and Greenwood, 1987; Barrett and Christophel, 1990). This structure is finely laminated with numerous laterally extensive, thin mats of leaves defining the layers. Preservation varies from densely packed mummified leaves, fruits, staminate flowers (Christophel and Greenwood, 1987), sporangiate fern fronds (*Lygodium* sp.) and leaf domatia with oribatid mites (O'Dowd *et al.*, in press) to an organic stain (occasionally preserving cuticle) revealing high detail of venation on a leached lighter matrix (Christophel and Greenwood, 1987; Figure 7.2 herein).

Barrett and Christophel (1990) collected macrofossils from the Golden Grove macroflora laterally along a single bedding plane for three vertically separated horizons. The separation between successive bedding planes in the Golden Grove clay lens was variable, although typical separation was in the order of a few centimetres, with two lower layers separated by about 2 cm and separated by about 50 cm from the upper layer sampled. Barrett and Christophel (1990) found significant differences in taxonomic membership and frequency between the upper and lower layers at Golden Grove; however, summing the leaves from the two lower layers and comparing this to the upper layer gave a different result (Figure 7.6). This procedure gave similar sample sizes (upper layer 433 leaves and lower layers 506 leaves). The same taxon was dominant (leaves of Elaeocarpaceae aff. *Sloanea*) laterally within and between each of the layers and the frequency and representation of other taxa changed only marginally between the lower layers and the upper layer. Some distinctive elements from the upper layer, such as aff. *Neorites* (Proteaceae) and 'taxon 25' (affinities unknown), were rare or absent in the two lower layers (Barrett and Christophel, 1990). The floras of the separate layers were nevertheless essentially the same. The mean size of leaves varied, with

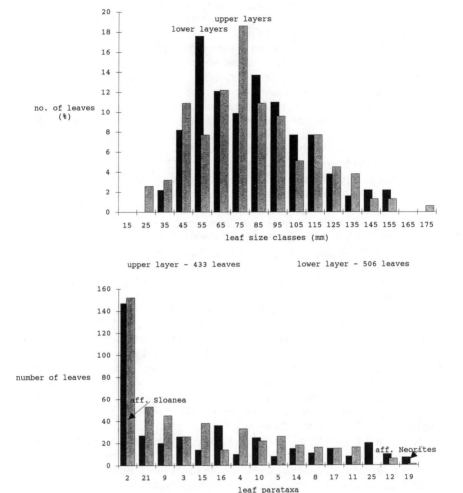

Figure 7.6 Comparison of leaf sizes (specimens) and taxonomic membership, lower leaf layers versus upper leaf layers, Eocene Golden Grove macroflora (adapted from Barrett, unpublished data; Christophel and Greenwood, 1987; Barrett and Christophel, 1990).

the upper layer yielding slightly larger leaves than the two lower layers (Figure 7.6). The LSI, based on the relative size of each taxon (Wolfe, 1979; Burnham, 1989), is nevertheless similar for the upper and lower layers (25 and 22.5).

It is likely that the separate organic-rich layers at Golden Grove represent deposition of litter over perhaps only one to a few generations of trees in the original standing vegetation, a time period of hundreds or tens of years to perhaps only between seasons (Barrett and Christophel, 1990). The small observed variation of taxonomic membership dominance between the three bedding planes may, therefore, reflect either seasonal changes in the leaf-rain contributing to the leaf assemblage, or changes in the composition of the source standing vegetation (Barrett and Christophel, 1990). Alternatively, the changes may reflect differences in the size and orientation of the source area sampled by the transporting medium. If the differences in the physiognomy of the leaves between the lower and upper layers are significant, then this would imply a climatic change between the times of deposition of these layers, thus supporting the interpretation of ecologically controlled change over longer time periods (hundreds of years). However, comparisons between litter samples within a single stream flowing through Complex Notophyll Vine Forest in Queensland have produced similar variation (Greenwood, 1987a; Christophel and Greenwood, 1989) and so it is likely that the differences observed in foliar physiognomy (Figure 7.6) are due to taphonomic, not ecologic, factors. It can be concluded, however, that within this single lens the same flora recurs in separate lamina.

At the Anglesea locality several lenses are fossiliferous. Plant remains in most of the lenses are typically mummified and can be extracted from the matrix intact by bulk disaggregation of the sediment. Each separate clay lens has been found to contain a characteristic macroflora and, to a lesser degree, a characteristic microflora (Christophel *et al.*, 1987). Overall macrofossil diversity is high with over 100 leaf and flower/fruit taxa recognized for the whole flora, although individual lenses typically have much lower diversity with 15–20 taxa, with a single lens containing 50 taxa. The analysis of the Golden Grove clay-lens macroflora (Barrett and Christophel, 1990; and above) suggests that the macroflora of each individual clay lens at Anglesea represents a discrete plant community growing within very close proximity to each of the channel infill deposits (Christophel *et al.*, 1987).

The floristic variation between the separate clay-lens macrofloras at Anglesea has been interpreted as reflecting the original vegetational mosaic of the floodplain (Christophel *et al.*, 1987; Christophel and Greenwood, 1988; 1989). In modern mesic mesothermal environments (such as subtropical and tropical rainforests), a mosaic of plant communities is often observed, reflecting edaphically and successionally controls on species membership. Each separate lens represents a different biofacies (see, for example, Taggert, 1988; Burnham, 1989). If the macrofloras of each clay lens had been lumped together (as a single Anglesea 'flora'), a quite different interpretation would have been produced of the palaeovegetation of the Anglesea area in the Eocene.

SUMMARY

Experiments with individual plant organs and modern vegetation have demonstrated that the leaf-rain potentially contributing to plant fossil beds reflects trees within only short distances of the area of deposition. Separate sedimentary facies in fluvial, paludal and lacustrine environments preserve plant macrofossil assemblages which reflect varying biases in the level of transport (autochthonous to allochthonous deposition) and hydrodynamic sorting (Figure 7.5). Different vegetation types within any landscape will have a varied proportional representation in these sedimentary facies, reflecting proximity to depositional sites, the mode of deposition of both plant parts and sediment, and the energy of transport. Each 'flora' present within an exposure of particular facies will represent a subsample of the total vegetational mosaic, in some cases strongly biased towards individual plant communities, in other cases containing elements from several communities.

In consequence of these observations, plant macrofossil studies of palaeovegetation must (where possible) sample from within discrete bedding planes and consider sedimentary facies when attempting floristic reconstructions of palaeovegetation. While the potential sources of bias are great, observations of modern plant fossil sedimentary analogues allows predictive models to be constructed that allow palaeovegetation reconstructions to account for sedimentary facies, biofacies and differential dispersal (and small-scale variation through seasonal effects?). Such applications of taphonomy are reliant on careful and systematic stratigraphic sampling and result in a finer resolution of the palaeocommunity. Previous approaches of treating single plant fossil localities as a 'flora' must be abandoned in favour of such an approach.

ACKNOWLEDGEMENTS

I thank D.C. Christophel, K.L. Johnson and A.I. Rowett for constructive criticisms on the text, and R.A. Spicer, D.K. Ferguson, C.R. Hill, D.J. Barrett and M.E. Collinson for many useful discussions on plant taphonomy. This chapter was written with the generous provision of facilities by the Botany Department, University of Adelaide.

NOTE

1. Leaf size index is defined as

$$LSI = [\% \text{ microphyll spp.} + 2(\% \text{ notophyll spp.}) +$$
$$3(\% \text{ mesophyll spp.}) - 100] \times O.5.$$

(see Wolfe, 1979; Burnham, 1989).

REFERENCES

Ambrose, G.J., Callen, R.A., Flint, R.B. and Lange, R.T., 1979, *Eucalyptus* fruits in stratigraphic context in Australia, *Nature*, **280** (5721): 387–9.

Anderson, J.M. and Swift, M.J., 1983, Decomposition in tropical forests. In S.L. Sutton, T.C. Whitmore, and A.C. Chadwick (eds), *Tropical rainforest: ecology and management, British Ecological Society Special Publication*, 2: 287–309.

Barrett, D.J. and Christophel, D.C., 1990, The spatial and temporal components of Australian Tertiary megafossil deposits. In J.G. Douglas and D.C. Christophel (eds), *Proceedings of the 3rd Conference of the International Organisation of Palaeobotanists, Melbourne, 1988.*

Bassinger, J.F. and Christophel, D.C., 1985, Fossil flowers and leaves of the Ebenaceae from the Eocene of southern Australia, *Canadian Journal of Botany*, **63** (10): 1825–43.

Birks, H.J.B. and Birks, H.H., 1980, *Quaternary palaeoecology*, Edward Arnold, London.

Blackburn, D.T., 1981, Tertiary megafossil flora of Maslin Bay, South Australia: numerical taxonomic study of selected leaves, *Alcheringa*, **5** (1): 9–28.

Brasell, H.M., Unwin, G.L. and Stocker, G.C., 1980, The quantity, temporal distribution and mineral-element content of litterfall in two forest types at two sites in tropical Australia, *Journal of Ecology*, **68** (1): 123–39.

Brenchley, P.J., 1990, Biofacies. In D.E.G. Briggs and P.R. Crowther (eds), *Palaeobiology: a synthesis*, Blackwell Scientific Publications, Oxford: 395–400.

Brünig, E.F., 1983, Vegetation structure and growth. In F.B. Golley (ed.), *Tropical rainforest ecosystems: structure and function, Ecosystems of the world*, **14A**, Elsevier, Amsterdam: 49–76.

Burnham, R.J., 1989, Relationships between standing vegetation and leaf litter in a paratropical forest: implications for paleobotany, *Review of Palaeobotany and Palynology*, **58** (1): 5–32.

Burnham, R. J., 1990, Some Late Eocene depositional environments of the coal-bearing Puget Group of western Washington, *International Journal of Coal Geology*, **15** (1): 27–51.

Burnham, R.J. and Spicer, R.A., 1986, Forest litter preserved by volcanic activity at El Chichón, Mexico: a potentially accurate record of the pre-eruption vegetation, *Palaios*, **1** (2): 158–61.

Cameron, C.C., Esterle, J.S. and Palmer, C.A., 1989, The geology, botany and chemistry of selected peat-forming environments from temperate and tropical latitudes. In P.C. Lyons and B. Alpern (eds), *Peat and coal: origin, facies, and depositional models, International Journal of Coal Geology*, **12**: 89–103.

Carpenter, R.J. and Horowitz, P., 1988, Leaf litter in two southern Tasmanian creeks and its relevance to palaeobotany, *Papers and Proceedings of the Royal Society of Tasmania*, **122** (2): 39–45.

Chaney, R.W., 1924, Quantitative studies of the Bridge Creek flora, *American Journal of Science*, **8** (2): 127–44.

Chaney, R.W. and Sanborn, E.I., 1933, The Goshen flora of west-central Oregon, *Contributions to Paleontology, Carnegie Institute of Washington, Publication* **439**: 1–103.

Christophel, D.C., 1981, Tertiary megafossil floras as indicators of floristic associations and palaeoclimate. In A. Keast (ed.), *Ecological biogeography of Australia*, Junk, The Hague: 379–90.

Christophel, D.C., 1984, Early Tertiary Proteaceae: the first floral evidence for the Musgraveinae, *Australian Journal of Botany*, **32** (2): 177–86.

Christophel, D.C., 1988, Evolution of the Australian flora through the Tertiary, *Plant Systematics and Evolution*, **162** (1): 63–78.

Christophel, D.C. and Greenwood, D.R., 1987, A megafossil flora from the Eocene of Golden Grove, South Australia, *Transactions of the Royal Society of South Australia*, **111** (3): 155–62.

Christophel, D.C. and Greenwood, D.R., 1988, A comparison of Australian tropical rainforest and Tertiary fossil leaf-beds. In R. Kitching (ed.), *The ecology of Australia's wet tropics, Proceedings of the Ecological Society of Australia*, **15**: 139–48.

Christophel, D.C. and Greenwood, D.R., 1989, Changes in climate and vegetation in Australia during the Tertiary, *Review of Palaeobotany and Palynology*, **58** (2): 95–109.

Christophel, D.C., Harris, W.K. and Syber, A.K., 1987, The Eocene flora of the Anglesea locality, Victoria, *Alcheringa*, **11** (2): 303–23.

Collinson, J.D., 1986, Alluvial sediments. In H.G. Reading (ed.), *Sedimentary environments and facies* (2nd edn), Blackwell Scientific Publications, Oxford: 20–62.

Collinson, M.E., 1983, Accumulations of fruits and seeds in three small sedimentary environments in southern England and their palaeoecological significance, *Annals of Botany*, **52** (4): 583–92.

Collinson, M.E., 1988, Freshwater macrophytes in palaeolimnology, *Review of Palaeobotany and Palynology*, **52** (4): 317–42.

Dolph, G.E. and Dilcher, D.L., 1979, Foliar physiognomic analysis as an aid in determining paleoclimate, *Palaeontographica*, **B170**: 151–72.

Drake, H. and Burrows, C.J., 1980, The influx of potential macrofossils into Lady Lake, north Westland, New Zealand, *New Zealand Journal of Botany*, **18** (2): 257–74.

Efremov, I.A., 1940, Taphonomy: a new branch of paleontology, *Pan American Geologist*, **74** (2): 81–93.

Ferguson, D.K., 1971, *The Miocene flora of Kreuzau, Western Germany: 1. The leaf remains*, North Holland, Amsterdam.

Ferguson, D.K., 1985, The origin of leaf-assemblages—new light on an old problem, *Review of Palaeobotany and Palynology*, **46** (1/2): 117–88.

Flores, R.M., 1981, Coal deposition in fluvial palaeoenvironments of the Paleocene Tongue River Member of the Fort Union Formation, Powder River area, Wyoming and Montana. In F.G. Ethridge and R.M. Flores (eds), *Modern and ancient nonmarine depositional environments, Special Publication of the Society of Economic Paleontologists and Mineralogists*, **31**: 169–90.

Frakes, L.A. and Francis, J.E., 1990, Cretaceous palaeoclimates. In R.N. Ginsburg and B. Beaudoin (eds), *Cretaceous resources, events and rhythms*, Kluwer Academic Publishers, The Netherlands: 273–87.

Gastaldo, R.A., 1986, Implications on the paleoecology of autochthonous Carboniferous lycopods in clastic sedimentary environments, *Palaeogeography, Palaeoclimatology, Palaeoecology*, **53** (2–4): 191–212.

Gastaldo, R.A., 1988, Conspectus of phytotaphonomy. In W.A. DiMichelle and S.L. Wing (eds), *Methods and applications of plant paleoecology, Paleontological Society Special Publication*, **3**: 14–28.

Gastaldo, R.A., Douglass, D.P. and McCarroll, S.M., 1987, Origin, characteristics and provenance of plant macrodetritus in a Holocene crevasse splay, Mobile delta, Alabama, *Palaios*, **2** (3): 229–40.

GreatRex, P.A., 1983, Interpretation of macrofossil assemblages from surface sampling of macroscopic plant remains in mire communities, *Journal of Ecology*, **71** (3): 773–91.

Greenwood, D.R., 1987a, *The foliar physiognomic analysis and taphonomy of leaf-beds derived from modern Australian Rainforest*, Ph.D. thesis, University of Adelaide.

Greenwood, D.R., 1987b, Early Tertiary Podocarpaceae: megafossils from the Eocene Anglesea locality, Victoria, Australia, *Australian Journal of Botany*, **35** (2): 111–33.

Greenwood, D.R., Callen, R.A. and Alley, N.F., 1990, Tertiary macrofloras and Tertiary stratigraphy of Poole Creek palaeochannel, Lake Eyre Basin, *Abstracts, 10th Australian Geological Congress, Hobart*.

Heath, G.W. and Arnold, M.K., 1966, Studies in leaf-litter breakdown. II. Breakdown rate of 'sun' and 'shade' leaves, *Pedobiologia*, **6** (3/4): 238–43.

Hickey, L.J., 1980, Paleocene stratigraphy and flora of the Clark's Fork basin. In P.D. Gingerich (ed.), *Early Cenozoic paleontology and stratigraphy of the Bighorn Basin, Wyoming, University of Michigan Papers in Paleontology*, **24**: 33–49.

Hill, R.S. and Gibson, N., 1986, Distribution of potential macrofossils in Lake Dobson, south central Tasmania, Australia, *Journal of Ecology*, **74** (2): 373–84.

Hill, R.S. and Macphail, M.K., 1983, Reconstruction of the Oligocene vegetation at Pioneer, northeast Tasmania, *Alcheringa*, **7** (2): 281–99.

Kemp, E.M., 1972, Reworked palynomorphs from the west Ice Shelf area, east Antarctica, and their possible geological and palaeoclimatological significance, *Marine Geology*, **13** (3): 145–57.

Knoll, A., 1985, Exceptional preservation of photosynthetic organisms in silicified carbonates and silicified peats. In H.B. Whittington and S.C. Morris (eds), *Extraordinary fossil biotas: their ecological and evolutionary significance, Philosophical Transactions of the Royal Society of London*, **B311**: 111–22.

Kovach, W.L. and Dilcher, D.L., 1984, Dispersed cuticles from the Eocene of North America, *Botanical Journal of the Linnean Society*, **88** (1): 63–104.

Lange, R.T., 1970, The Maslin Bay flora, South Australia, 2. The assemblage of fossils, *Neues Jahrbuch für Geologie und Paläontologie Monatshefte*, **1970** (8): 486–90.

Lange, R.T., 1976, Fossil epiphyllous 'germlings', their living equivalents and their palaeohabitat value, *Neues Jahrbuch für Geologie und Paläontologie Abhandlungen*, **151** (2): 142–65.

Lange, R.T., 1978, Southern Australian Tertiary epiphyllous fungi, modern equivalents in the Australasian region, and habitat indicator value, *Canadian Journal of Botany*, **56** (5): 532–41.

Lange, R.T., 1982, Australian Tertiary vegetation, evidence and interpretation. In J.M.G. Smith (ed.), *A history of Australasian vegetation*, McGraw-Hill, Sydney: 44–89.

Luly, J., Sluiter, I.R. and Kershaw, A.P., 1980, Pollen studies of Tertiary brown coals: preliminary analyses of lithotypes within the Latrobe Valley, Victoria, *Monash Publications in Geography*, **23**: 1–78.

Macphail, M.K. and Hope, G.S., 1985, Late Holocene mire development in montane southeastern Australia: a sensitive climatic indicator, *Search*, **15** (11–12): 344–8.

Martin, H.A., 1978, Evolution of the Australian flora and vegetation through the Tertiary: evidence from pollen, *Alcheringa*, **2** (2): 181–202.

Martin, H.A., 1981, The Tertiary flora. In A. Keast (ed.), *Ecological biogeography of Australia*, Junk, The Hague: 391–406.

Martin, H.A., 1982, Changing Cenozoic barriers and the Australian palaeobotanical record, *Annals of the Missouri Botanical Gardens*, **69**: 625–67.

McQueen, D.R., 1969, Macroscopic plant remains in Recent lake sediments, *Tuatara*, **17** (1): 13–19.

Moore, P.D., 1989, The ecology of peat-forming processes: a review. In P.C. Lyons and B. Alpern (eds), *Peat and coal: origin, facies, and depositional models*, *International Journal of Coal Geology*, **12**: 89–103.

Moore, R.C., 1949, Meaning of facies, *Geological Society of America Memoir*, **38**: 1–34.

O'Dowd, D.J., Brew, C.R., Christophel, D.C. and Norton, R.A., (in press), Evidence for 40 million year-old mite-plant associations from the Eocene of southern Australia, *Science*.

Peters, M.D. and Christophel, D.C., 1978, *Austrosequoia wintonensis*, a new taxodiaceous cone from Queensland, Australia, *Canadian Journal of Botany*, **56** (24): 3119–28.

Potter, F.W. and Dilcher, D.L., 1980, Biostratigraphic analyses of Eocene clay deposits in Henry County, Tennessee. In D.L. Dilcher and T.N. Taylor (eds), *Biostratigraphy of fossil plants, successional and paleoecological analyses*, Dowden, Hutchinson and Ross, Pennsylvania: 211–25.

Raymond, A., 1987, Interpreting ancient swamp communities: can we see the forest in the peat?, *Review of Palaeobotany and Palynology*, **52** (2/3): 217–31.

Reading, H.G., 1986, Facies. In H.G. Reading (ed.), *Sedimentary environments and facies* (2nd edn), Blackwell Scientific, Oxford: 4–19.

Rogers, R.W. and Barnes, A., 1986, Leaf demography of the rainforest shrub *Wilkiea macrophylla* and its implications for the ecology of foliicolous lichens, *Australian Journal of Ecology*, **11** (4): 341–6.

Roth, J.L. and Dilcher, D.L., 1978, Some considerations in leaf size and margin analysis of fossil leaves, *Courier Forschungsinstitut Senckenberg*, **30**: 165–71.

Scheihing, M. and Pfefferkorn, H., 1984, The taphonomy of land plants in the Orinoco delta: a model for the incorporation of plant parts in clastic sediments of Late Carboniferous Age of Euramerica, *Review of Palaeobotany and Palynology*, **41** (3/4): 205–40.

Schopf, J.M., 1975, Modes of fossil preservation, *Review of Palaeobotany and Palynology*, **20** (1): 27–53.

Scott, A.C., 1990, Anatomical preservation of plants. In D.E.G. Briggs and P.R. Crowther (eds), *Palaeobiology: a synthesis*, Blackwell Scientific, Oxford: 263–6.

Seilacher, A., 1990, Taphonomy of Fossil-Lagerstätten. In D.E.G. Briggs and P.R. Crowther (eds), *Palaeobiology: a synthesis*, Blackwell Scientific, Oxford: 266–70.

Spain, A.V., 1984, Litterfall and the standing crop of litter in three tropical Australian Rainforests, *Journal of Ecology*, **72** (3): 947–61.

Spicer, R.A., 1980, The importance of depositional sorting to the biostratigraphy of plant megafossils. In D.L. Dilcher and T.N. Taylor (eds), *Biostratigraphy of fossil plants, successional and paleoecological analyses*, Dowden, Hutchinson and Ross, Pennsylvania: 171–83.

Spicer, R.A., 1981, The sorting and deposition of allochthonous plant material in a modern environment at Silwood Lake, Silwood Park, Berkshire, England, *US Geological Survey Professional Paper*, **1143**: 1–77.

Spicer, R.A., 1988, Quantitative sampling of plant megafossil assemblages. In W.A. DiMichelle and S.L. Wing (eds), *Methods and applications of plant paleoecology, Paleontological Society Special Publication*, **3**: 29–51.

Spicer, R.A., 1989, The formation and interpretation of plant fossil assemblages, *Advances in Botanical Research*, **16**: 96–191.

Spicer, R.A., 1990, Fossils as environmental indicators: climate from plants. In D.E.G. Briggs and P.R. Crowther (eds), *Palaeobiology: a synthesis*, Blackwell Scientific, Oxford: 401–3.

Spicer, R.A. and Wolfe, J.A., 1987, Plant taphonomy of late Holocene deposits in Trinity (Clair Engle) Lake, northern California, *Paleobiology*, **13** (2): 227–45.

Stocker, G.C., Thompson, W.A., Irvine, A.K., Fitzsimon, J.D. and Thomas, P.R., (in press), Rainforest litterfall in tropical Australia. 1. Annual and site patterns, *Journal of Ecology*.

Taggert, R.E., 1988, The effect of vegetation heterogeneity on short stratigraphic sequences. In W.A. DiMichelle and S.L. Wing (eds), *Methods and applications of plant paleoecology, Paleontological Society Special Publication*, **3**: 147–71.

Thomas, B.A., 1990, Rules of nomenclature: disarticulated plant fossils. In D.E.G. Briggs and P.R. Crowther (eds), *Palaeobiology: a synthesis*, Blackwell Scientific, Oxford: 421–3.

Thomas, B.A. and Spicer, R.A., 1987, *The evolution and palaeobiology of land plants*, Croom Helm, London.

Truswell, E.M. and Harris, W.K., 1982, The Cainozoic palaeobotanical record in arid Australia: fossil evidence for the origin of an arid-adapted flora. In W.R. Barker and P.J.M. Greenslade (eds), *Evolution of the flora and fauna of arid Australia*. Peacock Publications, South Australia: 67–76.

Upchurch, G.R., Jr., 1989, Terrestrial environmental changes and extinction patterns at the Cretaceous–Tertiary boundary, North America. In S.K. Donovan (ed.), *Mass extinctions: processes and evidence*, Belhaven Press, London: 195–216.

Webb, L.J., 1959, A physiognomic classification of Australian rainforests, *Journal of Ecology*, **47** (3): 551–70.

Wilson, M.V.H., 1980, Eocene lake environments: depth and distance from shore variation in fish, insect, and plant assemblages, *Palaeogeography, Palaeoclimatology, Palaeoecology*, **32** (1): 21–44.

Wing, S.L., 1984, Relation of paleovegetation to geometry and cyclicity of some fluvial carbonaceous deposits, *Journal of Sedimentary Petrology*, **54** (1): 52–66.

Wing, S.L., 1987, Eocene and Oligocene floras and vegetation of the Rocky Mountains, *Annals of the Missouri Botanical Gardens*, **74**: 748–84.

Wing, S.L., 1988, Depositional environments of plant bearing sediments. In W.A. DiMichelle and S.L. Wing (eds), *Methods and applications of plant paleoecology, Paleontological Society Special Publication*, **3**: 1–11.

Wolfe, J.A., 1979, Temperature parameters of humid to mesic forests of eastern Asia and relation to forests of other regions of the Northern Hemisphere and Australasia, *US Geological Survey Professional Paper*, **1106**: 1–37.

Wolfe, J.A., 1985, The distribution of major vegetational types during the Tertiary. In E.T. Sundquist and W.S. Broecker (eds), *The carbon cycle and atmospheric CO_2, natural variations Archaen to present*, American Geophysical Union, *Geophysical Monograph*, **32**: 357–75.

Wolfe, J.A., 1990, Palaeobotanical evidence for a marked temperature increase following the Cretaceous/Tertiary boundary, *Nature*, **343** (6254): 153–6.

Wolfe, J.A. and Schorn, H.E., 1989, Paleoecologic, paleoclimatic, and evolutionary significance of the Oligocene Creede flora, Colorado, *Paleobiology*, **15** (2): 180–98.

Wolfe, J.A. and Upchurch, G.R., Jr., 1987, North American nonmarine climates and vegetation during the Late Cretaceous, *Palaeogeography, Palaeoclimatology, Palaeoecology*, **61** (1): 33–77.

Chapter 8

THE TAPHONOMY OF FORAMINIFERA IN MODERN CARBONATE ENVIRONMENTS:

Implications for the formation of foraminiferal assemblages

Ronald E. Martin and W. David Liddell

INTRODUCTION

Foraminifera are much more distinctive in size, shape, wall composition and microstructure than most other sediment grains. They may therefore provide taphonomic signatures indicative of the preservational consequences of Earth processes (Worsley *et al.*, 1986). Indeed, foraminifera may well prove to be the *Drosophila* of taphonomy (Prothero and Lazarus, 1980); the group is represented by over 3600 genera ranging from the Cambrian to the Recent and occurring in hypersaline lagoons to abyssal depths, as well as in the epipelagic zone of the open ocean. Foraminifera also exhibit a variety of test compositions, microstructures and morphologies (Tappan and Loeblich, 1988). 'Similar morphological features of test size, shape, chamber arrangement, chamber subdivision, and apertural modifications have appeared repeatedly in unrelated groups, as similar restrictions were imposed by their unicellular nature, the necessity of facilitating cytoplasmic movement and intercameral communication, method of test construction, mode of feeding, and habitat' (Tappan and Loeblich, 1988; see also Brasier, 1986). Homeomorphy (Bandy, 1964) within the foraminifera provides fertile ground for testing hypotheses of evolutionary palaeoecology against secular preservation resulting from the changing backdrop of ocean palaeochemistry, circulation, sedimentation and bioerosion (adaptive test traits in life versus traits 'selected' during preservation; see, for example, Wetmore, 1987).

Micropalaeontological analyses of well samples were conducted as early as 1878 (Howe, 1959) and were renewed in earnest after the First World War by J.A. Cushman, J.J. Galloway, and J.A. Udden (Stuckey, 1978) as petroleum companies established industrial laboratories on the US Gulf Coast (Owen, 1975) and in southern California (Kleinpell, 1971). Finally, in a landmark paper published in 1925, Esther Applin of Rio Bravo, Alva Ellisor with Humble (now Exxon) and Hedwig Kniker with the Texas Company (Texaco) revolutionized industrial micropaleontology by demonstrating the utility of benthic foraminifera in subsurface correlation of Texas and Louisiana (Applin *et al.*, 1925). A decade later, a standardized zonation based on benthic foraminiferal data was established for the Gulf Coast (Israelsky *et al.*, 1933), Kleinpell (1933) published his classic *Miocene stratigraphy of California* and Natland (1933) demonstrated the group's potential in defining the bathymetry of modern offshore California basins and, therefore, of ancient petroliferous basins buried onshore.

Nevertheless, it was not until after the Second World War that micro-palaeontologists seriously began to examine the distribution of living forami-nifera in relation to their subfossil counterparts. The advent of the rose Bengal staining technique (Walton, 1952; see criticism of the method by LeCalvez and Cesana, 1972; Martin and Steinker, 1973; Bernhard, 1988) opened the gates to a flood of papers comparing distributions of living and dead foraminifera (primarily in terrigenous environments) that has persisted to the present (summarized in Phleger, 1960; Murray, 1973; 1976; Culver, 1987). Most recently, the value of larger symbiont-bearing foraminifera in palaeoenvironmental interpretation of tropical carbonate sediments has been demonstrated (Frost and Langenheim, 1974; Chaproniere, 1975; Reiss and Hottinger, 1984; Hallock and Glenn, 1985; 1986).

Despite numerous distributional surveys, however, comparative taphono-mic studies of benthic foraminifera are still sorely lacking (Behrensmeyer and Kidwell, 1985). We know almost nothing about information loss in the transition from living foraminiferal populations through incipient fossil assemblages of the surface mixed layer (Martin and Wright, 1988; Denne and Sen Gupta, 1989; Loubere, 1989) to the deeper preservation zone. This omission is all the more glaring considering the few autecologic studies that have been conducted on growth rates, sediment production and population dynamics of foraminifera, mostly on selected larger reef-dwelling genera (for example, *Amphistegina*, *Archaias*; Muller, 1974; Hallock *et al.*, 1986a; 1986b). Moreover, the sheer abundance of these larger genera in sediment assemblages (up to about 50 per cent at Discovery Bay, Jamaica; Martin and Liddell, 1988) may obscure subtler bathymetric (and taphonomic) gradients based on smaller reef-dwelling species (Martin and Liddell, 1989; see also Schröder *et al.*, 1987; Sen Gupta *et al.*, 1987). Nor do we have much information about relative preservation in carbonate (Brasier, 1975a; 1975b; Buzas *et al.*, 1977; Martin, 1986; Martin and Liddell, 1988; Martin and Wright, 1988; Kotler *et al.*, 1989) versus terrigenous shelf sediments (see, for example, Murray and Wright, 1970; Collen and Burgess, 1979; Douglas *et al.*, 1980; Khusid, 1984; Smith, 1987; Goldstein, 1988; Goldstein and Barker, 1988; Goldstein *et al.*, 1989; Murray, 1989), much less continental slope and deep sea environments (Douglas *et al.*, 1980; Douglas and Woodruff, 1981). Unlike most of the distributional studies of modern foraminifera, the majority of previous studies of foraminiferal taphonomy have been restricted to either certain aspects of foraminiferal preservation or to single species (a notable exception is the study of foraminifera of the California continental borderland by Douglas *et al.*, 1980). In contrast, the investigations discussed herein represent an integrated approach that utilizes experimental analyses of abrasion and dissolution resistance, *in situ* dissolution and bioerosion, and determination of settling and traction threshold velocities to formulate and test hypotheses about information loss in the transition from living foraminiferal populations to incipient fossil assemblages.

TAPHONOMIC STUDIES OF MODERN FORAMINIFERA FROM CARBONATE ENVIRONMENTS

Jamaican taphofacies model

Modern carbonate environments provide ideal settings for the evaluation of the effect of taphonomic processes on the formation of subfossil assemblages of foraminifera. Like their terrigenous counterparts (see, for example, Brett and Baird, 1986; Kidwell *et al.*, 1986; Speyer and Brett, 1988; Fürsich and Flessa, 1987; Meldahl and Flessa, 1989), they recur throughout the rock record and exhibit pronounced depth-related gradients in water turbulence, sedimentation rate, organic carbon content, water and pore-fluid chemistry of sediments, bioerosion and bioturbation, and corresponding taphonomic gradients in post-mortem abrasion, dissolution and transport (Buxton and Pedley, 1989; Graus and Macintyre, 1989). Based on previous studies at Discovery Bay, Jamaica (summarized in Liddell *et al.*, 1984; see also Liddell *et al.*, 1987; Martin and Liddell, 1988; 1989) and elsewhere (Martin, 1986; Martin and Wright, 1988), we have developed a taphofacies model (Liddell and Martin, 1989) based upon the predicted degree of abrasion, dissolution, bioerosion and transport of foraminiferal tests along taphonomic gradients (Figure 8.1). The model has allowed us to develop and test hypotheses about the taphonomic behaviour of foraminiferal tests via laboratory and field experiments. It also provides an initial theoretical framework for testing the influence of ecologic and taphonomic constraints on test morphology, evolution, and preservation through time.

Taphofacies I represents a low-energy setting with low transport potential (that is, sheltered lagoon of inner reef tract or below wave base with a low slope angle). Assemblages consist of a diverse mixture of autochthonous abrasion-resistant and non-resistant tests of small to large miliolids, peneroplids, and soritids (Martin and Liddell, 1988; Martin and Wright, 1988) that include differing shapes and sizes (different susceptibilities to transport or destruction).

Although active carbonate precipitation has been reported for shallow-water sites at Jamaica (Land and Goreau, 1970; Pigott and Land, 1986), it is conceivable that surface sediment pore waters of the quiet-water backreef lagoon may become undersaturated with respect to calcium carbonate. Quiet-water environments are characterized by an abundant burrowing infauna, which rework sulphide particles and abundant organic matter to the surface mixed layer. Here the particles are oxidized to produce acids, which lower pore-water pH (Aller, 1982) and which may etch foraminiferal test surfaces (Murray and Wright, 1970; Cottey and Hallock, 1988; Murray, 1989; Martin, personal observations). Aller (1982) and Davies *et al.* (1989; see also Alexandersson, 1979) maintain that in terrigenous sediments (Long Island Sound and shallow Texas Bays, respectively) rapid transfer of calcareous shells from the surface dissolution zone (Aller, 1982; 'taphonomically-active zone' of Davies *et al.*, 1989) into the preservation zone (Aller, 1982) is necessary for the preservation of assemblages. Pore waters of the preservation zone are characterized by alkalinity build-up and are saturated with respect to calcium carbonate (Ginsburg, 1957; Berner *et al.*, 1970; Aller, 1982).

Bioerosion may also be an important factor in test destruction in quiet-water environments (Swinchatt, 1965; May and Perkins, 1979; Poulicek *et al.*, 1981; Peebles and Lewis, 1988). Bioerosion of foraminiferal tests by microscopic agents appears to be concentrated in quiet-water sediments (Swinchatt, 1965; Alexandersson, 1972; May and Perkins, 1979; Kloos, 1982; Goldstein and Barker, 1988; Peebles and Lewis, 1988; see also Poulicek *et al.*, 1981). Tests of certain foraminiferal species occasionally exhibit trails or borings near their periphery which may weaken the test and hasten its destruction.

High-energy outer back reef (taphofacies II) and shallow (10–15 m) terrace (taphofacies III) assemblages are dominated by abrasion-resistant species. Outer backreef assemblages consist of lags of both abrasion-resistant autochthonous (for example, *Archaias angulatus*) and allochthonous (for example, *Amphistegina gibbosa*) species that have been transported over the reef crest by storms (Martin, 1986; Martin and Liddell, 1988; Martin and Wright, 1988). Shallow terrace (taphofacies III) assemblages are dominated by the robust *Amphistegina gibbosa*; small or less-resistant species have presumably been transported elsewhere (for example, into the lagoon) or destroyed by abrasion.

Taphofacies IV (upper terrace: 5–10 m) and V (forereef slope: 30–75 m) are similar to taphofacies III and I, respectively, but differ in that tests with high susceptibility to transport are winnowed from assemblages with a resultant decrease in diversity. Because sediments contain abundant organic matter, dissolution in surface sediment layers may also be important on the forereef slope.

Taphofacies VI (island slope), located adjacent to a type III, IV, or V setting, is characterized by a diverse mixture of allochthonous and autochthonous species, including planktonic components; diversity is high due to mixing and low post-mortem abrasion.

Laboratory and field studies

Abrasion
Despite our use of a variety of species that differ in test shape, composition, and structure (Table 8.1), there is little difference between species in their

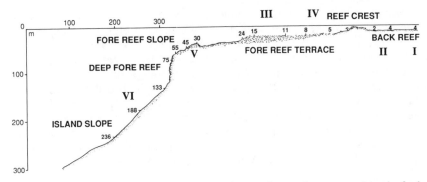

Figure 8.1 Profile of Discovery Bay showing major reef zones and taphofacies (Roman numerals). Depths in metres.

resistance to abrasive test reduction (Kotler *et al.*, 1989; Figure 8.2 herein). Even after 1000 hours of abrasion, which corresponds roughly to several hundred kilometres of movement (depending upon species and assumptions about wavelength, periodicity, etc.; see transport, below), most foraminifera are still easily recognizable to the species level. *Halimeda* platelets also show little wear. Abrasion-induced test surface features appear to reflect primarily test wall thickness (for example, thin wall in *Archaias angulatus*; thick in *Amphistegina gibbosa*; Figures 8.2A and 8.2B). Only the relatively thin-walled, fragile and encrusting species *Planorbulina acervalis* (Figure 8.2C) and *Sorites marginalis* are easily destroyed. Our results suggest that abrasion alone is often ineffective in test destruction in carbonate sediments, even in high-energy settings such as the shallow outer backreef and shallow forereef (taphofacies II and IV, respectively; Figure 8.1).

This is in marked contrast to experimental studies conducted on foraminiferal abrasion in terrigenous sediments. Driscoll and Weltin (1973; see also Chave, 1964) found that bivalves and gastropods were most rapidly abraded in very coarse quartz sand, less so in very fine sand, and least in medium sand. Moberly (1968) found that smaller foraminifera undergo a more rapid percentage weight loss than larger specimens when abraded in a mixture of carbonate and detrital grains, and Miller and Ellison (1982) demonstrated that certain agglutinated foraminifera are destroyed much more quickly than a calcareous species when agitated with glass beads. Cottey and Hallock (1988) found that the outer window-like test layer of *Archaias angulatus* was more rapidly abraded in quartz sand than carbonate sediment.

Dissolution

Unlike abrasion, foraminiferal species differ considerably in their relative resistance to dissolution (Figure 8.3). At the end of the dissolution experiments (about 80 hours duration), only two specimens of *Quinqueloculina tricarinata* and five of *Amphistegina gibbosa* were retrieved; all others had dissolved. Retrieved specimens were badly corroded and in no way resembled the majority of the tens of thousands of specimens from Discovery Bay

Table 8.1 *Morphology of common Jamaican foraminifera used in laboratory studies*

Species	Test microstructure	Test shape	Wall thickness	Size	Relative weight	Comments
Amphistegina gibbosa	Calcareous, perforate	Lenticular	Thick	Small–large (1–2 mm)	Low–high (size-dependent)	Dominant on forereef terrace, apparently abrasion-resistant
Archaias angulatus	Calcareous, imperforate	Lenticular	Thin, but reinforced by pillars	Small–large (1–2 mm)	Low–high (size-dependent)	Dominant in backreef, apparently abrasion-resistant
Asterigerina carinata	Calcareous, perforate	Conical	Thick, with heavy umbilical plug	Small (<0.5 mm)	High for size	Common on forereef terrace
Bigenerina irregularis	Agglutinated	Curvilinear	Thin, lightly cemented	Small (<0.5 mm)	Low	Common on forereef slope
Cyclorbiculina compressa	Calcareous, imperforate	Discoidal	Thin, but reinforced by pillars	Small–large (1–2 mm)	Low–high (size-dependent)	Most abundant in backreef, apparently abrasion-resistant
Discorbis rosea	Calcareous, perforate	Spherical	Thick	Small (<0.5 mm)	High for size	Most abundant on forereef terrace, apparently abrasion-resistant
Globigerinoides quadrilobatus	Calcareous, perforate	Globular	Thin, fragile	Small (<0.5 mm)	Low	Most abundant on forereef slope and deeper environments, delicate
Orbulina universa	Calcareous, perforate	Spherical	Thin, fragile	Small (<0.5 mm)	Low	Most abundant on forereef slope and deeper environments, delicate
Planorbulina acervalis	Calcareous, perforate	Discoidal	Thin, fragile	Medium (>0.5 mm)	Low	Common in backreef
Peneroplis proteus	Calcareous, imperforate	Discoidal	Thin	Small–large (0.5–2 mm)	Low	Common in backreef
Quinqueloculina spp.	Calcareous, imperforate	Spindle	Thick	Small (<0.5 mm)	Variable	Ubiquitous backreef to forereef slope
Sorites marginalis	Calcareous, imperforate	Discoidal	Thin, fragile	Medium (>0.5 mm)	Low	Common in backreef

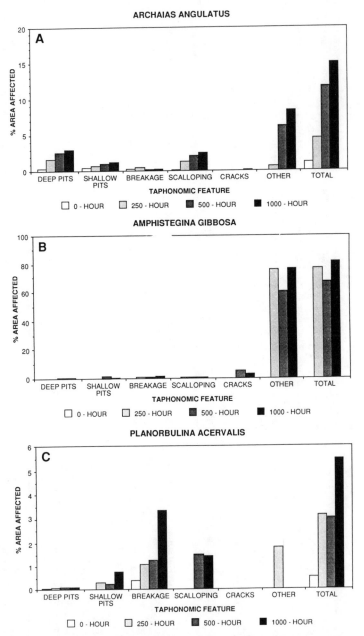

Figure 8.2　Test surface features produced experimentally by abrasion of foraminiferal tests. Ten intact specimens of each of the most common species of
foraminifera at Discovery Bay (Table 8.1), as well as platelets of the calcareous
green alga *Halimeda*, were subjected to abrasion in medium-size (mean diameter about 0.5 mm) carbonate sand in artificial seawater (pH about 8.0) on a

surface sediments examined by us over a seven-year period (Martin and Liddell, 1988; 1989). In this experiment, wall ultrastructure seems to have had little effect: *Quinqueloculina tricarinata* and *Amphistegina gibbosa* possess calcareous imperforate (suborder Miliolina) and perforate (suborder Rotaliina) tests, respectively, but both are also thick-walled.

Corliss and Honjo (1981) performed an experiment similar to ours. They suspended specimens of *Amphistegina* and characteristic bathyal and abyssal species (for example, *Gyroidina* spp.) on moorings in carbonate-undersaturated waters of the Pacific Ocean for two months. Interestingly, they found *Amphistegina* to be much less resistant to dissolution than the deep-dwelling species, although all species possess calcareous perforate tests.

In contrast to the results of our laboratory experiment, only the thin, window-like outer test layer of *Archaias angulatus* is usually corroded on specimens examined from organic-rich fine-grained surface sediments (our taphofacies I; see also Cushman, 1930, Plates 16, 17). Test surfaces of other species in field samples show only minor etching (indicated by relatively dull lustre). Also, specimens of intact species that had been glued to stakes (made of PVC pipe), and buried beneath the sediment at different depths for periods of three to six months, show only minor loss of surface lustre. Exposure of intact tests to running seawater of approximately pH 7.7 for up to four months produced similar results. This is in marked contrast to reports of rapid skeletal dissolution in modern terrigenous sediments (Alexandersson, 1972; 1976; 1979; Fitzgerald *et al.*, 1979; Bottjer and Douglas, 1984; Davies *et al.*, 1989).

The increased abundance of agglutinated foraminifera in taphofacies V (forereef slope; Martin and Liddell, 1988) would seem to indicate increased dissolution in sediments, but increased abundance of intact dissolution-susceptible planktonic foraminifera here (Martin and Liddell, 1988) is a contraindication of dissolution. The abundance of agglutinated foraminifera in sediment is known to be accentuated by differential dissolution of calcareous species (Murray, 1989), although this is highly variable between environments (Smith, 1987). Some agglutinated species are very fragile and are easily destroyed by deposit feeders (Douglas *et al.*, 1980; Douglas and Carlos, 1986) or bacterial degradation of the organic cement (Goldstein and Barker, 1988). In the case of Discovery Bay, increased abundances of agglutinated taxa on the forereef slope most likely indicate populations which live infaunally (Goldstein, 1988; Tappan and Loeblich, 1988).

---

rotary shaker at 150 rpm (sufficient to cause movement of sand without suspension). Histograms represent average percentage surface area (digitized from scanning electron micrographs) represented by each surface feature for each species. (A) *Archaias angulatus* (suborder Miliolina); (B) *Amphistegina gibbosa* (suborder Rotaliina); (C) *Planorbulina acervalis* (suborder Rotaliina). Scalloping = scalloping of test margin. Other = miscellaneous test features, including removal of outer test layer, obliteration of test aperture, loss of detail of sutures and exposure of chamber interior.

NORMALIZED DISSOLUTION RATE VS TIME

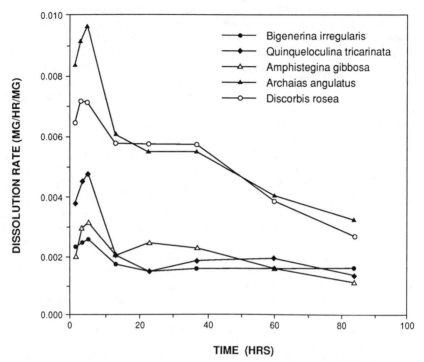

Figure 8.3   Rates of dissolution for five foraminiferal species. A separate group of five intact specimens of each of five species was weighed and then exposed to calcium-free seawater in a fluidized bed reactor (Chou and Wollast, 1984) for approximately 80 hours. Graph represents dissolution rates for each species normalized to the respective initial weight of each group of specimens. Each point represents the mean weight loss (milligrams per hour per milligram original test weight) for each group of specimens as calculated by measuring Ca^{2+} released through time in the reactor by flame atomic absorption spectrophotometry.

Bioerosion

Biological destruction of foraminiferal tests occurs as the result of the activities of a variety of micro-organisms, especially cyanobacteria (blue-green algae), bacteria, and fungi (Golubic *et al.*, 1975; 1984), but bioerosion has received even less attention than have abrasion, dissolution and transport, even though the process is important in the formation of fine-grained sediment (Perkins and Halsey, 1971). Bioerosion of tests by microscopic agents appears to be concentrated in quiet-water sediments (Swinchatt, 1965; Alexandersson, 1972; May and Perkins, 1979; Kloos, 1982; Goldstein and Barker, 1988; Peebles and Lewis, 1988; see also Poulicek *et al.*, 1981). Patterns produced by microborers ('endoliths') are dependent on both the boring organism (genetic) and the substrate (environment) involved (Rooney and Perkins, 1972; Perkins and Tsentas, 1976). In many cases, microborers appear to avoid pores and other cavities in the test (see, for example, Cottey and Hallock, 1988), and are often concentrated near the periphery of the test, where they may weaken the shell and accelerate its breakage.

In our studies, to date, five specimens each of *Archaias angulatus* (suborder Miliolina) and *Amphistegina gibbosa* (suborder Rotaliina) have been exposed to potential bioeroders above the sediment–water interface (SWI) in backreef waters. The intensity of bioerosion is remarkable, with the majority of specimens of *Archaias angulatus* exposed in the water column being completely, or nearly so, destroyed by bioerosion—particularly by microboring filamentous algae (Figure 8.4). Of course, tests completely exposed in the water column might be expected to show accelerated bioerosion relative to specimens in the surface sediment which are periodically buried (protected) and exhumed (exposed to bioeroders). *Amphistegina gibbosa* exhibits some damage due to bioeroders, but the tests are still recognizable to species. These results parallel the findings of Peebles and Lewis (1988), who also found a greater frequency of microboring in tests of miliolinid than of rotalinid foraminifera from shallow (3 m) forereef sediment samples from San Salvador in the Bahamas.

Bioerosion of fossil hardparts is not confined to microbes, however. Both selective and indiscriminate ingestion of foraminiferal tests by a wide variety of invertebrates has been reported (Sliter, 1971; Lipps and Ronan, 1974; Mageau and Walker, 1976; Buzas and Carle, 1979; Hickman and Lipps, 1983), which may produce characteristic punctures and gouges. Etching of the test surface by low-pH fluids of the digestive tract may also occur, although reports of test dissolution via biological agents vary from complete dissolution of calcareous tests to little or no effect (see, for example, Walker, 1971). Mageau and Walker (1976) report that agglutinated foraminifera were unaffected by ingestion, but that calcareous tests were often dissolved. Buzas (1978; 1982) noted indiscriminate ingestion of foraminifera by a wide variety of deposit feeders, and concluded that living foraminiferal population densities are controlled by predation. Lipps (1988) noted cropping of foraminifera by fish on Eniwetok Atoll and suggested that persistent cropping would exert selective pressure on foraminifera towards small size and rapid reproduction, and, therefore, perhaps sediment production, as well.

Transport
Settling velocities and movement thresholds of intact foraminiferal tests were measured in a flume and settling tube under controlled laboratory conditions (Figure 8.5). We found an approximately linear relationship between settling velocity and minimum test intercept for the foraminiferal morphotypes tested (Cunningham *et al.*, 1989; see also Maiklem, 1968). In addition, effective test density in water may also be used to predict settling velocity. Settling motion varies from straight path (*Discorbis rosea*; approximately spherical test) through spiral (*Bigenerina*; curvilinear) to side-to-side (*Sorites*; disc-shaped). Movement threshold velocities are largely determined by the distance which tests project above the bottom, which, in turn, is a function of shape,

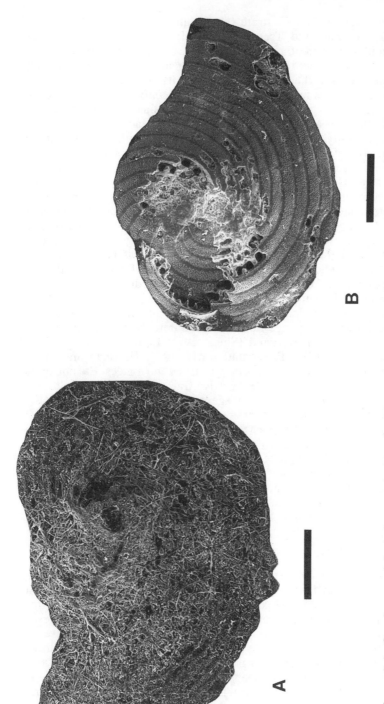

Figure 8.4   Specimens of *Archaias angulatus* exposed to bioeroders for a three-month interval. Specimens were cemented to PVC pipe and positioned 10 cm above the sediment water interface. The experiments were located at a depth of 2 m in the lagoon at Discovery Bay, Jamaica. (A) Heavily bored specimen. Note erosion of test margin, removal of test wall, and numerous filaments of microborers. (B) Note relatively large borings and removal of test wall. Damage is most likely due to invertebrate borers and, possibly, vertebrate grazers. Both scale bars represent 250 μm.

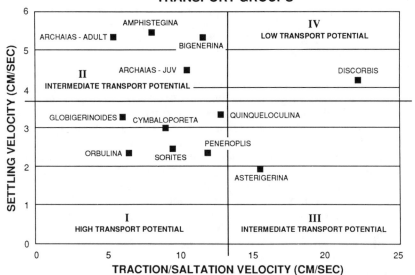

Figure 8.5   Foraminiferal transport groups. Settling velocity is plotted against lateral movement thresholds (fixed 0.5 mm sand substrate). Settling tube experiments were conducted in a 0.15 m diameter by 1.80 m length clear acrylic tube filled with distilled water at a temperature of 22°C. Prior to performing the settling experiments, all specimens were vacuum-immersed in water in order to remove air from the test chambers. After vacuum-immersion, specimens were transferred in water via pipettes to the settling column. Single specimens were allowed to fall for 45 cm prior to initiation of timing in order to allow for acceleration. Specimens were timed to 0.01 s using an electronic timer while falling over a distance of 100 cm, and the type of settling motion exhibited (spiral, straight, oscillation from side to side, etc.) was noted during descent.

Flume experiments were conducted in a 3.0 m × 0.3 m × 0.6 m (length, width, depth) flume, and flow velocities were measured with an electromagnetic flowmeter. Velocities were recorded to the nearest 0.01 m s^{-1}. At the beginning of each experimental run, a single specimen was placed upon a 10 × 10 cm platform located 15 cm above the bottom of the flume. Flow was then initiated within the flume and velocities slowly increased until movement of the foraminiferal test occurred. The sensor for the flowmeter was located at the same level in the water column and immediately behind the specimens. Initial orientation of the test, current velocity at the time of first movement (movement threshold), and the type of movement (traction, saltation, suspension) were noted. If initial movement was by traction or saltation, flow velocities were increased in an attempt to determine suspension thresholds. In order to test the influence of substrate type upon movement thresholds, specimen platforms with differing surface textures were also employed. Platforms utilized were smooth, coated with cemented fine (0.147 mm) sand or coated with cemented coarser (0.589 mm) sand.

orientation, overall test size and substrate grain size (see also Kontrovitz *et al.*, 1978). In order accurately to predict movement thresholds, various shapes, such as sphere (*Orbulina*), disc (*Amphistegina*) and cylinder (*Bigenerina*), must be isolated (Table 8.1). The type of movement initiated varies from rolling through tumbling to saltation and suspension.

Plotting foraminiferal settling velocities against traction/saltation velocities results in delineation of four 'transport groups' (Figure 8.5; see also Grabert, 1971; Schafer and Young, 1977; Hale *et al.*, 1985). Transport group I is characterized by high transport potential (low traction and low settling velocities); that is, species are easily set into suspension, but settle slowly. This group is typified by both planktonic genera tested (*Globigerinoides*, *Orbulina*) and by two benthic genera (*Peneroplis*, *Sorites*). The two planktonic genera have low-density tests (and spines in life) as an adaptation for floating; the two benthic genera have flattened tests with thin outer walls as an adaptation for housing symbiotic algae (Table 8.1). The planktonic genera are most prominent in taphofacies V–VI (open waters above fore reef slope-island slope; Martin and Liddell, 1988), while the benthic genera are characteristic of quiet-water backreef environments where wave energy is at a minimum.

Transport group II is characterized by genera which are easily moved, but which settle rapidly. Both *Archaias* and *Amphistegina* have dense, lenticular tests and are found in greatest abundance in wave-swept outer backreef (taphofacies II) and shallow forereef (taphofacies III–IV) environments, respectively, where they contribute to lag deposits (Martin, 1986; Martin and Wright, 1988; Martin and Liddell, 1989). Test shape has also been shown to influence fecundity in *Amphistegina*, which in turn affects its contribution to subfossil assemblages (Hallock, 1979; Hallock *et al.*, 1986b).

Transport group III is represented by *Asterigerina carinata* (cone-shaped), which plots to the left of the field. This species is most prominent in shallow terrace lag deposits, but is frequently swept into outer backreef sediments by storms (Martin and Wright, 1988). Based on our experiments, it is less easily moved than other species, but once in suspension it settles relatively slowly. Interestingly, the cone-shape of *Asterigerina* (Table 8.1) approximates to that of the planktonic foraminifer *Globorotalia truncatulinoides*.

Discorbis rosea is the sole representative of transport group IV. Like *Asterigerina*, this species is also most numerous in shallow terrace sediments (Martin and Liddell, 1988), but because of its spherical shape (decreased resistance to settling) it settles more rapidly than *Asterigerina*, which may in part explain its much greater abundance (relative to *Asterigerina*) in terrace sediments.

DISCUSSION AND CONCLUSIONS

Carbonate and shell-poor siliciclastic regimes appear to differ fundamentally in the taphonomic constraints they place on our interpretation of the palaeoecology, biostratigraphy and evolution of ancient micro-organisms.

Preservation in shelfal carbonate sediments appears to mimic that of condensed intervals of siliciclastic environments, or high-productivity communities like oyster banks, in which shell-rich layers create an environment favourable to their preservation by buffering pore waters of the surface mixed layer against dissolution (Kidwell, 1989). Moreover, shell-rich layers may provide opportunities for the establishment of living foraminiferal populations through 'taphonomic feedback' (Kidwell, 1986). Conversely, carbonate and shell-poor terrigenous regimes should differ in the intensity of taphonomic processes affecting incipient fossil assemblages. Rates of carbonate destruction, especially through dissolution, are predicted to be much higher in shell-poor terrigenous sediments than in carbonate deposits (Alexandersson, 1976; Davies *et al.*, 1989; Kotler *et al.*, 1989). Carbonate and terrigenous sedimentary regimes also differ in the continuity and rate of sedimentation. Although sedimentation in carbonate environments may be exceedingly rapid in the short term (Milliman, 1974), long-term carbonate accumulation tends to be slower (and more discontinuous) than in terrigenous environments (Wilson, 1975; Schindel, 1980). Obviously, these considerations bear upon phylogenetic reconstructions, sampling strategies, and biostratigraphic and palaeoenvironmental resolution (see, for example, Schindel, 1982; Signor and Lipps, 1982). For example, Ross and Ross (1985) recognized more than fifty late Palaeozoic eustatic transgressive-regressive depositional sequences (of 2 million years average duration) that are correlative world-wide with established fusulinid zones, the lineages of which have been the subject of extensive phylogenetic studies (Ross, 1982); they concluded that it is the fossil record that is punctuated, not evolution.

Carbonate and terrigenous sedimentary regimes may also differ fundamentally in the relative importance of biological destruction of tests. Because of lesser amounts of carbonate grains (tests) available for infestation by bioeroders in siliciclastic sediments, we predict that foraminiferal tests will be more intensively colonized and bored in carbonate environments. The intensity of bioerosion may also vary from high at the SWI to low within the sediment (compare results of field studies of test destruction above and below the SWI). Also, species-specific differences in intensity of bioerosion (for example, *Archaias angulatus* versus *Amphistegina gibbosa*) may reflect a higher-level (subordinal) difference in susceptibility to bioerosion (see also Peebles and Lewis, 1988). Future studies will incorporate a wide variety of species (see, for example, Table 8.1) in an effort to determine the relative influence of test microstructure (crystal arrangement) and test architecture (wall thickness, pillars, sculpture) on susceptibility to bioerosion. We must also evaluate the relative importance of microbial bioerosion (for example, by algae and fungi) versus bioerosive activities of invertebrate deposit feeders and vertebrates (see, for example, Mageau and Walker, 1976; Lipps, 1988) along bathymetric gradients and through geologic time (see, for example, Zeff and Perkins, 1979; Campbell, 1983).

Unfortunately, our knowledge of carbonate budgets and the relative rates of skeletal breakdown and transport is meager at best (for example, Chave *et al.*, 1972; Land, 1979; Land and Moore, 1980; Scoffin et al., 1980). We are also ignorant of 'the relative importance of biological versus physicochemical processes in the generation of carbonate sequences, even in modern shelf-reefal settings' (Wilkinson and Walker, 1989; see also Berger and Keir, 1984). These matters demand our attention if we are to decipher the subtle taphonomic signals hidden in the rock record. The breakdown of calcareous algae (such as *Halimeda* and *Penicillus*) into submicroscopic needles has been reported as the prime mechanism in the formation of carbonate muds (Neumann and Land, 1975; see also Steinen *et al.*, 1988; Shinn *et al.*, 1989; Steinen and Tennet, 1990). Because of their small size, the needles are 'supersoluble' (Bathurst, 1975) and redissolve, possibly with implications for atmospheric CO_2 and global temperature change (Walter and Burton, 1990; Walter *et al.*, 1990; see also MacKenzie and Sabine, 1990). Although foraminifera typically comprise only a few per cent of modern and ancient shallow-water carbonate sediments (Boss and Liddell, 1987), they may still produce upwards of several hundred grams per square metre per year in carbonate environments in the case of larger genera like *Amphistegina* and *Archaias* (Hallock *et al.*, 1986a; 1986b). In fact, in the terrigenous continental borderland of southern California, foraminifera are the largest source of calcium carbonate (Smith, 1971).

Species inhabiting shallow carbonate and terrigenous shelves may differ, however, in their susceptibility to destruction. The field experiments of Corliss and Honjo (1981) suggest that relatively subtle differences in test microstructure and architecture may have a pronounced effect on differential preservation (for example, *Amphistegina* versus deep-sea species; see also Walter and Morse, 1984; Walter, 1985; Henrich and Wefer, 1986), and that, perhaps, reef-dwelling species are inherently less preservable than those inhabiting terrigenous environments. Indeed, differences in test structure of deep-sea species appear to reflect not only modern gradients in calcium carbonate availability (see, for example, Bremer and Lohmann, 1982; see also Greiner, 1969) but also ancient ones (Gradstein and Berggren, 1981; Tjalsma and Lohmann, 1983). Thus, scenarios that attribute the demise of larger, symbiont-bearing foraminifera (such as fusulinids) to nutrient excess (Hallock and Schlager, 1986; Hallock, 1987; 1988; Tappan and Loeblich, 1988) may be compounded by selective destruction of reef-dwelling species. Obviously, more data are needed to confirm or refute this hypothesis. What is needed is a taphofacies model for terrigenous sediments, analogous to the one we have constructed for carbonate environments, that will allow us to generate testable hypotheses about differential foraminiferal preservation in siliciclastic regimes.

Indeed, there may be taphonomic 'grades' (Brett and Baird, 1986; Speyer and Brett, 1988) and 'signatures' (Brandt, 1989) recorded by foraminiferal assemblages and test surface features (Flessa and Brown, 1983; Martin, 1986; Cutler, 1987; Cottey and Hallock, 1988; Kotler *et al.*, 1989), respectively, that would allow us to distinguish the differing sedimentological and geochemical conditions (and time-scales) of shell accumulation in carbonate, terrigenous and mixed carbonate-siliciclastic regimes (see, for example, Aigner, 1982a; 1982b; Kidwell, 1988; Kidwell *et al.*, 1986). Taphonomic grades (and signatures) of fossil assemblages are no doubt related to diagenetic potential, that is, the extent of alteration which sediment will undergo during burial and cementation (Schlanger and Douglas, 1974). As originally applied by

Schlanger and Douglas (1974) to pelagic carbonates, diagenetic potential is a function of the depth to the Calcite Compensation Depth (the depth below which carbonate-free sediments accumulate; Kennett, 1982), fertility (sediment production) of surface waters, and the size distribution of microfossil remains. Conversely, diagenetic potential is of import to our interpretations of the sedimentological, geochemical and climatic phenomena that are inextricably intertwined with evidence for tempo and mode in evolution and episodes of mass extinction. Obviously, the concept of diagenetic potential may be applied to shallow-water carbonates and siliciclastic sediments. We may view the latitudinal transition zone between terrigenous and tropical carbonate environments as a sort of 'latitudinal lysocline' that has moved across the surface of the Earth in response to perturbations of the CO_2–bicarbonate–carbonate equilibrium of the ocean–continent–atmosphere system, just as the deep-sea lysocline (the depth below which significant calcium carbonate dissolution first occurs; Berger, 1970) has shifted due to similar perturbations. Positive and negative 'chemo-climatic feedback' (Berger, 1982) between carbonate and terrigenous shelf regimes may have been critical in the evolution and differential preservation of fossil biotas. Eventually, by integrating both laboratory and field studies of foraminiferal taphonomy, and by drawing upon ancient examples, we should arrive at a better understanding of the distinctive geohistorical and biohistorical signatures recorded in fossil assemblages and sediments of the Earth's crust (Worsley *et al.*, 1986).

ACKNOWLEDGEMENTS

Our investigations of foraminiferal taphonomy at Discovery Bay have been supported by National Science Foundation (Stratigraphy and Paleontology) Grant Number EAR-8815997 to REM and Number EAR-8815876 to WDL. Thanks to W.J. Ullman and S. Welch of the College of Marine Studies, University of Delaware, for running the dissolution reactor experiments, and for their counsel on calcium carbonate dissolution. Our deep appreciation to Karl Flessa, Sue Kidwell, and Carl Brett for moral support, and to Steve Donovan and the publisher, Belhaven Press, for the invitation to publish this paper. Barbara Broge drafted figures and Pat Musa carefully and cheerfully typed numerous versions of the manuscript.

REFERENCES

Aigner, T., 1982a, Event-stratification in nummulite accumulations and in shell beds from the Eocene of Egypt. In G. Einsele and A. Seilacher (eds), *Cyclic and event stratification*, Springer-Verlag, Berlin: 248–65.

Aigner, T., 1982b, Biofabrics as dynamic indicators in nummulite accumulations, *Journal of Sedimentary Petrology*, **55** (1): 131–4.

Alexandersson, E.T., 1972, Micritization of carbonate particles: process of precipitation and dissolution in modern shallow-marine sediments, *Bulletin Geological Institute University Uppsala*, new series, **3** (7): 201–36.

Alexandersson, E.T., 1976, Actual and anticipated petrographic effects of carbonate undersaturation in shallow seawater, *Nature*, **262** (5570): 653–7.

Alexandersson, E.T., 1979, Marine maceration of skeletal carbonate in the Skaggerak, North Sea, *Sedimentology*, **26** (6): 845–52.

Aller, R.C., 1982, Carbonate dissolution in nearshore terrigenous muds: the role of physical and biological reworking, *Journal of Geology*, **90** (1): 79–95.

Applin, E.R., Ellisor, A.E. and Kniker, H.T., 1925, Subsurface stratigraphy of the coastal plain of Texas and Louisiana, *American Association of Petroleum Geologists Bulletin*, **9** (1): 79–122.

Bandy, O.L., 1964, General correlation of foraminiferal wall structure with environment. In J. Imbrie and N. Newell (eds), *Approaches to paleoecology*, Wiley, New York: 75–90.

Bathurst, R.G.C., 1975, *Carbonate sediments and their diagenesis*, Elsevier, Amsterdam.

Behrensmeyer, A.K. and Kidwell, S.M., 1985, Taphonomy's contribution to paleobiology, *Paleobiology*, **11** (1): 105–19.

Berger, W.H., 1970, Planktonic foraminifera: selective dissolution and the lysocline, *Marine Geology*, **8** (2): 111–38.

Berger, W.H., 1982, Deep-sea stratigraphy: Cenozoic climate steps and the search for chemo-climatic feedback. In G. Einsele and A. Seilacher (eds), *Cyclic and event stratification*, Springer-Verlag, Berlin: 121–57.

Berger, W.H. and Keir, R.S., 1984, Glacial-Holocene changes in atmospheric CO_2 and the deep-sea record. In J.E. Hansen and T. Takahashi (eds), *Climate processes and climate sensitivity, American Geophysical Union Geophysical Monograph 29, Maurice Ewing Volume*, **5**: 337–51.

Berner, R.A., Scott, M.R. and Thomlinson, C., 1970, Carbonate alkalinity in the pore waters of anoxic marine sediments, *Limnology and Oceanography*, **15** (4): 544–9.

Bernhard, J.M., 1988, Postmortem vital staining in benthic foraminifera: duration and importance in population and distributional studies, *Journal of Foraminiferal Research*, **18** (2): 143–6.

Boss, S.K. and Liddell, W.D., 1987, Back reef and fore reef analogs in the Pleistocene of North Jamaica: implications for facies recognition and sediment flux in fossil reefs, *Palaios*, **2** (3): 219–28.

Bottjer, D.J. and Douglas, R.G., 1984, Comparative taphonomy of molluscs and foraminifera in paleoenvironmental interpretation of west coast active margin strata, *Geological Society of America Abstracts with Programs*, **16** (3): 126.

Brandt, D.S., 1989, Taphonomic grades as a classification for fossiliferous assemblages and implications for paleoecology, *Palaios*, **4** (4): 303–9.

Braiser, M.D., 1975a, Ecology of Recent sediment-dwelling and phytal foraminifera from the lagoons of Barbuda, West Indies, *Journal of Foraminiferal Research*, **5** (1): 42–62.

Braiser, M.D., 1975b, The ecology and distribution of Recent Foraminifera from the reefs and shoals around Barbuda, West Indies, *Journal of Foraminiferal Research*, **5** (2): 193–210.

Braisier, M.D., 1986. Form, function, and evolution in benthic and planktic foraminiferal test architecture. In B.S.C. Leadbeater and R. Riding (eds), *Biomineralization in lower plants and animals, Systematic Association Special Volume*, **30**: 251–68.

Bremer, M.L. and Lohmann, G.P., 1982, Evidence for primary control of the distribution of certain Atlantic Ocean benthic foraminifera by degree of carbonate saturation, *Deep-Sea Research*, **29** (8A): 987–98.

Brett, C.E. and Baird, G.C., 1986, Comparative taphonomy: a key to paleoenvironmental interpretation based on fossil preservation, *Palaios*, **1** (3): 207–27.

Buxton, M.W.N. and Pedley, H.M., 1989, A standardized model for Tethyan Tertiary carbonate ramps, *Journal of the Geological Society of London*, **146** (6): 746–8.

Buzas, M.A., 1978, Foraminifera as prey for benthic deposit feeders: results of predator exclusion experiments, *Journal of Marine Research*, **36** (4): 617–25.

Buzas, M.A., 1982, Regulation of Foraminiferal densities by predation in the Indian River, Florida, *Journal of Foraminiferal Research*, **12** (1): 66–71.

Buzas, M.A. and Carle., K.J., 1979, Predators of Foraminifera in the Indian River, Florida, *Journal of Foraminiferal Research*, **9**(4): 336–40.

Buzas, M.A., Smith, R.K. and Beem, K.A., 1977, Ecology and systematics of Foraminifera in two *Thalassia* habitats, Jamaica, West Indies, *Smithsonian Contributions to Paleobiology*, **31**: 1–139.

Campbell, S.E., 1983, The modern distribution and geological history of calcium carbonate boring microorganisms. In P. Westbroek and E.W. de Jong (eds), *Biomineralization and biological metal accumulation: biological and geological perspectives*, D. Reidel, Dordrecht: 99–104.

Chaproniere, G.C.H., 1975, Paleoecology of Oligo-Miocene larger Foraminiferida, Australia, *Alcheringa*, **1** (1): 37–58.

Chave, K.E., 1964, Skeletal durability and preservation. In J. Imbrie and N. Newell (eds), *Approaches to paleoecology*, Wiley, New York: 377–87.

Chave, K.E., Smith, S.V. and Roy, K.J., 1972, Carbonate production by coral reefs, *Marine Geology*, **12** (1): 123–40.

Chou, L. and Wollast, R., 1984, Study of the weathering of albite at room temperature and pressure with a fluidized bed reactor, *Geochimica et Cosmochimica Acta*, **48**: 2205–17.

Collen, J.D. and Burgess, C.J., 1979, Calcite dissolution, overgrowth and recrystallization in the benthic foraminiferal genus *Notorotalia*, *Journal of Paleontology*, **53** (6): 1343–53.

Corliss, B.H. and Honjo, S., 1981, Dissolution of deep-sea benthonic foraminifera, *Micropaleontology*, **27** (4): 356–78.

Cottey, T.L. and Hallock, P., 1988, Test surface degradation in *Archaias angulatus*, *Journal of Foraminiferal Research*, **18** (3): 187–202.

Culver, S.J., 1987, Foraminifera. In T.W. Broadhead (ed.), *Fossil prokaryotes and protists: notes for a short course*, *University of Tennessee Studies in Geology*, **18**: 169–212.

Cunningham, J.L., Liddell, W.D. and Martin, R.E., 1989, Hydraulic properties of foraminifera: implications for foraminiferal taphonomy, *Geological Society of America Abstracts with Programs*, **21** (6): A46.

Cushman, J.A., 1930, The foraminifera of the Atlantic Ocean, Part 7. Nonionidae, Camerinidae, Peneroplidae and Alveolinellidae, *US National Museum Bulletin*, **104** (7): 1–79.

Cutler, A.H., 1987, Surface textures of shells as taphonomic indicators. In K.W. Flessa (ed.), *Paleoecology and taphonomy of Recent to Pleistocene intertidal deposits*, *Gulf of California, Paleontological Society Special Publication*, **2**: 164–76.

Davies, D.J., Powell, E.N. and Stanton, R.J., 1989, Relative rates of shell dissolution and net sediment accumulation—a commentary: can shell beds form by the gradual accumulation of biogenic debris on the sea floor?, *Lethaia*, **22** (2): 207–12.

Denne, R.A. and Sen Gupta, B.K., 1989, Effects of taphonomy and habitat on the record of benthic foraminifera in modern sediments, *Palaios*, **4** (5): 414–23.

Douglas, R.G. and Carlos, A., 1986, Modern foraminiferal taphonomy and its paleoecological and biostratigraphic implications, *Fourth North American Paleontological Congress, Abstracts*: 12.

Douglas, R.G., Liestman, J., Walch, C., Blake, G. and Cotton, M.L., 1980, The transition from live to sediment assemblages in benthic foraminifera from the southern California borderland. In M.E. Field, A.H. Bouma, I.P. Colburn, R.G. Douglas, and J.C. Ingle (eds), *Quaternary depositional environments of the Pacific coast, Society of Economic Paleontologists and Mineralogists, Pacific Section Symposium*, **4**: 257–80.

Douglas, R.G. and Woodruff, F., 1981, Deep-sea benthic foraminifera. In C. Emiliani (ed.), *The oceanic lithosphere, the sea*, Wiley Interscience, New York, **7**: 1233–1327.

Driscoll, E.G. and Weltin, T.P., 1973, Sedimentary parameters as factors in abrasive shell reduction, *Palaeogeography, Palaeoclimatology, Palaeoecology*, **13** (4): 275–88.

Fitzgerald, M.G., Parmenter, C.M. and Milliman, J.D., 1979, Particulate calcium carbonate in New England shelf waters: result of shell degradation and resuspension, *Sedimentology*, **26** (6): 853–7.

Flessa, K.W. and Brown, T.J., 1983, Selective solution of macroinvertebrate calcareous hard parts: a laboratory study, *Lethaia*, **16** (3): 193–205.

Frost, S.H. and Langenheim, R.L., 1974, *Cenozoic reef biofacies: Tertiary larger foraminifera and scleractinian corals from Chiapas, Mexico*, Northern Illinois University Press, DeKalb.

Fürsich, F.T. and Flessa, K.W., 1987, Taphonomy of tidal flat molluscs in the northern Gulf of California: paleoenvironmental analysis despite the perils of preservation, *Palaios*, **2** (6): 543–59.

Ginsburg, R.N., 1957, Early diagenesis and lithification of shallow-water carbonate sediments in south Florida. In R.J. LeBlanc and J.G. Breeding (eds), *Regional aspects of carbonate deposition, Society of Economic Paleontologists and Mineralogists Special Publication*, **5**: 80–100.

Goldstein, S.T., 1988, Foraminifera of relict salt marsh deposits, St. Catherines Island, Georgia: taphonomic implications, *Palaios*, **3** (3): 327–34.

Goldstein, S.T. and Barker, W.W., 1988, Test ultrastructure and taphonomy of the monothalamous agglutinated foraminifer *Cribrothalammina*, n. gen., *alba* (Heron-Allen and Earland), *Journal of Foraminiferal Research*, **18** (2): 130–6.

Goldstein, S.T. , Watkins, G.T. and Bagwell, E.C., 1989, Infaunal salt marsh foraminifera: effects on total foraminiferal assemblages in subsurface marsh sediments, *Geological Society of America Abstracts with Programs*, **21** (6): A62.

Golubic, S., Campbell, S.E., Drobne, K., Cameron, B., Balsam, W.L., Cimerman, F. and DuBois, L., 1984, Microbial endoliths: a benthic overprint in the sedimentary record and a paleobathymetric cross-reference with foraminifera, *Journal of Paleontology*, **58** (2): 351–61.

Golubic, S., Perkins, R.D. and Lucas, K.J., 1975, Boring microorganisms in carbonate substrates. In Frey, R., (ed.), *The study of trace fossils*, Springer-Verlag, New York: 229–59.

Grabert, B., 1971, Zur Eignung von Foraminiferen als Indikatoren für Sandwanderung, *Deutsche Hydrographische Zeitschrift*, **24** (1): 1–14.

Gradstein, F.M. and Berggren, W.A., 1981, Flysch-type agglutinated foraminifera and the Maestrichtian to Paleogene history of the Labrador and North Seas, *Marine Micropaleontology*, **6** (3): 211–68.

Graus, R.R. and Macintyre, I.G., 1989, The zonation by patterns of Caribbean coral reefs as controlled by wave and light energy input, bathymetric setting and reef morphology: computer simulation experiments, *Coral Reefs*, **8** (1): 9–18.

Greiner, G.O.G., 1969, Recent benthic foraminifera: environmental factors controlling their distribution, *Nature*, **223** (5202): 168–70.

Hale, W.R., Snyder, S.W. and Kontrovitz, M., 1985, Modern benthic foraminifera of the Washington continental shelf: assessment of post mortem transport based on estimated traction velocities, *Geological Society of America Abstracts with Programs*, **17** (7): 601.

Hallock, P., 1979, Trends in test shape with depth in large, symbiont bearing foraminifera, *Journal of Foraminiferal Research*, **9** (1): 61–9.

Hallock, P., 1987, Fluctuations in the trophic resource continuum: a factor in global diversity cycles?, *Paleoceanography*, **2** (5): 457–71.

Hallock, P., 1988, The role of nutrient availability in bioerosion: consequences to carbonate buildups, *Palaeogeography, Palaeoclimatology, Palaeoecology*, **63** (1–3): 275–91.

Hallock, P., Cottey, T.L., Forward, L.B. and Halas, J., 1986a, Population biology and sediment production of *Archaias angulatus* (Foraminiferida) in Largo Sound, Florida, *Journal of Foraminiferal Research*, **16** (1): 1–8.

Hallock, P., Forward, L.B. and Hansen, H.J., 1986b, Environmental influence of test shape in *Amphistegina*, *Journal of Foraminiferal Research*, **16** (3): 224–31.

Hallock, P. and Glenn, E.C., 1985, Numerical analysis of foraminiferal assemblages: a tool for recognizing depositional facies in Lower Miocene reef complexes, *Journal of Paleontology*, **59** (6): 1382–94.

Hallock, P. and Glenn, E.C., 1986, Larger foraminifera: a tool for paleoenvironmental analysis of Cenozoic carbonate depositional facies, *Palaios*, **1** (1): 55–64.

Hallock, P. and Schlager, W., 1986, Nutrient excess and the demise of coral reefs and carbonate platforms, *Palaios*, **1** (4): 389–98.

Henrich, R. and Wefer, G., 1986, Dissolution of biogenic carbonates: effects of skeletal structure, *Marine Geology*, **71** (3/4): 341–62.

Hickman, C.S. and Lipps, J.H., 1983, Foraminiferivory: selective ingestion of foraminifera and test alterations produced by the neogastropod *Olivella*, *Journal of Foraminiferal Research*, **13** (2): 108–14.

Howe, H.V., 1959, Fifty years of micropaleontology, *Journal of Paleontology*, **33** (3): 511–15.

Israelsky, M.C., Brehm, R.C., Hanna, M.A., Miller, J.C. and Rolshausen, F.W., 1933, Coastal plain stratigraphic nomenclature, *American Association of Petroleum Geologists Bulletin*, **17** (12): 1535–6.

Kennett, J.P., 1982, *Marine geology*, Prentice Hall, Englewood Cliffs, New Jersey.

Khusid, T.A., 1984, On dissolution of carbonate tests of benthic foraminifera on the Pacific shelf of South America, *Oceanology*, **24** (1): 85–91.

Kidwell, S.M., 1986, Taphonomic feedback in Miocene assemblages: testing the role of dead hardparts in benthic communities, *Palaios*, **1** (3): 239–55.

Kidwell, S.M., 1988, Archaic vs. modern shell concentrations: Phanerozoic trends in nature of the fossil record, *Society of Economic Paleontologists and Mineralogists Midyear Meeting, Columbus, Ohio, Abstracts with Program*: 880.

Kidwell, S., 1989, Stratigraphic condensation of marine transgressive records: origin of major shell deposits in the Miocene of Maryland, *Journal of Geology,* **97** (1): 1–24.

Kidwell, S.M., Fürsich, F.T. and Aigner, T., 1986, Conceptual framework for the analysis and classification of fossil concentrations, *Palaios*, **1** (3): 228–38.

Kleinpell, R.M., 1933, *Miocene stratigraphy of California*, American Association of

Petroleum Geologists, Tulsa, Oklahoma.

Kleinpell, R.M., 1971, California's early 'oilbug' profession, *Journal of the West*, **10** (1): 72–101.

Kloos, D.P., 1982, Destruction of tests of the foraminifer *Sorites orbiculus* by endolithic microorganisms in a lagoon on Curaçao (Netherlands Antilles), *Geologie en Mijnbouw*, **61** (2): 201–5.

Kontrovitz, M., Snyder, S.W. and Brown, R.J., 1978, A flume study of the movement of foraminifera tests, *Palaeogeography, Palaeoclimatology, Palaeoecology*, **23** (1/2): 141–50.

Kotler, E., Martin, R.E. and Liddell, W.D., 1989, Experimental analysis of abrasion-resistance of modern reef dwelling foraminifera from Discovery Bay, Jamaica, *Geological Society of America Abstracts with Programs*, **21** (6): A46.

Land, L.S., 1979, The fate of reef-derived sediment on the North Jamaican island slope, *Marine Geology*, **29** (1/4): 55–71.

Land, L.S. and Goreau, T.F., 1970, Submarine lithification of Jamaican reefs, *Journal of Sedimentary Petrology*, **40** (1): 457–62.

Land, L.S. and Moore, C.H., 1980, Lithification, micritization and syndepositional diagenesis of biolithites on the Jamaican island slope, *Journal of Sedimentary Petrology*, **50** (2): 357–70.

LeCalvez, Y. and Cesana, D., 1972, Detection de l'état de vie chez les foraminifères, *Annales of Paleontology*, **58** (2): 129–34.

Liddell, W.D., Boss, S.K., Nelson, C.V. and Martin, R.E., 1987, Sedimentological and foraminiferal characterization of shelf and slope environments (1–234 m), north Jamaica. In H.A. Curran (ed.), *Proceedings of the Third Symposium on the Geology of the Bahamas*: 91–8.

Liddell, W.D. and Martin, R.E., 1989, Taphofacies in modern carbonate environments: implications for formation of foraminiferal sediment assemblages, *28th International Geological Congress, Washington DC, 9–19 July, Abstracts*, **2**: 299.

Liddell, W.D., Ohlhorst, S.L. and Coates, A.G., 1984, *Modern and ancient carbonate environments of Jamaica* [Sedimenta X], University of Miami Comparative Sedimentology Laboratory, Miami.

Lipps, J.H., 1988, Predation on foraminifera by coral reef fish: taphonomic and evolutionary implications, *Palaios*, **3** (3): 315–26.

Lipps, J.H. and Ronan, T.E., 1974, Predation on foraminifera by the polychaete worm, *Diopatra*, *Journal of Foraminiferal Research*, **4** (2): 139–43.

Loubere, P., 1989, Bioturbation and sedimentation rate control of benthic microfossil taxon abundances in surface sediments: a theoretical approach to the analysis of species microhabits, *Marine Micropaleontology*, **14** (4): 317–25.

MacKenzie, F.T. and Sabine, C.L., 1990, Early diagenesis of benthically derived carbonates and their importance in the carbon cycle, *American Association of Petroleum Geologists Bulletin*, **74**: 711.

Mageau, N.C. and Walker, D.A., 1976, Effects of ingestion of foraminifera by larger invertebrates. In C.T. Schafer and B.R. Pelletier (eds), *First International Symposium on Benthic Foraminifera of Continental Margins. Part A. Ecology and Biology, Maritime Sediments Special Publication*, **1**: 89–105.

Maiklem, W.R., 1968, Some hydraulic properties of bioclastic carbonate grains, *Sedimentology*, **10** (2): 101–9.

Martin, R.E., 1986, Habitat and distribution of the foraminifer *Archaias angulatus* (Fichtel and Moll) (Miliolina, Soritidae), *Journal of Foraminiferal Research*, **16** (3): 201–6.

Martin, R.E. and Liddell, W.D., 1988, Foraminiferal depth zonation on a north coast fringing reef (0–75 m), Discovery Bay, Jamaica, *Palaios*, **3** (3): 298–314.

Martin, R.E. and Liddell, W.D., 1989, Relation of counting methods to taphonomic gradients and information content of foraminiferal sediment assemblages, *Marine Micropaleontology*, **15** (1/2): 67–89.

Martin, R.E. and Steinker, D.C., 1973, Evaluation of techniques for recognition of living foraminifera, *Compass of Sigma Gamma Epsilon*, **50** (4): 26–30.

Martin, R.E. and Wright, R.C., 1988, Information loss in the transition from life to death assemblages of foraminifera in back reef environments, Key Largo, Florida, *Journal of Paleontology*, **62** (3): 399–410.

May, J.A. and Perkins, R.D., 1979, Endolithic infestation of carbonate substrates below the sediment–water interface, *Journal of Sedimentary Petrology*, **49** (2): 357–78.

Meldahl, K.H. and Flessa, K.W., 1989, Taphonomic pathways and comparative biofacies and taphofacies in a Recent intertidal/shallow shelf environment, *Lethaia*, **23** (1): 43–60.

Miller, D.J. and Ellison, R.L., 1982, The relationship of foraminifera and submarine topography on the New Jersey–Delaware continental shelf, *Geological Society of America Bulletin*, **93** (3): 239–45.

Milliman, J.D., 1974, *Marine carbonates*, Springer-Verlag, New York.

Moberly, R., 1968, Loss of Hawaiian littoral sand, *Journal of Sedimentary Petrology*, **38** (1): 17–34.

Muller, P.H., 1974, Sediment production and population biology of the benthic foraminifer *Amphistegina madagascariensis*, *Limnology and Oceanography*, **19** (5): 802–9.

Murray, J.W., 1973, *Distribution and ecology of living benthic foraminiferids*, Crane, Russak, New York.

Murray, J.W., 1976, Comparative studies of living and dead benthic foraminiferal distributions. In R. Hedley and C.G. Adams (eds), *Foraminifera*, Academic Press, New York, **2**: 45–109.

Murray, J.W., 1989, Syndepositional dissolution of calcarous foraminifera in modern shallow-water sediments, *Marine Micropaleontology*, **15** (1/2): 117–21.

Murray, J.W. and Wright, C.A., 1970, Surface textures of calcareous foraminiferids, *Paleontology*, **13** (2): 184–7.

Natland, M.L., 1933, The temperature- and depth-distribution of some recent and fossil foraminifera in the southern California region, *Scripps Institute of Oceanography Bulletin*, **3**: 225–30.

Neumann, A.C. and Land, L.S., 1975, Lime mud deposition and calcareous algae in the Bight of Abaco, Bahamas: a budget, *Journal of Sedimentary Petrology*, **45** (4): 763–86.

Owen, E.W., 1975, Trek of the oil finders: a history of exploration for petroleum, *American Association of Petroleum Geologists Memoir*, **6**: 1–647.

Peebles, M.W. and Lewis, R.D., 1988, Differential infestation of shallow-water benthic foraminifera by microboring organisms: possible biases in preservation potential, *Palaios*, **3** (3): 345–51.

Perkins, R.D. and Halsey, S.D., 1971, Geologic significance of microboring fungi and algae in Carolina shelf sediments, *Journal of Sedimentary Petrology*, **41** (3): 843–53.

Perkins, R.D. and Tsentas, C.I., 1976, Microbial infestation of carbonate substrates planted on the St. Croix shelf, West Indies, *Geological Society of America Bulletin*, **87** (11): 1615–28.

Phleger, F.B., 1960, *Ecology and distribution of recent foraminifera*, Johns Hopkins Press, Baltimore, Maryland.

Pigott, J.D. and Land, L.S., 1986, Interstitial water chemistry of Jamaican reef sediment: sulfate reduction and submarine cementation, *Marine Chemistry*, **19** (4): 355–78.

Poulicek, M., Jaspar-Versali, M.F. and Goffinet, G., 1981, Étude expérimentale de la dégradation des coquilles de mollusques au niveau des sèdiments marins, *Bulletin de la Société Royale Scièntifique de Liège*, **50** (11–12): 512–18.

Prothero, D.R. and Lazarus, D.B., 1980, Planktonic microfossils and the recognition of ancestors, *Systematic Zoology*, **29** (2): 119–29.

Reiss, Z. and Hottinger, L., 1984, *The Gulf of Aqaba: ecological micropaleontology*, Springer-Verlag, Berlin.

Rooney, W.S. and Perkins, R.D., 1972, Distribution and geologic significance of microboring organisms within sediments of the Arlington Reef Complex, Australia, *Geological Society of America Bulletin*, **83** (4): 1139–50.

Ross, C.A., 1982, Paleozoic foraminifera-fusulinids. In T.W. Broadhead (ed.), *Foraminifera: notes for a short course, University of Tennessee Studies in Geology*, **6**: 163–76.

Ross, C.A. and Ross, J.R.P., 1985, Late Paleozoic depositional sequences are synchronous and worldwide, *Geology*, **13** (3): 194–7.

Schafer, C.T. and Young, J.A., 1977, Experiments on mobility and transportability of some nearshore benthonic foraminifera species, Report of Activities, Part C, *Geological Survey of Canada Paper*, **77–1C**: 27–31.

Schindel, D.E., 1980, Microstratigraphic sampling and the limits of paleontologic resolution, *Paleobiology*, **6** (4): 408–26.

Schindel, D.E., 1982, Resolution analysis: a new approach to the gaps in the fossil record, *Paleobiology*, **8** (4): 340–53.

Schlanger, S.O. and Douglas, R.G., 1974, Pelagic ooze–chalk–limestone transition and its implications for marine stratigraphy. In K.J. Hsü and H.C. Jenkyns (eds), *Pelagic sediments: on land and under the sea, International Association of Sedimentologists Special Publication*, **1**: 117–48.

Schröder, C.J., Scott, D.B. and Medioli, F.S., 1987, Can smaller benthic foraminifera be ignored in paleoenvironmental analyses?, *Journal of Foraminiferal Research*, **17** (2): 101–5.

Scoffin, P.W., Stearn, C.W., Barcker, D., Frydl, P., Hawkins, C.M., Hunter, I.G. and MacGeadry, J.K., 1980, Calcium carbonate budget of a fringing reef on the west coast of Barbados, Part II—Erosion, sediments and internal structure, *Bulletin of Marine Science*, **30** (2): 475–508.

Sen Gupta, B.K., Shin, I.C. and Wendler, S.T., 1987, Relevance of specimen size in distribution studies of deep-sea benthic foraminifera, *Palaios*, **2** (4): 332–8.

Shinn, E.A., Steinen, R.P., Lidz, B.H. and Swart, P.K., 1989, Whitings, a sedimentologic dilemma, *Journal of Sedimentary Petrology*, **59** (1): 147–61.

Signor, P.W. and Lipps, J.H., 1982, Sampling bias, gradual extinction patterns and catastrophes in the fossil record, *Geological Society of America Special Paper*, **190**: 291–6.

Sliter, W.V., 1971, Predation on benthic foraminifers, *Journal of Foraminiferal Research*, **1** (1): 20–9.

Smith, R.K., 1987, Fossilization potential in modern shallow-water benthic foraminiferal assemblages, *Journal of Foraminiferal Research*, **17** (2): 117–22.

Smith, S.V., 1971, Budget of calcium carbonate, southern California continental borderland, *Journal of Sedimentary Petrology*, **41** (3): 798–808.

Speyer, S.E. and Brett, C.E., 1988, Taphofacies models for epeiric seas: Middle Paleozoic examples, *Palaeogeography, Palaeoclimatology, Palaeoecology*, **63** (1–3): 225–62.

Steinen, R.P., Swart, P.K., Shinn, E.A. and Lidz, B.H., 1988, Bahamian lime mud: the algae didn't do it, *Geological Society of America, Abstracts with Programs*, **20** (7): A209.

Steinen, R. and Tennet, P., 1990, Origin of fine-grained Holocene shallow marine carbonate sediment on the Florida–Bahama platform, *American Association of Petroleum Geologists*, **74** (5): 771.

Stuckey, C.W., 1978, Milestones in Gulf Coast economic micropaleontology, *Gulf Coast Association of Geological Societies Transactions*, **28**: 621–5.

Swinchatt, J.P., 1965, Significance of constituent composition, texture, and skeletal breakdown in some recent carbonate sediments, *Journal of Sedimentary Petrology*, **35** (1): 71–90.

Tappan, H. and Loeblich, A.R., 1988, Foraminiferal evolution, diversification, and extinction, *Journal of Paleontology*, **62** (5): 695–714.

Tjalsma, R.C. and Lohmann, G.P., 1983, Paleocene-Eocene bathyal and abyssal benthic foraminifera from the Atlantic Ocean, *Micropaleontology Special Publication*, **4**: 1–90.

Walker, D.A., 1971, Etching of the test surface of benthonic foraminifera due to ingestion by the gastropod *Littorina littorea* Linné, *Canadian Journal of Earth Science*, **8** (11): 1487–91.

Walter, L.M., 1985, Relative reactivity of skeletal carbonates during dissolution: implications for diagenesis. In N. Schneidermann and P.M. Harris (eds), *Carbonate cements, Society of Economic Paleontologists and Mineralogists Special Publication*, **36**: 3–16.

Walter, L.M., Bonnell, L. and Patterson, W.P., 1990, Syndepositional dissolution of shallow marine carbonates, *American Association of Petroleum Geologists Bulletin*, **74**: 787.

Walter, L.M. and Burton, E.A., 1990, Dissolution of recent platform carbonate sediments in marine pore fluids, *American Journal of Science*, **290** (6): 601–43.

Walter, L.M. and Morse, J.W., 1984, Reactive surface area of skeletal carbonates during dissolution: effect of grain size, *Journal of Sedimentary Petrology*, **54** (4): 1081–90.

Walton, W.R., 1952, Techniques for recognition of living foraminifera, *Contributions of the Cushman Foundation for Foraminiferal Research*, **3** (2): 56–60.

Wetmore, K.L., 1987, Correlations between test strength, morphology and habitat in some benthic foraminifera from the coast of Washington, *Journal of Foraminiferal Research*, **17** (1): 1–13.

Wilkinson, B.H. and Walker, J.C.G., 1989, Phanerozoic cycling of sedimentary carbonate, *American Journal of Science*, **289** (4): 525–48.

Wilson, J.L., 1975, *Carbonate facies in geologic history*, Springer-Verlag, New York.

Worsley, T.R., Nance, R.D. and Moody, J.B., 1986, Tectonic cycles and the history of the Earth's biogeochemical and paleoceanographic record, *Paleoceanography*, **1** (3): 233–64.

Zeff, M.L. and Perkins, R.D., 1979, Microbial alteration of Bahamian deep-sea carbonates, *Sedimentology*, **26** (1): 175–201.

Chapter 9

TRILOBITE TAPHONOMY: A BASIS FOR COMPARATIVE STUDIES OF ARTHROPOD PRESERVATION, FUNCTIONAL ANATOMY AND BEHAVIOUR

Stephen E. Speyer

INTRODUCTION

Trilobites, without doubt, are the best-represented arthropod group in the fossil record. The fact that their remains are restricted to Palaeozoic strata makes this preservation record all the more remarkable. In spite of an extensive database of occurrence information, our understanding of trilobites and other fossil arthropods as once living animals remains limited. Quite simply, we are not utilizing the available data to its full potential. During the past decade and a half, palaeoecology has experienced a renaissance of a sort, brought on largely by a Westernization of concepts and ideas that have flourished since before the turn of the century. The mostly German school of actualism (Aktuopaläontologie), the subject of books and many papers by such notables as S. Wiegelt, R. Richter, and W. Schäfer, gave birth to the conceptual basis of 'taphonomy' (*sensu* Efremov, 1940) as it is presently applied. In short, taphonomy is a summary of the preservational history of a fossil assemblage. The ultimate goal of palaeoecology is to provide testable models regarding the dynamic relationships among once living animals, and between animals and their environments. Comparative taphonomy provides a database of contrasting patterns which are correlated to these environmental and biological phenomena. Thus, by understanding the nature of our data-base (the fossil record) we realize the limitations of our data, and we equip ourselves better for tackling the problems and questions best suited to these data.

Trilobite remains occur in great abundance across many facies throughout the Palaeozoic stratigraphic record. Comparative taphonomy of these remains provides an important source of information regarding palaeoenvironmental conditions and trilobite palaeobiology. Trilobites offer extraordinary opportunities for research in comparative taphonomy because: their exo-skeleton consists of heavily calcified cuticle and is, therefore, readily pre-served; they inhabited a wide range of benthic habitats represented by a diversity of stratigraphic facies; as vagrant benthos, they were capable of a wide range of behaviours and behavioural attitudes that are potentially preservable; they span a significant portion of geologic time, thus commend-ing them for integrated studies of taphonomy, evolution and palaeoecology; and in conjunction with this last point, they provide a substantial database for comparison with taphonomic grades and preservation modes documented among other fossil, as well as modern (extant) arthropod, groups (for example, Chelicerates and Crustacea). Fossil arthropods, including trilobites, exhibit particular patterns of stratigraphic occurrence that reflect preservatio-nal, evolutionary and ecological constraints.

This chapter will explore the relationships among these constraints and discuss differences in the preservational history of varied arthropod groups as they appear throughout geologic time. Trilobites provide a convenient, though not so obvious, standard against which the reliability of ecological and environmental inferences regarding fossil arthropod deposits may be com-pared. Comparisons of this sort may help evaluate the reliability of diversity data upon which so much emphasis and interest are presently placed (see Sepkoski, 1984). Studies of arthropod taphonomy, taphonomic resolution and diversity of preservation modes across Phanerozoic time and among marine facies provide data useful in evaluating the impact of evolution (in its broadest definition) on taphonomic criteria used in differentiating facies and deciphering form–function relationships.

THE FOSSIL RECORD OF TRILOBITES

Plotnick (1990) outlined a straightforward history of arthropod preservation that includes: death (or ecdysis) and the introduction of the cuticle as a 'sedimentary particle' (*sensu* Seilacher, 1973) into the sedimentary system; incipient decay of the soft-part anatomy by bacterial and necrolytic agents; disarticulation of the tergite and the sternite, and dissociation of individual and groups of sclerites; chemical and biological disintegration of the sclerites themselves as a function of environmental dissolution and/or bioerosion; and diagenetic alteration of whatever remains. The integrity of cuticle preser-vation is directly related to the length of time exuvial or carcass remains are processed in this idealized taphonomic history. Remains that are buried shortly after ecdysis or death are more likely to be well preserved than those remains that are continually processed at the surface. Therefore, the amount of preservation bias incurred is a function of duration of residence in the hypothetical flowchart which, in turn, is a consequence of the relationship between environmental setting and event sedimentology (Figure 9.1).

Trilobites are abundantly preserved in virtually all marine facies spanning the entire Palaeozoic. This impressive record is due largely to the construc-tion of the trilobite cuticle which is primarily calcitic with varied amounts of

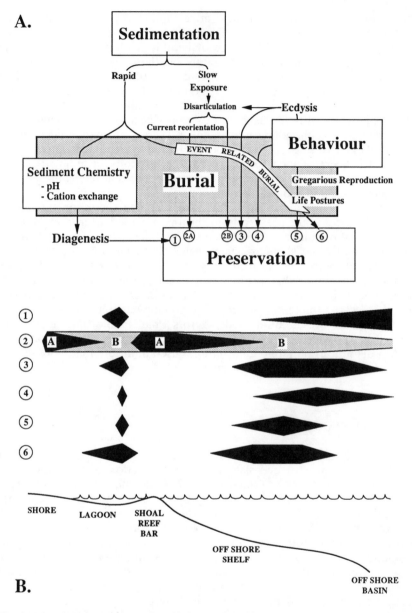

Figure 9.1 (A) Flowchart summary of arthropod taphonomy based, in part, upon Speyer (1987) and Speyer and Brett (1986a). Extrinsic factors (sedimentation rate, sediment chemistry) influence the duration of surface exposure and period of time potential fossils are subjected to destructive taphonomic processes (disarticulation, fragmentation). Intrinsic factors (behaviour) modify the manner of preservation and may (as with ecdysis) affect the relative integrity of preservation.

phosphatic material. Although the specific biostratinomic and diagenetic details vary between facies, the simple presence and persistence of trilobite remains attests to the robustness of the generic trilobite skeleton. The taphonomy and mode of preservation of trilobite remains are moderated by four basic parameters: chemistry (both environmental and skeletal); sedimentology (including aspects of current energy, sedimentation rate and, indirectly, bioturbation); morphology and Bauplan (including inherent hydrodynamic properties and structural strength); and behavioural attributes (including mode of life, style of moulting, specialized behaviours, etc.). The relationships among these four parameters and the generalized taphonomic history compose a complex mosaic that is roughly represented in Figure 9.1.

Importantly, extrinsic factors (environmental attributes such as current energy and mode, rate of sedimentation, and sediment geochemistry) directly affect the integrity of preservation of intrinsic attributes (biological characteristics including behavioural and skeletal characteristics). Moreover, complex feedback relationships exist between morphology and environmental selection, environmental setting and preservation, and behaviour and environmental circumstance. These relationships and their taphonomic significance are explored in later sections of this chapter.

Arthropod cuticle and preservation potential

Given that trilobites lived in a wide range of environments and that these environmental settings are characterized by particular taphonomic attributes, it stands to reason that the amount and calibre of data as well as the degree of palaeoecologic resolution afforded in these settings also varies. The most fundamental limitation on preservation potential is the composition of the skeleton. Differences in dissolution vulnerability and overall constructional strength determine a first-order separation of preserved from non-preserved

---

Six modes of occurrence are recognized. (1) Diagenetic enhancement including carbonate, pyrite, silica and phosphate syngenetic mineralization. (2) Disarticulated remains may be predominantly concave-down (2A) or randomly orientated (2B). (3) Intact moult ensembles with corresponding sclerites in association. (4) Enrolled bodies indicating a behavioural response to rapid burial and/or toxicity. (5) Varied aspects of behaviour including gregarious reproduction, synchronized ecdysis and particular body postures. (6) Solitary articulated bodies.

(B) Enviromental distribution of modes of occurrence detailed in Figure 9.1A. These modes of occurrence represent specific pathways on the illustrated flowchart and are distributed, non-randomly, according to environmental conditions. The association of particular modes of arthropod preservation reflects a particular environmental setting and may be used to infer palaeoenvironmental circumstances.

remains (see Flessa and Brown, 1982; Speyer and Brett, 1988a). The preservation of mineralized remains is a function of original skeletal chemistry, sediment composition (siliciclastic versus carbonate; grain size), and episodicity of burial (see Figure 9.1).

Arthropod exoskeletons are composed of cuticle that shows a wide range of compositions and preservation potentials. Arthropod cuticle consists of three structural components: the protein, chitin and mineralized layers. These layers roughly correspond to the epicuticle (outermost layer) and the endocuticle and innermost exocuticle (together composing the procuticle). Although terminologies vary among workers (see Dalingwater and Mutvei, 1989, for review), differences in the degree to which the layers are developed and mineralized directly affect the preservation potential of the organism's remains (see Figure 9.2). The physiological relationships between these layers in cross-section have been discussed by numerous workers and the interested reader is directed to Neville's (1975) excellent review. From the standpoint of evolutionary morphology, however, it is interesting to note that the biology of these layers, though they themselves are variably developed among the diversity of arthropod taxa, is tightly integrated and the genetic relationship between layers is cyclically manifest following each ecdysis.

Modern marine arthropod groups display the full range from non-mineralized, protein- and chitin-base cuticles (marine chelicerates, many isopods, amphipods and some penaeid shrimps; see Figure 9.2) through partially mineralized cuticles, the calcite integrated into an organic framework within the endocuticle (most decapod crustaceans), to well-mineralized carapaces with minimal protein or chitin (most ostracodes). Trilobite cuticle, unlike typical crustacean (ostrocodes being a notable exception) and chelicerate cuticle, is heavily calcified. It is not surprising that the degree of mineralization is strongly correlated to cuticle preservation. Allison (1986; 1988), Plotnick (1986), and Plotnick *et al.* (1988) discussed the preservation potential of varied crustaceans in terms of bacterial decomposition of the protein and chitinous constituents of the arthropod cuticle.

Trilobites apparently possessed a mostly calcitic exoskeleton, perhaps with a thin outer phosphatic layer (see Tiegler and Towe, 1975; McAllister and Brand, 1989; Wilmot and Fallick, 1989). Although unlike normal crustacean cuticle, which is characteristically protein- and chitin-rich, the trilobite exoskeleton is remarkably similar to the ostracod carapace (Towe, 1975). Towe (1973) and Dalingwater and Miller (1977) argued that the trilobite cuticle was extensively calcified, requiring a minimal organic base (such as characterizes the typical marine decapod crustacean). Wilmot and Fallick (1989) argued that the outer phosphatic layer, recognized by Tiegler and Towe (1975), is a secondary diagenetic artefact. Speyer (1987), however, identified a 'brassy veneer' on numerous articulated trilobites from a particularly fine-grained siliclastic facies that he believed to represent the altered remnants of an outer phosphatic layer (see Speyer, 1987, figure 7C). The existence of such a layer is consistent with chemical conditions that theoretically lead to early silica diagenesis and cuticle replacement (see Aller, 1983).

Differences in the construction of arthropod cuticle render the different arthropod groups as more or less likely to be preserved (Figure 9.2). Most significantly, the relative proportion of chitin and other protein-based components within the cuticle directly influences the biasing effects of decomposing and chitinivorous bacteria (see Benton, 1935; Campbell and Williams, 1955; Plotnick, 1986). Trilobites, because of pervasive cuticular mineralization, appear not to have had a significant organic matrix and were probably conserved from such biodegradation (McAllister and Brand, 1989). Although extraordinary circumstances of preservation will circumvent normal taphonomic biases (for example, preclude extensive loss of chitin; see Allison, 1988), the distribution of arthropod taxa across a spectrum of stratigraphic facies indicates that preservation predictably varies according to gross environmental setting. The pattern of distribution is further, and more fundamentally, influenced by the distributional dynamics of the living organism. Obviously, ecological constraints on environmental distribution are the primary controls on where a fossil organism might occur regardless of the circumstances of preservation.

Evolutionary innovations also result in taphonomically significant consequences (Figure 9.3). Although the phylogenetic relationships among the marine arthropods (Crustacea, Chelicerata, Trilobita) remain debated, several physiological and morphological developments in post-Palaeozoic groups importantly influence their preservation potential. Plotnick (1990) pointed out that calcification of appendages and overall external structure is a post-Palaeozoic phenomenon, citing the advent of crabs, lobsters and related decapod Crustacea. Although calcification of the appendages, in conjunction with an unprecedented specialization of anterior appendages, resulted in important ecological innovations (for example, predation and mobility), these groups clearly lack the robust tergite exhibited by most trilobites (certain Burgess Shale examples, *Naraoia*, *Tegopelte*, and *Liwia*, are possible exceptions; see Whittington, 1977; 1985; Dzik and Lendzion, 1988). The very nature of mineralization and cuticle construction appears to differ (Figure 9.2). Whereas calcification of trilobite cuticle is pervasive across the dorsal tergite and remains unchanged throughout the moult cycle (see Speyer, 1985), calcification in crustaceans occurs relatively late in ontogeny, is rigidly controlled by a pre-existing organic matrix and systematically accumulates during a post-ecdysial period, and is unevenly distributed across the exoskeleton. Finally, most decapods with a significantly mineralized cuticle display a degree of pre-ecdysial resorption that, although metabolically efficient, may limit the preservation potential of discarded exuviae in certain environmental settings. Rathbun (1935, p. 1) summarized the crustacean exoskeleton as 'thin, fragile, and easily destroyed, so that the remains consist largely of chelae or parts of chelae, as these are usually thicker than the carapace and the other appendages'.

Metabolic resorption of the mineral constituents of crustacean cuticle should, theoretically, be reflected in the manner of preservation exhibited by exuvial versus carcass remains (Figure 9.2). Feldman and Tshudy (1987) differentiated between exuviae and corpse cuticle among specimens of a Late Cretaceous nephropid decapod (*Haploparia stokesi*) on the basis of cuticle ultrastructure. Moult remains lacked prominent endocuticular laminations, whereas carcass remains were identified by the presence of such laminations.

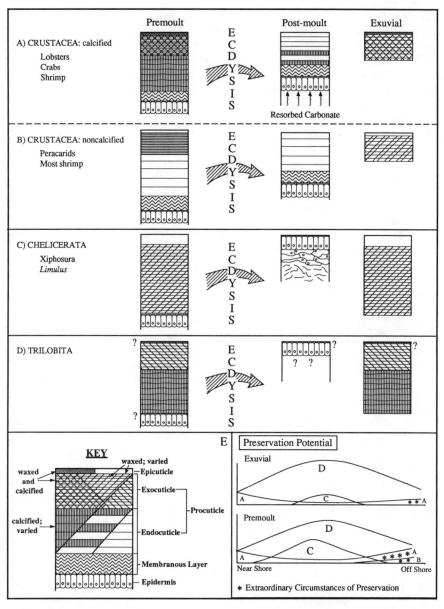

Figure 9.2 Arthropod preservation fundamentally depends upon the preservation potential of cuticle in various environmental situations. Cuticle composition varies among taxa and within a particular taxon according to life-history phenomena (that is, exuviation). Four typical marine arthropod groups are compared with respect to premoult (or intermoult), post-moult and exuvial construction and composition (based, in part, on Aitken, 1980; Barnes, 1980; Neville, 1975; Plotnick, 1990; Tiegler and Towe, 1975; Wainwright *et al.*, 1976).

In all cases, however, the remains were entirely or partially articulated, signifying rapid post-mortem or post-ecdysial burial. This appears to have been a critical stage in the taphonomic history of these remains, particularly in the face of abundant evidence for scavenging cephalopods noted for efficient deterioration of lobster and other crustacean exuviae (see also Schäfer, 1972). Indirectly, these data suggest that entirely dissociated cuticle (particularly moulted cuticle) is prone to taphonomic loss in most environmental settings. This basic model is corroborated by actualistic experiments conducted by Allison (1986) and Plotnick (1986) upon shrimp carcasses under varied scavenging and depositional circumstances.

Plotnick *et al.* (1988) demonstrated that the robust rock crab *Panopeus* sp. is remarkably resistant to preservation bias, a function, perhaps, of the severe and energetic environments it inhabits. The relative influence of various taphonomic agents, however, changes predictably through ontogenetic and physiological time (for example, moult cycle). For example, the chelae of

---

(A) Calcified Crustacea, including most decapod groups, display a complex ecdysial physiology involving pre-exuvial resorption of the carbonate matrix. Post-moult individuals reconstruct their endocuticle and exocuticle layers by laying down a carbonate matrix on a pre-formed framework of chitin. Discarded exuviae are partially calcified.

(B) Non-calcified Crustacea (such as peracarid groups and many shrimps) display cuticles that are variably waxed and tanned. Ecdysis does not involve resorption; the tanned cuticle must be completely reconstructed after exuvia has been released. Endocuticle and exocuticle are hardened to varying degrees according to sclerite adaptation and phylogenetic history. Ultrastructure is basically the same as calcified forms, but lacks the carbonate matrix laid down within the chitin frame.

(C) Chelicerates (Merostomata; Xiphosura) possess the thickest recorded cuticle among extant marine arthropods. Cuticle is entirely composed of layered chitin which is hardened and strengthened by a high wax that is tanned through a complex series of metabolic pathways (see Neville, 1975). During ecdysis the entire hardened cuticle is released and must be reconstructed during the subsequent intermoult period. Preservation potential of immediately post-moult individuals is, therefore, very low. Exuviae and pre-moult bodies theoretically display very similar, if not equivalent, potentials for preservation.

(D) Trilobites represent the most stable and, therefore, preservable of all marine arthropods because of a thick and heavily calcified cuticle. An extensive fossil record across many environmental facies indicates that pre-moult cuticle and discarded exuviae are taphonomically identical. Speyer (1985) reported soft-shell trilobites from carbonate-enhanced mudstone strata. These display a wrinkled surface texture, distinctly different from typically thick and heavy cuticle among articulated and disarticulated specimens from the same beds. Calcification does not appear to be organized as in decapod Crustacea (see (A)); that is, carbonate was not laid down on a chitinous framework.

(E) Variations among marine arthropod cuticles dictate particular patterns of preservation according to facies distributions. Refer also to Figures 9.1 and 9.4 (and corresponding text) for further information.

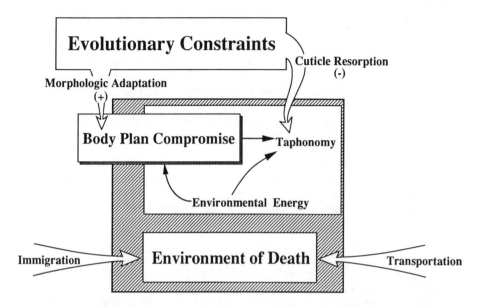

Figure 9.3 Summary chart detailing the theoretical relationships among taphonomy, environmental energy (as a proxy for gross environmental conditions)
and evolutionary constraints (including morphologic compromise and physio-
logic innovation). Decapod Crustacea possess a heavily calcified cuticle that is
moulted after some proportion of the carbonate matrix has been resorbed to be
used later in reconstruction. This reduces the preservation potential of crustacean exuviae. In contrast, many arthropods have specialized to particular
niches and environmental conditions that have necessitated changes in body
plan to accommodate severe physico-chemical situations; these accommo-
dations favourably influence preservation.

   The means by which a given organism made its way into a particular environment of death (preservation) is not a trivial consideration. Immigrants will
probably display a higher potential for preservation than will a transported
carcass, because the timing of burial following death is lengthened due to
transport and particular ecological adaptations may favour survival of the
immigrant, minimizing taphonomic destruction.

most crabs, lobsters and shrimp represent the most resistant sclerites owing to
their thickened calcitic structure (necessary for effective breakage and predation). Although the carapace and appendages may be disarticulated, dissolved and lost from the prospective fossil record, the chelae are almost
invariably present (North Sea siliciclastic sediments: Schäfer, 1972; siliciclastic sediments in the Miocene Chesapeake Group (Choptank Formation) of
the western shore of Chesapeake Bay, Maryland: personal observation).

   The preservation potential of arthropod remains is thus a consequence of
numerous, interrelated factors. These include: ecologic adaptation (taxa
inhabiting a particularly severe environmental setting will possess skeletal

attributes fit for survival); physiological innovations and variations (skeletal resorption significantly diminishes the preservation potential of moulted cuticle); ontogenetic maturity (immature crustaceans possess a poorly or non-calcified cuticle; skeletal strength increases as a function of body mass); and environment of death (environmental and biological constraints on cuticle preservation, dissolution and biodegradation) (see Figure 9.3).

Taphofacies and environmental inference

It is well known from an extensive literature that comparative taphonomy provides data useful in constructing taphofacies which, in turn, represent environmental settings. These taphofacies represent environment-proxies useful in evaluating the distributional biology of preserved assemblages while providing an independent index of bias and resolution. Taphofacies are facies descriptions that incorporate taphonomic data as the primary determinant in discriminating palaeoenvironmental settings. Speyer and Brett (1986a; 1986b) introduced this concept utilizing trilobite data from the Middle Devonian Hamilton Group of New York state. The decision to use trilobite data in this study was not altogether arbitrary. Trilobites are abundant in virtually all facies of the Hamilton Group, but, more importantly, trilobites possess numerous morphological and skeletal attributes that make them particularly well suited for comparative taphonomy. Multi-element skeletons (skeletons with numerous articulating and disarticulating parts) provide a particularly sensitive index of post-mortem disturbance that can be balanced against duration of surface exposure. In addition, the cephala and pygidia of many taxa react as do other concavo-convex elements in a current-dominated system, that is, assume a hydrodynamically stable orientation in response to persistent (laminar) current flow (see Speyer, 1987). Table 9.1 summarizes the variety of possible postures and taphonomic conditions that any given trilobite might assume during its taphonomic history.

The environmental distribution of fossil crustacean and, by inference, chelicerate remains, in contrast to the distribution of trilobite remains, is preservationally very restricted. The various peracarid crustaceans and other non-calcified groups are not preserved except under very unusual circumstances. Thus, these taxa represent an end-member condition opposite the trilobite, that of non-preservation (see Figure 9.4). Rathburn (1935), in her compendium, listed only one isopod genus and commented on the unlikelihood of preservation (see also Wieder and Feldman, 1989). The vast majority of 'complete' crabs, lobsters and related Decapoda are preserved in syngenetic concretions and nodules. These are typically carbonate, but are likely to represent a diversity of complex taphonomic and diagenetic histories. Speyer and Brett (1988a; in press; Speyer, in press) provide overviews of environmental situations and sedimentary circumstances that favour early diagenetic phenomena. In short, deep-water (or limited circulation) epeiric environments concentrate a variety of chemical species, including calcium (shelly remains) and phosphate (faecal pellets).

Table 9.1 *Modes of occurrence and taphonomic conditions among trilobite remains and assemblages (see Speyer and Brett, 1986a; Speyer, 1987)*

DISARTICULATED REMAINS

General: Disarticulated remains indicate periods of exposure associated with surface scavenging and/or current agitation. Post-burial disarticulation may result from intrastratal bioturbation.

MANNER OF OCCURENCE

1. Disarticulated skeletons with preferred convex-up orientations among concavo-convex elements.
2. Disarticulated skeletons with preferred concave-up orientations among concavo-convex elements.
3. Disarticulated skeletons show no preferred orientation.
4. Fragmentation of disarticulated sclerites.

TAPHONOMIC CONDITIONS

1. Persistent surface currents; not usually associated with burial.

2. Settling postures after rapid, turbid deposition; scavenger reorientation.

3. Rapid, mass deposition; long-term accumulation in low-energy environments.
4. Episodic or persistent agitation in high-energy environments.

ARTICULATED REMAINS

General: Articulated trilobite remains indicate that rapid burial and conservation from destructive agents (currents, bioturbation). Important index for recognizing event-deposited strata.

MANNER OF OCCURRENCE

1. Outstretched carcass remains.

2. Enrolled carcass remains.

3. Intact moult remains; moult ensembles and configurations.

TAPHONOMIC CONDITIONS

1. Rapid burial without pre-burial disturbance; overturned bodies indicate settling and/or behaviour.
2. Pre-burial disturbance (toxicity, turbidity) followed by deep burial.
3. Burial without current agitation; predominance indicates shallow burial.

OTHER CONSIDERATIONS

General: Various other conditions of preservation reflect specific taphonomic circumstances. These are, likewise, useful in evaluating the history of preservation of trilobite assemblages.

MANNER OF OCCURRENCE

1. Diagenetic mineralization; silica, pyrite, carbonate, phosphate.

2. Cuticle colour and thickness.

TAPHONOMIC CONDITIONS

1. Decay initiates chemical pathway according to background circumstances; indicates carcass remains.
2. Degree to which cuticle carbonate matrix has been dissolved and/or modified indicates ambient pH and level of carbonate saturation.

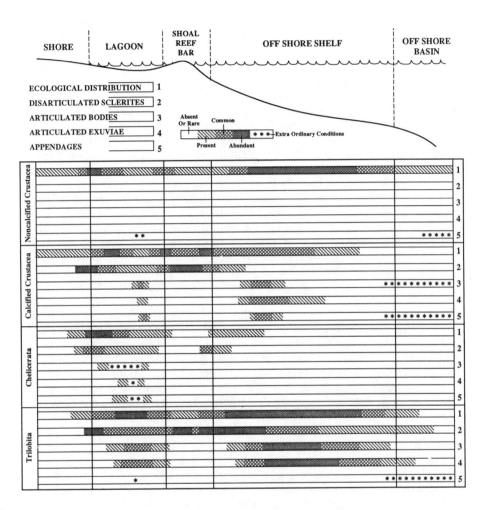

Figure 9.4 Comparative taphonomy of marine arthropod remains across a hypothetical series of environments. Preservation potential of arthropod remains varies according to environmental conditions (see Fig. 9.1), cuticle composition (see Fig. 9.2), and ecological distribution. The distribution of disarticulated sclerites, articulated bodies, articulated moult remains and the rare occurrence of preserved appendages correlates to particular environmental gradients. Therefore, fossil assemblages may be related to specific environmental settings on the basis of modes of occurrence of remains. Such comparisons provide data important in considerations of evolutionary taphonomy, megabias and other changes in ecological distribution and/or stratigraphic resolution that derive from important evolutionary innovations. (Based, in part, on Bowen *et al.*, 1979; Coull and Bell, 1983; Rudloe, 1979; 1980; Speyer and Brett, 1986a; 1986b.)

Waage (1964) and Bishop (1981; 1986) discussed differing models for the genesis of Cretaceous carbonate concretions containing extraordinary decapods (Fox Hills and Pierre Shale Formations, South Dakota). Both models, however, emphasize rapid burial, stratigraphic condensation along a sediment-starved substrate, organic concentration and depressed oxygenation. These conditions dictate a deep-water setting well below normal- and, perhaps, storm-wave bases. Rathburn (1935) provided an abbreviated listing of stratigraphic occurrences of fossil Crustacea discussed in her monograph. These details, coupled with brief discussions of distribution for each of her taxa, indicate that these examples were also preserved in concretionary layers or diagnosed on the basis of fragmentary material only (typically chelae remains). The predominance of examples from various Cretaceous and Eocene marls and clays corroborates the hypothesis that intact preservation required rapid burial and low-agitation conditions. Disarticulated and fragmented chelae and similarly robust skeletal remains occur in abundance in many other environmental facies, particularly where organics had not significantly concentrated, oxygen was apparently not limited, current agitation was persistent, and burrowers and predators were locally abundant. Figure 9.1 summarizes the distribution of fossil remains and modes of occurrence across an idealized mosaic of environmental and taphonomic settings.

An important concern in palaeoecology is the integrity of fossil remains present in a particular assemblage; that is, do the remains represent individuals that lived in the area of occurrence? This, of course, is the critical question at the foundation of any palaeoecological or palaeobiological study and has been addressed in numerous papers including those of Boucot (1953), Johnson (1960), Schäfer (1972) and Brouwer (1966), among many others. Westrop (1986), in particular, has considered the problem of palaeoecological disinformation due to size-segregated trilobite taxa. Larger taxa were preferentially concentrated in winnowed lags while smaller taxa were carried downslope and accumulated elsewhere, thereby dissociating an original 'biocoenosis' (*sensu* Wasmund, 1926). Similar problems may plague studies of population dynamics that rely upon relative size (instar) frequency data. Shifts in the relative proportion of a particular size class may well be the result of persistent or episodic current sorting. These effects, however, may be recognized and the amount of misinformation minimized providing a more comprehensive evaluation of taphonomic attributes. Speyer and Brett (1986a; 1988a) place a great deal of emphasis upon articulation of multi-element skeletons (for example, arthropods and echinoderms) as a sensitive measure of current-related bias, an idea corroborated by Fortey (1975) and Meyer *et al*. (1990). Long-term exposure in low-energy environmental settings, however, also results in an assemblage dominated by disarticulated remains. Surface features and relative skeletal corrosion provide measures of surface residence time and may facilitate a more complete understanding of conditions leading up to preservation (Brett and Baird, 1986; Speyer, 1987).

Disarticulated trilobite remains (typically cephala, pygidia and thoracic segments) offer a source of data that are conveniently converted into biostratinomic indices that assess the relative influence of various environmental parameters (for example, exposure time, current energy and tempo, and bioturbation; see Table 9.1). Speyer and Brett (1986a) used convexity and articulation indices to diagnose current and burial regimes among the mosaic of taphofacies they identified in Middle Devonian strata. These criteria, together with indices that measure relative abundance of articulated remains (moult versus enrolled or outstretched carcass), provided the basis for inferring gross environmental conditions. Given that specific taphonomic traits deductively invoke common sense relationships (for example, predominantly convex-upward orientations indicate persistent background currents; see Speyer and Brett, 1988a), opportunities exist for widespread comparisons of trilobite (or other group) taphonomy across varied environmental facies and among numerous stratigraphic and temporal situations.

Articulated trilobite remains also provide data important in evaluating the environmental setting. The simple preservation of articulated remains indicates special conditions, usually involving rapid, episodic burial (Speyer, 1987). Persistently high background sedimentation may also conserve remains from post-mortem disturbance and facilitate intact preservation. There are three basic ways in which articulated arthropods (trilobites) occur: as outstretched carcass remains; as enrolled carcass remains; or as intact moult remains (exuviae). Although these modes of occurrence represent specific biological aspects of the trilobite (discussed below), a predominance of one or another also suggests particular attributes of the sedimentary environment (see Table 9.1).

Distributional ecology and migration

Comparative taphonomy and environmental inference based upon taphofacies provides a powerful and informative tool for the study of community and population ecology. The most useful application of taphonomically differentiated environments can be said to lie in the identification of distributional patterns of fauna among facies delineated on the basis of modes of fossil preservation. Such studies enhance the credibility of so-called biofacies by establishing the integrity of deposits from which the fossils are taken.

Speyer and Brett (1988b) utilized a conservative taphofacies model to determine the environmental distribution of size among *Phacops rana* populations in the Middle Devonian Hamilton Group. Although data from one facies were discounted because of potential bias, taphonomic characteristics argued that data from all other facies were viable. Size-frequency comparisons among disarticulated cephala and pygidia were used to establish that sorting currents had little influence on the deposit taphonomy. The distribution of articulated remains further corroborated the integrity of studied deposits. Taphonomic data were used to interpret the environment of deposition and to argue the ecological integrity of studied trilobite assemblages. *Phacops* 'populations' show a variable size frequency that is predictably distributed among the environmental facies comprising the Hamilton Group. These organize into three separate and statistically distinct trilobite size-facies categories that suggest that *Phacops*, like many long-lived Crustacea and Chelicerata, migrate among environments during life.

'Evolutionary taphonomy' and megabias

Hesselbo (1987) recognized that comprehensive models regarding trilobite taphonomy and taphofacies, particularly those erected on the basis of sclerite biostratinomy, are hampered by varied morphologies present on trilobites, the result of differing taxonomic affinities and evolutionary circumstance. Therefore, it is not surprising that dorso-ventrally flattened *Dikelocephalus* sclerites show a biostratinomy that differs from that exhibited by strongly concavo-convex cephala and pygidia of *Phacops* (Speyer, 1987). This rather obvious point underscores the need to integrate morphologic form and structure with taphonomic process and pattern; this is one aspect of 'evolutionary taphonomy' discussed by Speyer (in press).

Figure 9.3, in part, summarizes the feedback relationship between form, function and biostratinomy. Simply stated, animals are well adapted for a particular niche defined by a specific range of interrelated environmental gradients. When an animal dies, the effects of environmental conditions on the remains are mediated, at least in part, by the structural compensations opted during life. Ironically, particularly specialized taxa will reflect taphonomies that corroborate their confined ecological distribution. More conservative taxa, displaying a broader ecological distribution, provide a particularly powerful tool in comparative taphonomy because the remains are distributed across a range of environmental conditions and display a correspondingly varied biostratinomic pattern.

In addition to the changes in taphonomic attributes that might be correlated to intrinsic factors (for example, morphological evolution), 'evolutionary taphonomy' also encompasses those changes in taphofacies resolution and identification that are due to changes in extrinsic phenomena such as bioturbation intensity or gross environmental character. Changes in the biosphere, the result of evolutionary innovation in a faunal constituent, may result in significant changes in the taphonomic signature which corresponds to the affected environmental system(s). Droser and Bottjer (1987) argued that bioturbation mode and intensity show marked shifts throughout the Phanerozoic; these shifts should have affected patterns of skeletal biostratinomy. Similarly, changes in the distribution of fauna and flora due to revolutions in particular modes of growth and styles of construction may significantly alter pre-existing environmental settings so as to generate new, but diagnostic, taphonomic signatures (for example, evolution of reefs, bioherms and similar organic build-ups).

Likewise, chemical and physiographic changes in a particular, perhaps generalized, environment will result in specific changes in the expression and distribution of diagnostic taphonomic signatures. Palaeogeographic and climatic constraints may result in particular periods of time characterized by broad preservational attributes. Behrensmeyer and Kidwell (1985) discussed 'megabiases' as global or widespread biases that are reflected in the stratigraphic record as enhanced preservational bias. The reverse is also likely true; that is, certain periods of time are typified by enhanced preservation potential (perhaps abundant and worldwide distribution of Konservat-Lagerstätten, as has been argued for the Middle Cambrian; see Conway

Morris, 1989). In order better to study these phenomena, it is imperative that broad taphonomic patterns be diagnosed and then evaluated in light of other, concurrent trends of global significance (such as particular sedimentologic, stratigraphic and climatic trends).

In short, skeletal morphology, evolutionary ecology (including moult physiology), and global patterns of climate and geographic change all bear independently and, to a degree, interdependently upon the temporal and spatial patterns of skeletal preservation (Figure 9.3). Trilobite remains provide a particularly informative focus for such research because they display a broad environmental and stratigraphic distribution, possess easily defined morphological innovations, possess tangible modern analogues with notable, but confined, differences, and are exceptionally abundant in most marine strata.

BEHAVIOURAL PALAEOBIOLOGY OF TRILOBITES

As previously stated, articulated arthropod remains indicate special circumstances of preservation and a particular taphonomic history. In addition, the ways in which articulated trilobite and other arthropod remains occur may offer otherwise untenable insights into aspects of organism behaviour and biology. These then can be represented as adaptive paradigms that may lie on an evolutionary continuum. This allows us to investigate the relationship between morphology and behaviour without being vulnerable to gross speculation; that is, our database lies in the tabulated matrix of observed manners of occurrence. These data, for example, permit a consideration of changes in moult and enrollment procedure, both among trilobite taxa and between higher taxa (for example, crustacean versus chelicerate versus trilobite). Thus, at one level we consider the phylogenetic context of behaviour and at another we may view the constructional constraints upon these behaviours.

The facies distribution of articulated arthropod (trilobite) remains may be widepread, but the abundance of such occurrences is certainly confined to a narrow range of stratigraphic facies (see Figure 9.4). Speyer and Brett (1986a) reported that even the most energetic of their inferred environments exhibit a taphofacies within which articulated trilobites and crinoids are present. Although the presence of articulated remains argues for episodically rapid burial, background reworking within normal- and subsequent storm-wave base diminishes the preservation potential of most buried carcasses and exuviae.

Speyer and Brett (1990) argued that, all other things being equal, biostratinomic preservation is maximally best in intermediate- to deep-water settings typified by episodically rapid burial followed by quiet post-burial conditions. In such a taphonomic setting, trilobite remains are abundantly articulated and occur in a diversity of ways that might correspond to biological and ecological behaviours.

Trilobites, like many modern marine arthropods, were vagrant and are likely to have exhibited a wide range of behaviours. Although many fixed and sessile organisms (such as crinoids) provide sensitive biostratinomic indices, the range of biologically relevant postures, attitudes and associations is severely limited. Arthropods (excepting the Cirripedes and parasitic groups) possess complex life histories and exhibit diverse behaviours that correspond to ecological and biological circumstances. Presumably, trilobites also exhibited a comparatively complex range of attitudes and behaviours. Frequency-of-occurrence data simply and straightforwardly indicate 'normal' and aberrant orientations. These, then, may be evaluated in light of modern arthropod phenomena, with the caveat that convergent form–function relationships are a sound basis for understanding fossilized postures and modes of occurrence.

Arthropods and, by inclusion, trilobites display several basic behaviours that are potentially represented in the fossil record: ecdysis (moulting); escape (including enrollment); social organization; and migration. A hierarchical organization of behavioural inferences according to reliability and viability of fossil data is provided in Boucot's (1990) comprehensive book. These range from incontrovertible evidence (so-called 'frozen behaviour') to entirely unreliable interpretations built solely upon speculation. This calibration of reliability is an important measure of confidence and makes us consider alternatives to our favourite models.

Ecdysis and moult behaviour

Trilobites, like most arthropods, moulted to accommodate changes in size and shape during development. Moulted exuviae and instar cuticle, however, are taphonomically undifferentiated, making it impossible to distinguish between carcass and moult remains in the fossil assemblage. This poses a unique problem for palaeoecological studies involving population estimates and individual counts. Quite simply, how many cephala represent individuals that died and how many represent moulting individuals that proceeded to the next age class (instar)? Speyer and Brett (1988a) reasoned that the total number of individuals represented by moulted and carcass remains provides a relative approximation of the number of individuals living at the time of assemblage formation. That is, the absolute number of living individuals and the relative number indicated by moult and carcass remains combined are not significantly different. Moreover, it is biologically well founded that arthropods decrease the number of moults per unit time through ontogeny. In other words, the young trilobite may moult frequently, but the older, aged individual may moult only rarely; the remains of the older segment of the population, therefore, are more likely to be represented by carcass remains than are the younger segments. These life-history patterns may be indirectly corroborated by size-frequency analysis of intact, undisturbed assemblages.

Moulting among trilobites and other fossil arthropods has received considerable attention (trilobites: see Henningsmoen, 1975; McNamara and Rudkin, 1984; Speyer, 1985; Whittington, 1990; other arthropods: see Schäfer, 1972; Bishop, 1986; Feldmann and Tshudy, 1987). Trilobites apparently did not resorb any portion of their mineralized cuticle prior to ecdysis and, therefore, moulted the entire calcified skeleton basically intact. The taphonomic consequence of this less than optimal physiology is enhanced preservation of exuviae. Differentiation among discarded exuviae and carcass remains, however, is tenuous and relies mostly upon biostratinomic differences among articulated remains.

Trilobite exuviae are identified on the basis of diagnostic orientations of sclerites; disarticulated remains are, thus far, left undifferentiated. Speyer (1990) discussed two basic modes of moult occurrence: taxa with fused cephala (for example, *Phacops*) dislocated the cephalon from the thoracopygidium and then discarded the two components separately (the occipital juncture, then, represents the point of emergence in most cases); and trilobites with articulated free cheeks appear to have dissociated the free cheeks from the cranidium and emerged through the resulting gap, without disrupting the cephalo-thoracic segment. Articulation of the thoracic assembly and the cephalo-thoracic segment is a strong argument for preservation prior to substantial disturbance.

Speyer and Brett (1986a) and Speyer (1985; 1987) discussed the taphonomy of deposits with a high relative proportion of moult remains (with respect to other types of articulated remains). These deposits display a wide range of taphonomic attributes, corroborating the idea that they represent long-term accumulations of moult remains; that is, those exuviae that were rapidly buried display considerable articulation, those that were exposed for long periods of time show little remnant articulation. Carcass remains (with all sclerites intact) are exceedingly rare, indicating that living individuals escaped from beneath thin sediment blankets. Speyer and Brett (1986a) noted that fixo-sessile taxa (arborescent tabulate corals, brachiopods, crinoids) in these same beds were preserved essentially intact. Logically, delicate articulations of crinoid columns would be entirely separated had currents or scavengers been significant taphonomic agents in this particular environmental facies. Therefore, it stands to reason that occurrences of articulated trilobite thoracopygidia (pygidium with attached and mostly complete thorax) must reflect a biological configuration and not a mechanical condition. Although conclusive, incontrovertible evidence is not likely to surface to confirm or disprove these inferences, taphonomic data provide powerful arguments regarding physico-chemical environmental conditions at the time and site of burial. Dislocation of a single dorsal joint (at the occipital suture or along the facial sutures) is unlikely to have been the result of indiscriminate scavenging or current disarticulation.

Enrollment and escape behaviours

Trilobite enrollment has been described for many groups, and enrollment-related structures (coaptative structures as per Clarkson and Henry, 1973) have figured importantly in systematic arguments and phylogenetic reconstruction (see Hammann, 1983; Henry, 1985). Interpretations of the biological and behavioural significance of trilobite enrollment rely largely upon morphologic, uniformitarian and taphonomic comparisons (see Speyer, 1988).

The most fundamental question concerns the relationship between enrollment, life and death of the trilobite. The tight integration of cephalic, pygidial and thoracic morphologies argues strongly against the notion that enrollment (in its strictest definition, with interlocked coaptative structures) was a general response to *rigor mortis* following death. However, it is clear that many trilobites died while enrolled; in many strata the number of enrolled trilobites exceeds the number of outstretched specimens (see Babcock and Speyer, 1987). This, again, is strong evidence that enrollment and death were biologically related. Moreover, certain taphofacies in Hamilton Group strata are characterized by an abundance of enrolled trilobites which display varying degrees of early pyrite diagenesis. Speyer (1987) related pyrite diagenesis to decaying softpart anatomy shortly after death; the enclosed nature of the microenvironment was theoretically critical to pyritization (see Dick and Brett, 1986; Brett *et al.*, 1990). Death, therefore, followed after enrollment.

The widespread distribution of enrolled trilobites within correlated strata places certain constraints upon our interpretation of enrollment-inducing phenomena (see Brandt Velbel, 1985; Speyer and Brett, 1986a; Babcock and Speyer, 1987). Although predation and the threat of predation are likely to have evoked enrollment, stratigraphic evidence for this correlation is absent; the enrolled trilobite either would have been eaten anyway or would have outrolled after evading predation. In either case preservation of the enrolled individual is unlikely. Natural experiments and observation indicate that the modern limulid *Xiphosura* enrolls as a generalized response towards environmental perturbations including toxic water conditions and increased turbidity, as well as predatory threat (see Fisher, 1977; Fields, 1982). In some instances, enrollment is associated with metabolic shutdown and anaerobiosis (see Fields, 1982). By uniformitarian analogy, it seems likely that trilobites also enrolled in response to widespread environmental catastrophes (such as storms and turbidity currents). Providing burial was sufficiently deep and/or event-related chemical changes persisted long enough, the trilobite would have died in an enrolled position (see Speyer, 1987; 1988).

Enrollment in fossil limulids has been reported by Fisher (1977) and Rolfe and Beckett (1985). Although most decapod crustaceans do not possess a body plan appropriate for enrollment, anaerobiosis has been documented and studied in lobsters, crabs and certain shrimps (see Brand, 1946; Bridges and Brand, 1980; McMahon and Wilkens, 1975; Oertzen, 1982). Enrollment and related responses to adverse chemical situations have also been documented in living and fossil oniscomorph millipedes (Hannibal and Feldman, 1981), isopods (Mead, 1978), and marine polychaetes and amphipods (Dean *et al.*, 1964). Although the phenomenon may be taxonomically and environmentally widespread, our fossil record is limited by preservation potential which, in turn, relates to cuticle composition and sedimentary conditions.

The natural taphonomic consequence of morphological and behavioural adaptations against adverse environmental conditions is prolonged lifespan; the simple fact that these adaptations were, in certain situations, unsuccessful (as indicated by the fossil record of enrolled limulaceans and trilobites) corroborates our interpretation of their significance (the exception proving the rule). The evolutionary significance of these developments may be indicated in a complexity of enrollment-related devices and behaviours versus taxonomic longevity correlation (presuming, of course, that functional morphology and enrollment procedure were tightly integrated as demonstrated in two phacopid taxa, *Greenops* and *Phacops*; see Speyer, 1988). Enrollment behaviour in these taxa was apparently a compromise between constructional morphology and varied functional constraints. Biostratinomic details provided an outline of two very different procedures that were best understood in terms of pronounced morphologic differences.

Phacopid trilobites exhibit a diversity of body plans (Baupläne) with a corresponding diversity of enrollment-related morphologies. However, it is clear that most, if not all, phacopid taxa enrolled and fostered the capacity for relatively complex interlocking structures (see Bergström, 1973). In constrast, asaphine, olenid and proetid trilobites all display enrollment, but show very limited diversification of enrollment-related structures. Although phacopid structures are well suited for maintaining an enrolled state (for example, vincular structures; see Speyer, 1988), these other groups are less likely to remain enrolled unless confining pressure by surrounding sediment ensures it. In fact, contraction across the dorso-ventral muscular system (see Campbell, 1975) may actually have favoured outrolling following death. Again, the construction of the organisms directly influences the preservation potential and affects the integrity of the fossil deposit. Enrollment-related morphologies and procedure directly affect the preservation potential of enrolled trilobites. Competent interlocking morphologies and/or intrastratal enrollment (cf. *Phacops*) would obviously have enhanced preservation potential, whereas less complex coaptation and/or surface enrollment (cf. *Greenops*) would have diminished preservation potential.

Social organization and gregarious behaviour

Extraordinary circumstances of preservation sometimes conserve extraordinary biological phenomena. Speyer and Brett (1985) discussed the preservation and interpretation of clustered trilobite assemblages which occur throughout Palaeozoic marine strata. Although the behavioural aspects of trilobite clusters may not be as comprehensive as the stratigraphic and taxonomic occurrence (see Brezinski, 1986), certain taphonomic criteria add credence to the thought that some represent gregarious associations. The precise nature of such gregarious associations, however, remains speculative, based almost entirely upon modern analogy.

Two types of clusters are present in the rock record; body clusters which comprise complete carcasses only, and moult clusters which are aggregates composed mostly of articulated moults (see also Henningsmoen, 1957). Speyer and Brett (1985) demonstrated that these two types of clustered assemblages co-occur along bedding planes in certain strata and, further, comprise size-frequency data that are statistically undifferentiated. These are taphonomic arguments that support *in situ* preservation and justify the exploration of behavioural correlations, between moult and body clusters. They reasoned that a genetic relationship existed, which commenced with synchronized ecdysis and, by uniformitarian inference, concluded with gregarious reproduction. Although largely built upon modern analogue comparison and, therefore, suffering from a lack of discreet, conclusive data, the relationship between moult clusters and body clusters (at least for the Hamilton Group examples) is firmly rooted in hard taphonomic data.

SUMMARY AND PROSPECTUS

The reliability of interpretations discussed in this chapter varies among examples according to the nature of the phenomenon investigated, the inherent preservation potential of the organism involved, and the range of environments across which the phenomenon and involved organism occur. The relationship among these variables is illustrated in Figure 9.4. It is important to understand that the resolution of our interpretations is predicated by numerous, interrelated factors. Speyer and Brett (1990) offered a calibration of palaeoecologic resolution according to the environmental distribution of obrution (rapid burial) deposits, but this general model is subject to the potential for preservation inherent in a particular skeletal type and composition. The typical trilobite possessed a robust, strongly calcitic exoskeleton composed of many parts (sclerites) which were readily disarticulated. Many trilobites indicate a generalized ecology and display a widespread environmental distribution. As such, the trilobite provides a convenient endmember against which the taphonomic integrity of deposits of other marine arthropods (such as decapod crustaceans and xiphosuran chelicerates) may be compared.

   Arthropod taphonomy, as illustrated by the taphonomy of trilobite remains, is an important and underutilized tool in the study of palaeobiological and palaeoecological questions. Comparative taphonomy of arthropod remains enhances our understanding of biomineralization patterns and cuticle composition through the extent of their fossil record. Furthermore, taphonomic studies provide important information regarding the diversity of behaviours and a source of data concerning constructional and functional attributes corresponding to those behaviours. These data ultimately provide new approaches to understanding the relationship between function and morphology in an evolutionary as well as ecological context.

REFERENCES

Aitken, D.E., 1980, Molting and growth. In J.S. Cobbs and B.F. Phillips (eds), *The biology and management of lobsters, Volume 1: Physiology and behavior*, Academic Press, New York: 91–163.

Aller, R.C., 1983, The importance of diffusive permeability of animal burrow linings in determining sediment chemistry, *Journal of Marine Research*, **41** (2): 299–322.

Allison, P.A., 1986, Soft-bodied animals in the fossil record: the role of decay in fragmentation during transport, *Geology*, **14** (12): 979–81.

Allison, P.A., 1988, The role of anoxia in the decay and mineralization of protein-aceous macro-fossils, *Paleobiology*, **14** (2): 139–54.

Babcock, L.E. and Speyer, S.E., 1987, Enrolled trilobites from the Alden Pyrite Bed, Ledyard Shale (Middle Devonian) of western New York, *Journal of Paleontology*, **61** (3): 539–48.

Barnes, R.D., 1980, *Invertebrate zoology*, Saunders College, Philadelphia.

Behrensmeyer, A.K. and Kidwell, S.M., 1985, Taphonomy's contribution to paleo-biology, *Paleobiology*, **11** (2): 105–19.

Benton, A.G., 1935, Chitinivorous bacteria, *Journal of Bacteriology*, **29** (4): 449–63.

Bergström, J., 1973, Organization, life and systematics of trilobites, *Fossils and Strata*, **2**: 1–69.

Bishop, G.A., 1981, Occurrence and fossilization of the *Dakotacancer* assemblage, Upper Cretaceous Pierre Shale, South Dakota. In J. Gray, A.J. Boucot and W.B.N. Berry (eds), *Communities of the past*, Hutchinson Ross, Stroudsburg, Pennsylvania: 383–414.

Bishop, G.A., 1986, Taphonomy of North American decapods, *Journal of Crustacean Biology*, **6** (3): 326–55.

Boucot, A.J., 1953, Life and death assemblages, *American Journal of Science*, **251** (1): 25–40.

Boucot, A.J., 1990, *Evolutionary paleobiology of behavior and coevolution*, Elsevier, Amsterdam.

Bowen, M.A., Smyth, P.O., Boesch, D.F. and Montfrans, J. van, 1979, Comparative biogeography of benthic macrocrustaceans of the middle Atlantic (U.S.A.) conti-nental shelf. In A.B. Williams (ed.), *Symposium on the composition and evolution of crustaceans in the cold and temperate waters of the World Ocean, Bulletin of the Biological Society of Washington*, **3**: 214–55.

Brand, T.F. von, 1946, Anaerobiosis in invertebrates, *Biodynamica Monographs*, **4**: 1–328.

Brandt Velbel, D., 1985, Ichnologic, taphonomic, and sedimentologic clues to the deposition of the Cincinnatian Shales (Upper Ordovician), Ohio, U.S.A. In H.A. Curran (ed.), *Biogenic structures: their use in interpreting depositional environ-ments, SEPM Special Publication*, **35**: 299–307.

Brett, C.E. and Baird, G.C., 1986, Comparative taphonomy: a key to paleoenviron-mental interpretation based on fossil preservation, *Palaios*, **1** (3): 207–27.

Brett, C.E., Miller, K.B. and Baird, G.C., 1990, A temporal hierarchy of paleo-ecologic processes within a Middle Devonian epeiric sea. In W. Miller III (ed.), *Paleocommunity temporal dynamics: the long-term development of multispecies assemblages, Paleontological Society Special Publication*, **5**: 178–209.

Brezinski, D., 1986, An opportunistic Upper Ordovician trilobite assemblage from Missouri, *Lethaia*, **19** (4): 315–25.

Bridges, C.R. and Brand, A.R., 1980, The effect of hypoxia on oxygen consumption and blood lactate levels in some marine Crustacea, *Comparative Biochemistry and Physiology*, **65** (A): 399–409.

Brouwer, A., 1966, *General palaeontology*, University of Chicago Press, Chicago.

Campbell, K.S.W., 1975, The functional morphology of phacopid trilobites: musculature and eyes, *Proceedings of the Royal Society of New South Wales*, **108**: 168–88.

Campbell, L.L. and Williams, O.B., 1955, A study of chitin-decomposing microorganisms of marine origin, *Journal of General Microbiology*, **5**: 894–905.

Clarkson, E.N.K. and Henry, J.-L., 1973, Structures coaptatives et enroulement chez quelques Trilobites Ordoviciens et Siluriens. *Lethaia*, **6** (2): 105–32.

Conway Morris, S., 1989, Burgess Shale faunas and the Cambrian explosion, *Science*, **246** (4928): 339–46.

Coull, B.C. and Bell, S.S., 1983, Biotic assemblages: populations and communities. In F.J. Vernberg and W.B. Vernberg (eds), *The biology of Crustacea, Volume 7: Behavior and ecology*, Academic Press, New York: 283–319.

Dalingwater, J.D. and Miller, J., 1977, The laminae and cuticular organization of the trilobite *Asaphus raniceps*, *Palaeontology*, **20** (1): 21–32.

Dalingwater, J.D. and Mutvei, J., 1989, Arthropod exoskeletons. In J.G. Carter (ed.), *Skeletal biomineralization: patterns, processes and evolutionary trends*, *American Geophysical Union, Short Course in Geology*, **5** (2): 83–96.

Dean, D., Rakin, J. and Hoffman, E., 1964, A note on the survival of polychaetes and amphipods in stored jars of sediment, *Journal of Paleontology*, **38** (3): 608–9.

Dick, V.B. and Brett, C.E., 1986, Petrology, taphonomy and sedimentary environments of pyritic fossil beds from the Hamilton Group (Middle Devonian) of western New York. In C.E. Brett (ed.), *Dynamic stratigraphy and depositional environments of the Hamilton Group (Middle Devonian) in New York state, part 1*, *New York State Museum Bulletin*, **457**: 102–38.

Droser, M.L. and Bottjer, D.J., 1987, Early Phanerozoic step-wise increase in bioturbation, *Geological Society of America Abstracts with Programs*, **19** (7): 647.

Dzik, J. and Lendzion, K., 1988, The oldest arthropods of the east European Platform, *Lethaia*, **21** (1): 29–38.

Efremov, J.A., 1940, Taphonomy: a new branch of paleontology, *Pan-American Geologist*, **74** (1): 81–93.

Feldmann, R.M. and Tshudy, D., 1987, Ultrastructure in cuticle from *Hoploparia stokesi* (Decapoda: Nephropidae) from the Lopez de Bertodano Formation (Late Cretaceous–Paleocene) of Seymour Island, Antarctica, *Journal of Paleontology*, **61** (6): 1194–1203.

Fields, J., 1982, Anaerobiosis in *Limulus*. In J. Bonaventura, C. Bonaventura and S. Tesh (eds), *Physiology and biology of horseshoe crabs*, Alan R. Liss, Inc.: 125–31.

Fisher, D., 1977, Mechanism and significance of enrollment in xiphosurans (Chelicerata, Merostomata), *Geological Society of America Abstracts with Programs*, **9** (7): 264–5.

Flessa, K.M. and Brown, T.J., 1982, Selective solution of macroinvertebrate calcareous hard parts: a laboratory study, *Lethaia*, **16** (2): 193–205.

Fortey, R., 1975, Early Ordovician trilobite communities, *Fossils and Strata*, **4**: 331–52.

Hammann, W., 1983, Calymenacea (Trilobita) aus dem Ordovizium von Spanien; ihre Biostratigraphie, Ökologie und Systematik, *Abhandlungen der Senkenbergisschen Naturforschenden Gesellschaft*, **542**: 1–177.

Hannibal, J.T. and Feldmann, R.M., 1981, Systematics and functional morphology of oniscomorph millipedes (Arthropoda: Diplopoda) from the Carboniferous of

North America, *Journal of Paleontology*, **55** (4): 730–46.

Henningsmoen, G., 1957, The trilobite family Olenidae, *Skrifter Utgitt av det Norske Videnskaps-Akademi i Oslo*, **1957** (1): 1–303.

Henningsmoen, G., 1975, Moulting in trilobites, *Fossils and Strata*, **4**: 179–200.

Henry, J.-L., 1985, New information on the coaptative devices in the Ordovician trilobites *Placoparia* and *Crozonaspis*, and its significance for their classification and phylogeny, *Transactions of the Royal Society of Edinburgh*, **76** (2/3): 319–24.

Hesselbo, S.P., 1987, The biostratinomy of *Dikelocephalus* sclerites: implications for the use of trilobite attitude data, *Palaios*, **2** (6): 605–8.

Johnson, R.G., 1960, Models and methods for analysis of the mode of formation of fossil assemblages, *Geological Society of America Bulletin*, **71** (7): 1075–86.

McAllister, J.E. and Brand, U., 1989, Primary and diagenetic microstructures in trilobites, *Lethaia*, **22** (1): 101–11.

McMahon, B.R. and Wilkens, J.L., 1975, Respiratory and circulatory responses to hypoxia in the lobster *Homarus americanus*, *Journal of Experimental Biology*, **62** (3): 637–55.

McNamara, K. and Rudkin, D., 1984, Techniques of trilobite exuviation, *Lethaia*, **17** (2): 153–73.

Mead, P.F., 1978, Analyse comparative de la séquence sexuelle chez les Isopodes Onniscoidea de la série lignée relation avec le développement de la volvation chez les Armadillidiidae et les Armadillidae, *Zeitschrift für Tierpsychologie*, **46** (1): 30–42.

Meyer, D.L., Ausich, W.I. and Terry, R., 1990, Comparative taphonomy of echinoderms in carbonate facies: Fort Payne Formation (Lower Mississippian) of Kentucky and Tennessee, *Palaios*, **4** (6): 533–52.

Neville, A.C., 1975, *Biology of the arthropod cuticle*, Springer-Verlag, Berlin.

Oertzen, J.-A. von, 1982, A comparative study of the respiratory responses of *Pomatoschistus microps* (Kroyer) and *Palaemon adspersus* (Rathke) to declining oxygen tension, *Ophelia*, **21** (1): 65–73.

Plotnick, R.E., 1986, Taphonomy of a modern shrimp: implications for the arthropod fossil record, *Palaios*, **1** (3): 286–93.

Plotnick, R.E., 1990, Paleobiology of arthropod cuticle, In D.G. Mikulic (ed.), *Arthropod paleobiology*, *Short Courses in Paleontology*, Paleontological Society, University of Tennessee, Knoxville: 177–96.

Plotnick, R.E., Baumiller, T. and Wetmore, K.L., 1988, Fossilization potential of the mud crab, *Panopeus* (Brachyura: Xanthidae) and temporal variability in crustacean taphonomy, *Palaeogeography, Palaeoclimatology, Palaeoecology*, **63** (1–3): 27–44.

Rathburn, M.J., 1935, Fossil Crustacea of the Atlantic and Gulf Coastal Plain, *Geological Society of America Special Paper*, **2**: 1–160.

Rolfe, W.D.I. and Beckett, E., 1985, Autecology of Silurian Xiphosurida, Scorpionida, Cirripedia, and Phyllocarida. In M.G. Bassett and J.D. Lawson (eds), *Autecology of Silurian organisms*, *Special Papers in Palaeontology*, **32**: 27–37.

Rudloe, A., 1979, *Limulus polyphemus*: a review of the ecologically significant literature. In E. Cohen (ed.), *Biomedical applications of the horseshoe crab (Limulidae)*, Alan R. Liss, Inc.

Rudloe, A., 1980, The breeding behavior and patterns of movement of horseshoe crabs, *Limulus polyphemus*, in the vicinity of breeding beeches in Apalachee Bay, Florida, *Estuaries*, **3** (2): 177–83.

Schäfer, W., 1972, *Ecology and palaeoecology of marine environments*, University of Chicago Press, Chicago.

Seilacher, A., 1973, Biostratinomy: the sedimentology of biologically standardized particles. In R. Ginsburg (ed.), *Evolving concepts in sedimentology, Johns Hopkins University Studies in Geology*, **21**: 159–77.

Sepkoski, J.J., Jr., 1984, A kinetic model of Phanerozoic taxonomic diversity. III. Post-Paleozoic families and mass extinction, *Paleobiology*, **10** (2): 246–67.

Speyer, S.E., 1985, Moulting in phacopid trilobites, *Transactions of the Royal Society of Edinburgh*, **76** (2/3): 239–54.

Speyer, S.E., 1987, Comparative taphonomy and palaeoecology of trilobite Lagerstätten, *Alcheringa*, **11** (2): 205–32.

Speyer, S.E., 1988, Biostratinomy and functional morphology of enrollment in two Middle Devonian trilobites, *Lethaia*, **21** (1): 121–38.

Speyer, S.E., 1990, Trilobite moult patterns. In A.J. Boucot, *Evolutionary paleobiology of behavior and coevolution*, Elsevier, Amsterdam: 491–7.

Speyer, S.E., in press, Comparative taphonomy and taphofacies: taphonomic approaches to marine paleoecology, *Acta Geologica Hispanica*.

Speyer, S.E. and Brett, C.E., 1985, Clustered trilobite assemblages in the Middle Devonian Hamilton Group, *Lethaia*, **18** (2): 85–103.

Speyer, S.E. and Brett, C.E., 1986a, Trilobite taphonomy and Middle Devonian taphofacies, *Palaios*, **1** (3): 312–27.

Speyer, S.E. and Brett, C.E., 1986b, Hamilton trilobite assemblages; biofacies patterns and inferred environmental controls, *Geological Society of America Abstracts with Programs*, **18** (1): 68.

Speyer, S.E. and Brett, C.E., 1988a, Taphofacies models for epeiric sea environments: Middle Paleozoic examples, *Palaeogeography, Palaeoclimatology, Palaeoecology*, **63** (1–3): 225–62.

Speyer, S.E. and Brett, C.E., 1988b, Paleobiology of size variation in Middle Devonian trilobites, *Geological Society of America Abstracts with Programs*, **20** (7): 47.

Speyer, S.E. and Brett, C.E., 1990, Environmental distribution of marine obrution deposits: facies models and paleoecological resolution, *Geological Society of America Abstracts with Programs*, **22** (7): 81.

Speyer, S.E. and Brett, C.E., in press, Taphofacies controls: background and episodic processes in fossil assemblage preservation. In P.A. Allison and D.E.G. Briggs (eds), *Taphonomy: releasing the data locked in the fossil record*, Plenum, London.

Tiegler, D.J. and Towe, K., 1975, Microstructure and composition of the trilobite exoskeleton, *Fossils and Strata*, **4**: 137–49.

Towe, K.M., 1973, Trilobite eyes: calcified lenses in vivo, *Science*, **179** (4077): 1007–9.

Towe, K.M., 1975, Do trilobites have a typical arthropod cuticle?, *Palaeontology*, **21** (2): 456–61.

Waage, K.M., 1964, Origin of repeated fossiliferous concretion layers in the Fox Hills Formation (Cretaceous, S.D.), *Kansas Geological Survey Bulletin*, **169**: 541–63.

Wainwright, S.A., Biggs, W.D., Currey, J.D. and Gosline, J.M., 1976, *Mechanical design in organisms*, Princeton University Press, Princeton, New Jersey.

Wasmund, E., 1926, Biocönose und Thanatocönose, *Archive Hydrobiologie*, **1**: 1–117.

Westrop, S.R., 1986, Taphonomic versus ecologic controls on taxonomic relative abundance patterns in tempestites, *Lethaia*, **19** (2): 123–32.

Wieder, R.W. and Feldman, R.M., 1989, *Palaega goedertorum*, a fossil isopod (Crustacea) from Late Eocene to early Miocene rocks of Washington state, *Journal of Paleontology*, **63** (1): 73–80.

Whittington, H.B., 1977, The Middle Cambrian trilobite *Naraoia*, Burgess Shale,

British Columbia, *Philosophical Transactions of the Royal Society of London*, **B280**: 409–43.

Whittington, H.B., 1985, *Tegopelte gigas*, a second soft-bodied trilobite from the Burgess Shale, Middle Cambrian, British Columbia, *Journal of Paleontology*, **59** (6): 1251–74.

Whittington, H.B., 1990, Articulation and exuviation in Cambrian trilobites, *Philosophical Transactions of the Royal Society of London*, **B299**: 27–46.

Wilmot, N.V. and Fallick, A.E., 1989, Original mineralogy of trilobite exoskeletons, *Palaeontology*, **32** (2): 297–304.

Chapter 10

ECTOCOCHLEATE CEPHALOPOD TAPHONOMY

William B. Boston and Royal H. Mapes

INTRODUCTION

Taphonomic studies of ectocochleate (externally shelled) cephalopods (nautiloids, ammonoids and bactritoids) have primarily dealt with the biostratinomic role of post-mortem drifting on the distribution of the shelly remains of these organisms outside of the geographic range and biologic habitat of the living populations. Most of these earlier studies attempted to ascertain the buoyancy parameters of the empty conch so as to establish a baseline susceptibility to post-mortem drift. While such studies allow a first approximation of the taphonomic pathways open to an individual conch, they generally do not assess the importance of the variety of physical and biological events that can modify the buoyancy parameters. Modification of initial conch buoyancy results in the occurrence of a majority of specimens of any taxon within the habitat of the living populations with minimal taphonomic transport. The occurrence of individual specimens of a taxon in taphofacies widely variant from the habitat of the living populations results from extensive taphonomic transport.

Analysis of an extensive collection of fossil ectocochleate cephalopod specimens from the Upper Palaeozoic strata of North America, collected primarily by R.H. Mapes, has provided the basis for the interpretation of the occurrence pattern for these fossils. The variable nature of the occurrence pattern appears to be a result of a variety of biostratinomic processes that have affected the various ectocochleate cephalopod morphoforms. This pattern contains extensive occurrences of ectocochleate cephalopod remains in lithic units that are interpreted as having been deposited in deeper-water epicontinental seas, and very localized occurrences of ectocochleate cephalopod remains in lithic units interpreted as very shallow-water deposits. The intermediate water-depth lithologies contain a modest number of ectococh-leate cephalopod occurrences scattered more or less randomly throughout the units.

MEANS OF INQUIRY

Examinations of ectocochleate cephalopod taphonomy can be divided into two general methods which utilize separate approaches and databases. The first method involves the study of modern *Nautilus* biology and related taphonomic processes, some of which are directly observable, that effect the distribution of *Nautilus* remains (see Saunders and Landman, 1987, for the most recent and complete summary available). The results of these studies can then be used as an analogue for the biologic and taphonomic processes that affected the distribution of extinct ectocochleate cephalopod remains. Such studies have come under critical questioning due to presumed disparities between *Nautilus* and the extinct forms of ectocochleate cephalopods.

The second generalized method utilizes a database acquired from the study of fossilized ectocochleate cephalopod conchs. This second means of inquiry also includes depositional and palaeoecological interpretations obtained from the strata from which these fossils were obtained. Calculations and experi-mental modelling of the buoyancy parameters of the fossil ectocochleate cephalopod conchs have been based on biometric data obtained from fossil-ized specimens.

Both of these methods have identified biostratinomic transportation, usually post-mortem drifting, as a primary controlling factor on the tapho-nomy of ectocochleate cephalopods, and the physical effects of the various shell morphologies on conch buoyancy conditions as being a limiting factor on biostratinomic transport. Additionally, physical events such as water infiltration and septa implosions, as well as biotic interactions such as dura-phagous predation and epizoan encrustations, have been identified as buoyancy-modifying processes which can effect biostratinomic transport and, hence, the taphonomic distribution of ectocochleate cephalopod remains.

Unfortunately, assessment of the taphonomic pathways open to these fossils is complicated by a lack of consensus on some of the important elements of the biostratinomic processes. These areas of contention include the buoyancy conditions of some of the extinct conch morphoforms, the application of the *Nautilus* analogue and buoyancy modification processes.

THE MODERN *NAUTILUS* ANALOGUE

The investigation of the taphonomy of extinct ectocochleate cephalopods has relied heavily on the utilization of modern *Nautilus* as an analogue. This reliance is based on the fact that *Nautilus* is one of only two extant forms of cephalopod with an exoskeletal shell, the other being *Argonauta* which develops an unseptate, paper-thin shell only during reproduction. *Spirula*, a form of squid with an internal shell, has also been utilized as an analogue for the extinct ectocochleate cephalopods because, upon removal of the soft tissues, the internal septate shell is similar to an ectocochleate cephalopod conch (Donovan, 1989).

The *Nautilus* conch is a relatively thick-walled, multi-chambered shell which the animal retains throughout its life. Thus, despite the unsatisfactory nature of *Nautilus* as an analogue for extinct ectocochleate cephalopod forms due to disparities in constructional morphology, *Nautilus* remains our extant model of last resort.

Nautilus buoyancy

Experimental *Nautilus* studies utilizing empty conchs, living animals and mathematical calculations based on conch biometrics have established the buoyancy parameters of *Nautilus* (Greenwald and Ward, 1987; Chamberlain, 1987; Ward, 1987). These studies have concluded that attainment of near-neutral buoyancy is essential for *Nautilus* to maintain its nektonic habits. *Nautilus* is an inefficient swimmer which utilizes a jet of water to propel itself (Chamberlain, 1987, among others). The jet of water is forced from the soft tissue through the hyponome, causing the animal to move in the opposite direction. Should a condition of excessive positive buoyancy occur, the animal would float nearly helplessly to the surface. A condition of excessive negative buoyancy would leave the *Nautilus* stranded on the sea floor, too heavy to be moved by the force of its jet propulsion. Studies of *Nautilus* have determined that the soft tissue and shell material are negatively buoyant (Chamberlain, 1987, p. 499). *Nautilus* counters this by displacing a volume of water with air which is contained within the chambers of the conch. The air within the conch is maintained at slightly less than 1 atm pressure and is less dense than the displaced volume of water. The air-filled conch is, therefore, positively buoyant and counteracts the negative buoyancy of the shell material and soft tissues. Chamberlain (1987) ascribed 90 per cent of the total dead weight of *Nautilus* to the shelly material. Variable fluid levels in the youngest chambers are used to fine-tune the buoyancy parameters to an approximately neutral level. These are varied by addition and removal of fluid within the chambers through the permeable connecting rings that make up part of the siphuncle, and which are connected to a wicking material which coats the chambers (Greenwald and Ward, 1987). These buoyancy conditions provide a basic framework for the analysis of taphonomic pathways available to *Nautilus* conchs.

Nautilus taphonomy

Upon removal of the soft tissue of *Nautilus* through the processes of putrefaction, predation or scavenging, which usually occurs within hours of the death of the animal (Chamberlain *et al.*, 1981), an undamaged conch will be positively buoyant and attempt to rise to the surface. This effect will be greatly enhanced if a portion of the body chamber has been removed, possibly through predation (Ward and Greenwald, 1982). Water will slowly flood the air-filled chambers of a conch, primarily through the siphuncle (Chamberlain *et al.*, 1981). The flooding of the buoyancy chambers will lower the buoyancy and, if sufficient water enters the conch, it will become negatively buoyant. A negatively buoyant conch will sink to the sea floor and may become incorporated in the sediment of the lithofacies and biofacies where the sinking takes place. Chamberlain *et al.* (1981) have calculated that the conch of a *Nautilus* that died in water deeper than 200–300 m, which lies within the interpreted preferred depth range of *Nautilus* (Saunders and Ward, 1987), will not reach the surface. This is due to water infiltrating the buoyancy chambers sufficiently to establish a condition of negative buoyancy. Extensive taphonomic transport by post-mortem drift is thus a rarity. Reyment (1958) has suggested that the sinking conchs may encounter a higher density layer within the water column that will impede the sinking and allow additional transport of the conch by current at depth. Observations of numerous *Nautilus* conchs at depths between 300 and 900 m, within and below the normal depth range of the *Nautilus* (Saunders and Ward, 1987 p. 161) during dives conducted during the CALSUB cruise off New Caledonia in 1989 (Roux, 1990) suggests that extensive taphonomic transport has not affected these specimens. The number of chambers that must be flooded to result in the conversion from a condition of positive to one of negative buoyancy is variable, depending on, for example, the maturity of the specimen, the condition of the body chamber and the physical integrity of he phragmocone (Chamberlain *et al.*, 1981).

The rare positively buoyant conch that has risen to the surface will be affected by wind and surface currents. These conchs may be incorporated into very shallow-water sediments if they strand at the shoreline, explaining the occurrence of *Nautilus* specimens in beach environments.

These analyses fail to deal with the effects of a variety of physical events and biotic interactions other than simple water infiltration through the siphuncle which may modify the buoyancy parameters of the conchs before, after or at the time of death of the *Nautilus*. This modification of buoyancy parameters results in additional taphonomic pathways which any conch can follow. Many of these same buoyancy modifying events and interactions appear to have affected the taphonomic pathways of fossil ectocochleate cephalopods as well.

FOSSIL ECTOCOCHLEATE CEPHALOPOD BUOYANCY STUDIES

The buoyancy parameters of fossil ectocochleate cephalopod conchs have generally been estimated by uniformitarian application of the study of *Nautilus* to the biometrics obtained from the variety of fossil shell forms. Analysis of the initial, unmodified buoyancy parameters of the various conch morphologies of extinct ectocochleate cephalopods is an important first step in the analysis of the various taphonomic pathways that were open to the shelly remains of these organisms. This is because a positively buoyant conch will start to rise to the surface and drift, while a negatively buoyant conch will begin to sink. The various fossil conch morphologies exhibit different initial buoyancy conditions. Study of the buoyancy parameters of extinct conch morphologies has been based on detailed study of fossilized conch material, mathematical calculations of buoyancy and limited attempts to model the extinct forms with materials that simulate the fossilized shells. These attempts to determine the flotational characteristics of these shell forms have been utilized as a proxy for differential susceptibility to nekroplanktonic dispersal.

Palaeontologists have assumed that the soft tissues and shell material of fossil cephalopods were negatively buoyant, and gas within the chambered phragmocone was used to offset this and obtain a condition of near-neutral buoyancy (Figure 10.1). This assumption mandates a nektonic habit, and points to a discrepancy between *Nautilus* and the fossil forms, in that both planktonic and benthic habits have been proposed for some forms of fossil ectocochleate cephalopods. The utilization of a planktonic habit would infer a condition of enhanced positive buoyancy, possibly through increased conch buoyancy, which would increase the propensity toward nekroplanktonic dispersal of the shelly remains. A benthic habit would infer a condition of enhanced negative buoyancy, possibly through decreased conch buoyancy, and, thus, a propensity away from nekroplanktonic dispersal of the shelly remains.

Further, the application of the results of taphonomic studies of *Nautilus* cannot be uniformly applied to the different constructional morphologies of the fossil cephalopods. These differences include septal spacing, folding, and mass, siphuncle construction, internal deposits (both cameral and siphuncular), external ornamentation and differential conch forms, such as orthocones, brevicones and those of the heteromorphic ammonoids. Nevertheless, some basic conclusions have been drawn about the buoyancy parameters of the various fossil ectocochleate cephalopod conch morphologies based on the biometrics and calculations obtained from the variety of their forms. This is because shell shape provides some buoyancy limitations and the size of the siphuncle tends to influence these parameters in conchs with similar morphologies. In those forms with similar shell shape and siphuncle size, internal deposits will significantly influence the buoyancy parameters of the conchs.

However, field-based occurrence data are in conflict with some of the theoretical flotation characteristics based on buoyancy calculations and experimental modelling of the various conch forms. This illustrates that factors other than initial conch buoyancy conditions affect the taphonomic distribution of the shelly remains of these organisms.

One long-standing problem which complicates the analysis of ectocochleate cephalopod taphonomy is that some of the taxonomic classifications of the various fossil cephalopods contain multiple conch morphologies. This necessitates that buoyancy studies be based on major conch morphologies, primarily overall shell shape and the general relative size of the siphuncle, rather than the buoyancy of the various taxonomic groups.

Planarspiral forms

This morphoform contains the vast majority of the nautiloids and ammonoids which, like *Nautilus*, have a planarspiral coil. None of this group contain internal deposits which negatively affected the buoyancy parameters of the

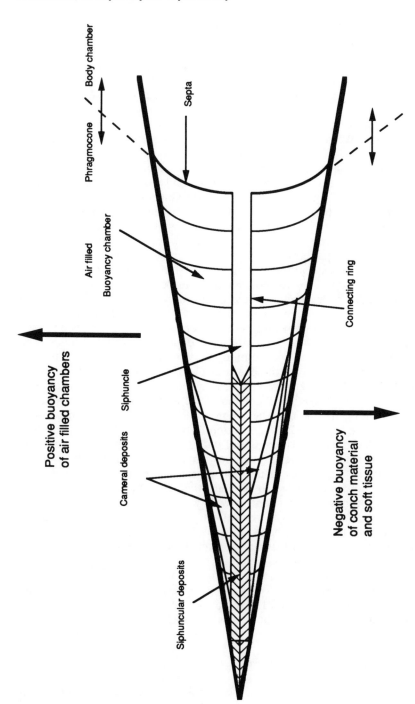

Figure 10.1 Morphology of an ectocochleate cephalopod: the siphuncle can be expanded, the deposits removed and the phragmocone coiled in variations of this basic plan.

conch. This has resulted in the direct application of the results of *Nautilus* taphonomic studies to the taphonomy of these fossil planarspiral forms, despite some pronounced differences between some of the fossil forms and *Nautilus*. While it is possible to assess the impact of any one or possibly several of these different variables acting in concert to affect the buoyancy conditions of the conch, the total combined impact of these variables is unknown.

One such difference is the type of material which makes up the siphuncle connecting ring. While the attributes of the various proposed compositions of the connecting ring material in the fossil ectocochleate cephalopods are the subject of continuing debate, the impact of the permeability of these materials is generally agreed upon. Decreased permeability of the connecting rings should have allowed the fossil conchs to maintain positive buoyancy longer than *Nautilus*, due to the slower rate of water entry into the buoyancy chambers. Conversely, increased permeability would decrease the relative ability of the fossil forms to maintain a condition of positive buoyancy.

The rate of water penetration into the buoyancy chambers of the ammonoids would be decreased by the higher calculated rupture strength of their connecting rings compared to those of the nautiloids (Westermann, 1973). These calculations have concluded that implosion would have occurred at far shallower depths in Palaeozoic nautiloids (a few tens of metres) than in the Palaeozoic ammonoids (a few hundreds of metres). This has also been taken to imply that the ammonoids lived at greater depths than the nautiloids. The greater depth tolerance of the ammonoids would have allowed a greater length of time to elapse from the death of the specimen until it floats to the surface. This would consequently have allowed time for the infiltration of more water into the buoyancy chambers of the ammonoids compared to the shallower-dwelling nautiloids. The ammonoids are also calculated to have had more relative septal mass. This is due to the increased number of septa associated with the increased number of phragmocone whorls. The increased relative septal mass and the larger body chamber negatively affected the buoyancy parameters of the conchs. The ammonoid conch wall was thinner and possibly more susceptible to damage than the thicker-walled shell of *Nautilus*. The thin ammonoid conch wall was, however, lighter than the thicker nautiloid conch wall. The ventral marginal placement of the siphuncle in the ammonoids might have been more susceptible to damage and thus may have increased the probability of buoyancy chamber flooding when compared with the central position of the siphuncle in most nautiloids. Despite these complications, and others, including the results of shell coiling compression, the planarspiral coiled morphoforms are generally viewed as being initially post-mortemly positively buoyant based on studies of *Nautilus* and modelling studies of an ammonoid conch form by Reyment (1980).

Straight and slightly curved shells with relatively large siphuncles

The members of this morphotype are members of the endoceratoid and actinoceratoid nautiloid lineages. The conchs are straight to slightly curved and all contain large siphuncles. The size of the siphuncle is very important to the flotational characteristics of such forms because a large siphuncle served to reduce the effective volume of the buoyancy chambers and acts as a more efficient conduit of water to these. Many of the cephalopods with this conch morphology utilized siphuncular and/or cameral deposits for what has been interpreted as ballast within the apical end of the phragmocone. These deposits would have allowed the animal to maintain a horizontal, rather than a vertical, orientation in the water (Furnish and Glenister, 1964). The internal deposits developed by this and other similar morphoforms negatively affected the buoyancy conditions of the conch. Thus, specimens of this morphoform with no siphuncular deposits would be marginal floaters due to the extensive volume of the siphuncle, which was relatively easy to flood, as well as the presumed relatively direct access of water to the buoyancy chambers. Those taxa that had siphuncular deposits may have floated somewhat better, because of the lack of easy water egress into the buoyancy chambers in the apical end of the phragmocone, despite the weight of these deposits. Those specimens with both siphuncular and cameral deposits seemingly would not float due to the greatly increased negative buoyancy of the conch that they caused. These flotation conclusions are drawn in part from work by Reyment (1958) involving plastic models of endocerid-like shells which were used to try to discern the floating characteristics of these shell forms.

The Reyment (1958) model suggested that the phragmocone of these forms (without body chamber material) floated low in the water. When Reyment added material to the chambers within the models to simulate cameral deposits, they sank, provided the large siphuncle was open to water influx. If the siphuncle was sealed so as to inhibit the influx of water the models would float. Reyment (1958) then added material to simulate the body chamber; the models sank when this additional weight exceeded 26 per cent of the length of the phragmocone. Based on these model experiments, it would seem that naturally occurring specimens with both siphuncular and cameral deposits must have been negatively buoyant (Teichert, 1964), especially if the body chambers were complete.

Straight or slightly curved shells with relatively small siphuncles

This morphoform primarily includes members of the orthocerid nautiloids, bactritoids and the straight-shelled ammonites. It also includes the deciduous apical phragmocone sections of some of the brevicone nautiloid forms.

The bactritoid and straight ammonite members of this morphoform would have been very buoyant due to a virtually complete lack of internal deposits within the phragmocone and also the narrow siphuncle, which restricted water influx to the buoyancy chambers. This resulted in an extremely buoyant phragmocone and has lead some authors to postulate a planktonic lifestyle for these forms.

The orthocerid nautiloids had a somewhat more varied siphuncular diameter and position, and almost invariably contained extensive siphuncular and cameral deposits. While these forms may have initially floated, even a small amount of water infiltration into the reduced number of buoyancy chambers was critical because of the mass of the deposits in the apical chambers.

The apical ends of some of the breviconic nautiloid forms were negatively buoyant because all of the chambers were deposit-filled. A physical separation of the deposit-filled apical end of an orthocerid phragmocone would have produced a physically similar fragment. These deposit-filled forms acted as a solid tube of calcite and/or aragonite which would have been extremely negatively buoyant. Some of the Palaeozoic orthocerid forms have been interpreted as hardly being able to float, based in part on the occurrence of these massive internal deposits (Teichert, 1964).

Reyment (1958) constructed a plastic model of this general morphoform to test the floatational calculations. The application of these conclusions are somewhat problematic in that Reyment (1958) did not report adding mass to the apical end of the phragmocone of his orthocerid-like model, which would have simulated the presence of the internal deposits. Reyment (1958, p. 120) pointedly indicated that he did add such weight to the endocerid shell models to simulate their internal deposits. Reyment (1958) also stated that the body chamber portion of the orthocerid-like model had to exceed approximately 50 per cent of the length of the phragmocone to cause the shell to sink, indicating considerable buoyancy within the phragmocone. The floating position of the Reyment (1958) orthocerid model was one of apex-up, which is inconsistent with the presence of deposit-filled apical chambers.

Brevicone shells

These forms include members of the several groups of nautiloids which have wide body chambers with a variety of forms and buoyancy-chamber configurations. The large mass of the body-chamber material and the small amounts of buoyancy-chamber volume severely limited the attainment of positive buoyancy. However, the constricted aperture on some brevicones has led some previous workers to invoke a planktonic lifestyle for these forms. A planktonic habitat would require a very buoyant conch leading to a high nekroplanktonic dispersal potential. Nekroplanktonic dispersal would have been enhanced by the very shallow depth in the water column at which the mortality of planktonic specimens would occur. However, as Teichert (1964) pointed out, the addition of ears and lobate projections to the aperture would have adversely affected buoyancy. A planktonic lifestyle would be limited to those forms which would have shed the deciduous deposit-filled apical end of the phragmocone. Those forms which did not shed the apical section may have been benthic and certainly would not have floated.

Helically coiled and other heteromorphic ammonoid shells

This morphoform includes some of the heteromorphic ammonites and rarer nautiloid forms. The interpretation of the buoyancy parameters of this morphoform is uncertain, because their interpreted lifestyle is controversial, planktonic and benthic habits having been invoked. The attendant interpretations of the buoyancy parameters and flotational conditions have been equally varied. The primary problem is the apparent improbability of a nektonic habit for this morphoform, because of the rotational forces produced by the eccentric hyponomically delivered water jet. Despite these interpretational problems, the conchs of this morphoform, lacking deposits, should have been buoyant enough to be nekroplanktonic if mortality occurred in water depths sufficient shallow for the conch to reach the surface before the flooding of the buoyancy chambers.

BUOYANCY-MODIFYING CONCH MORPHOLOGY CONDITIONS

It was noted by Reyment (1958) that, within the various cephalopod shell morphologies, physical conditions existed that would have directly influenced the calculated buoyancy conditions of the conch.

Presence of body-chamber material

Body-chamber material is dead weight and detrimental to the attainment of positive buoyancy (Reyment, 1958). Trueman (1941) stated that the body chamber of *Dactylioceras* weighed two to three times as much as the phragmocone. The modelling experiments by Reyment (1958) amply illustrated that the presence of body-chamber material contributed to negative buoyancy of the conch and was therefore a negative factor in the propensity towards nekroplanktonic dispersal. Chamberlain *et al.* (1981) attributed the occurrence of *Nautilus* specimens in beach environments and floating on the surface to the loss of body-chamber material, possibly due to predation.

Presence of ornamentation

Reyment (1958) indicated that the presence of ornamentation on the shell of ectocochleate cephalopods influenced buoyancy parameters. If the ornamentation was solid and thick-walled, it increased the weight of the conch and produced decreased buoyancy. If ornamentation was thin-walled and hollow, it would have increased the positive buoyancy of the conch by increasing buoyancy-chamber volume.

BUOYANCY-MODIFYING EVENTS AND INTERACTIONS

In addition to water infiltration, decreased buoyancy did, and will, result if the physical integrity of a conch was violated by either physical or biotic interactions. While the physical events and biotic interactions which modify buoyancy could occur in a variety of environments, the facies-related occurrences of these fossils seem to indicate that the majority of individuals inhabiting a specific environment suffered the same taphonomic history (Scott, 1940; Boardman *et al.*, 1984).

Physical events

The physical events, other than simple buoyancy-chamber flooding, interpreted to be buoyancy-modifying events include implosion at depth, heating and cooling of a floating conch and damage by physical abrasion or chemical dissolution. All of these events could have resulted in the flooding of the buoyancy chambers and the attainment of negative buoyancy.

Implosion at depth

Calculation and experimentation utilizing *Nautilus* indicate that there is a maximum depth at which septa will withstand hydrostatic pressure (about 750 m; Ward, 1987). Below this depth the septa fail and the conch will implode. The catastrophic failure of the conch results in negatively buoyant shelly fragments which will sink and be deposited below the site of implosion. Exceeding the implosion depth could result from the sinking of an ectocochleate cephalopod before the loss of the soft tissue while overall negative buoyancy was maintained. Modification of the buoyancy parameters and sinking of a previously buoyant conch which has been transported into an area with water depths in excess of the implosion depth could also result in collapse. All of the possible scenarios are dependent on at least one chamber containing some remnant air, so as to set up differential hydrostatic pressure in excess of septa strength. If water has free access to the buoyancy chambers and they all fill with water, the hydrostatic pressure will be equalized on both sides of the septa and the conch can remain intact at great depth. It is doubtful that implosion would have occurred in the case of deposit-filled conch segments due to the additional buttressing of the conch wall by these deposits. At and below implosion depths, dissolution may affect the fossilization potential of the calcareous conch material.

Diurnal heating and cooling

The buoyancy parameters of a floating conch can be modified by diurnal heating and cooling (Reyment, 1958). The air in the buoyancy chambers expands due to increased daytime temperatures, slightly increasing the internal pressure. Since the connecting rings of the siphuncle are permeable, the increased pressure is bled off into the siphuncle. During the nightly cooling of the air in the buoyancy chambers, partial vacuum occurs. This may draw water into the buoyancy chambers.

Physical abrasion and chemical dissolution

If the conch comes into contact with hard substrates such as coral, or should the floating conch encounter the breaker line on an exposed shoreline, then physical abrasion may result in damage to the phragmocone and flooding of the buoyancy chambers.

Perforation of *Nautilus* conchs at sites of pre-existing abnormalities, which are sites of shell repair, might result from dissolution due to the lack of external periostracum cover. The incorporation of a repaired portion of shell into the chambered phragmocone during ontogenetic growth may increase the potential for flooding at a later date.

Biologic interactions

A variety of biologic interactions can influence the taphonomic history of an individual conch. The two primary biologic controls are predation and epizoan encrustation. These various interactions provide a mechanism by which multiple taphonomic pathways can exist.

Duraphagous predation
Nautilus faces a number of potential predators including sharks, other fishes and cephalopods. Direct observational evidence indicates that the effects of duraphagous predators plays a significant role in the taphonomic distributions of *Nautilus* remains. Saunders *et al*. (1987) have published a set of stunning photographs that show a triggerfish attacking a *Nautilus*. While the attack might have been precipitated by the release of many captured specimens of *Nautilus* at the site, the photographs illustrate the shell breakage that would be expected to accompany such an onslaught in a natural setting. Saunders *et al*. (1987) also reported an attack on another *Nautilus* specimen by a grouper which had pinned the cephalopod against a reef face. In the case of the triggerfish attack the wounded *Nautilus* might have survived, but the grouper mortally wounded the cephalopod. Ward (1987, p. 163) stated that two live *Nautilus* specimens tethered to a moderately deep (50 m) portion of a reef face during daylight hours were subjected to predation, with the soft tissue being consumed by an unknown predator. Breakage of the phragmocone during such attacks would facilitate the encroachment of water into the buoyancy chambers and the attainment of negative buoyancy. Chamberlain *et al*. (1981) also related the occurrence of *Nautilus* mandibles in the stomach of a shark.

Several workers have detailed evidence of shell-penetrating boring predation of *Nautilus* by *Octopus*. Saunders *et al*. (1987) illustrated the sites of shell penetration borings on *Nautilus* conchs. Some of the sites are in the phragmocone region, which would allow water to enter the buoyancy chambers. While such attacks would not be immediately fatal, the lack of shell-secreting mantle in these regions would limit the possibility of repair. A healthy *Nautilus* might have been able to maintain its natural buoyancy condition, but a sick or injured specimen could have suffered complications related to buoyancy. Upon the death of the animal, shell penetrative boring in the phragmocone region would have facilitated the early attainment of negative buoyancy.

Evidence of predation on fossil ectocochleate cephalopods has been reported in Kauffman and Kesling (1960), who documented tooth marks left by an attack by a mosasaur on a *Placenticeous* ammonite. Mapes and Hansen

(1984) similarly identified the tooth marks of a *Symmorium reniforme* on a
?*Domatoceras* nautiloid conch. Westermann (1985) symbolized the 'sudden
death' of various cephalopod morphoforms with an attack by a 'monster fish'
which led to the occurrence of nautiloid specimens in water deep enough to
implode the septa. Westermann (1985) also attributed the fragmentation of
additional specimens to 'macropredation'. Mapes (1987, p. 526) published
photographic evidence of cephalopod mandibles stacked like ice-cream cones
in a phosphatic nodule. The stacking of these mandibles seesm to defy
explanation by random processes and is probably the result of a cephalopod-
eating predator's biologic digestive processes concentrating the undigestible
material (that is, cephalopod mandibles) in the middle of faecal material. The
presence of cephalopod conch material with shark remains was presented as
indirect evidence of a predator–prey relationship by Zangerl and Richardson
(1963). Thus, even a conch calculated to be strongly positively buoyant with
an associated propensity toward nekroplanktonic dispersal could be deposited at depth through the process of duraphagous predation (Chamberlain
et al., 1981).

Epizoan encrustation
Even if a conch has risen to the surface and is floating, epizoan attachments
may increase the mass of the conch until a buoyancy shift occurs and the shell
is dragged down to the sea floor under its own weight coupled with the mass
of epizoans. Epizoan attachments are more common on post-Palaeozoic
cephalopods, including recent epizoan attachments on *Nautilus* (Boston *et al.*,
1988). Ward (1987, p. 167) stated that *N. macromphalus* and *N. scrobiculates*
both have an open umbilicus which is invariably encrusted with epizoans.
Ward (1987, p. 168) further commented that such encrustations are usually
not significant enough to influence the swimming or buoyancy of *Nautilus*.
Meischner (1968), however, noted that epizoans attached to a *Ceratites
semipartitus* specimen did influence the buoyancy of the specimen and possibly contributed to the mortality of that cephalopod.

UPPER PALAEOZOIC OCCURRENCES

The occurrence of ectocochleate cephalopod remains in Upper Palaeozoic
strata, primarily the Pennsylvanian of North America, provides field data
from multiple depositional facies. The depositional environmental interpretations used in these studies have been drawn primarily from the work of
Heckel and Baesemann (1975) and Heckel (1977). These workers interpreted
lithic units within the vertically repetitive Pennsylvanian lithologies as having
been deposited in water depths ranging from shallow (exposed strandline) to
deep (epicontinental-basinal). The vertical repetitiveness of the strata was
interpreted to have resulted from transgressions and regressions associated
with glacial eustatic sea-level changes. These lithologies contain facies that
are associated with intergradational ectocochleate cephalopod accumulations
(Figure 10.2). The accumulations are interpreted as relating to the prevalent

taphonomic processes associated with the varying depositional environments (Figure 10.3).

Deep-water accumulations

The occurrences of ectocochleate cephalopods associated with deep water facies were best documented by Boardman *et al.* (1984), who designated a faunal community succession. Two palaeoecological communities were established; the shallower of these deep-water communities is divided into two subcommunities based in part on the occurrence of different ectocochleate cephalopods. These communities are associated with the core shale interval within the Heckel depositional model. Studies of additional faunal elements collected from these lithologies have shown close agreement with the Heckel lithostratigraphic and Boardman biocommunity models. This has provided independent confirmation of the deep-water cephalopod occurrence.

Boardman *et al.* (1984) interpreted the community succession as being related to varying dissolved oxygen content in the water at depth. Tasch (1953) suggested that the distribution of the shelly elements in these facies represents sorting by currents at depth. However, the occurrence of multiple size fractions of shelly debris in these facies appears to rule out extensive reworking. Thus, these occurrences associated with deep water facies are interpreted to be the result of minimal biostratinomic transport.

Caneyella–Dunbarella–ammonoid–radiolarian community
Within the Boardman *et al.* (1984) succession the community associated with the deepest water facies is the *Caneyella–Dunbarella*–ammonoid–radiolarian community. This community is interpreted to have been deposited in anoxic (O_2 less than 0.1 ml l^{-1}) conditions. Ammonoids dominate the ectocochleate cephalopod portion of this community and represent specimens that swam or drifted into the area. Predation by fishes, including sharks, higher in the water column would facilitate the accumulation of ectocochleate cephalopod debris in these benthically inhospitable environs. In some cases the sinking to the substrate must have been relatively rapid because some phosphatic nodules contain ectocochleate cephalopod specimens with *in situ* mandibles and/or radulae (Mapes, 1987). The presence of these organs *in situ* demonstrates that little, if any, post-mortem transportation has affected these specimens.

Trepospira–Sinuitina–ammonoid–*Anthraconelio* community
The *Trepospira–Sinuitina*–ammonoid–*Anthraconelio* community is interpreted to have been deposited in slightly shallower water than the *Caneyella–Dunbarella*–ammonoid–radiolarian community. It is subdivided into the *Sinuitina*–juvenile ammonoid–*Anthraconelio* subcommunity deposited in dysaerobic facies (O_2 between 1.0 and 0.1 ml l^{-1}) and the *Trepospira*–mature ammonoid–*Anthraconelio* subcommunity deposited in well-oxygenated water (O_2 more than 1.0 ml l^{-1}).

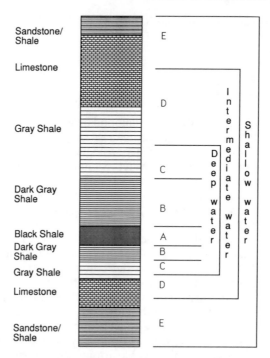

Figure 10.2 Water depth-related facies and the associated ectococohleate cephalopod accumulations (A–E, also see Figure 10.3) in the vertically repetitive lithic cycles of Pennsylvanian age of North America.

Sinuitina–juvenile ammonoid–*Anthraconelio* subcommunity

The taphonomy of the ectocochleate cephalopod specimens in the *Sinuitina*–juvenile ammonoid–*Anthraconelio* subcommunity is interpreted to have been controlled by the restricted mobility of the juvenile ammonoids. The mobility of these forms is interpreted to have been restricted by the relatively massive amount of conch material that these forms possessed, as well as the cameral fluid that must have been present in the phragmocone prior to hatching. Additionally, many of the juvenile specimens preserve complete body chambers. Boardman *et al.* (1984) attributed the occurrence of these specimens to periodic depletion of the dissolved oxygen content of the water, resulting in mass kills of the semimobile juveniles. The more mobile adult and subadult specimens that were more neutrally buoyant would have been able to flee these conditions.

Most of the microscopic bactritoid forms documented by Mapes (1979) have been collected almost exclusively from this subcommunity. Many of these specimens are juvenile forms that exhibit some crushing of the buoyancy chambers. Even the adults of this morphoform are rarely collected from shallow or intermediate water-depth lithologies, despite the very positively buoyant condition of the depositless phragmocones. This might be related to the depth at which bactritoids have been interpreted to have laid

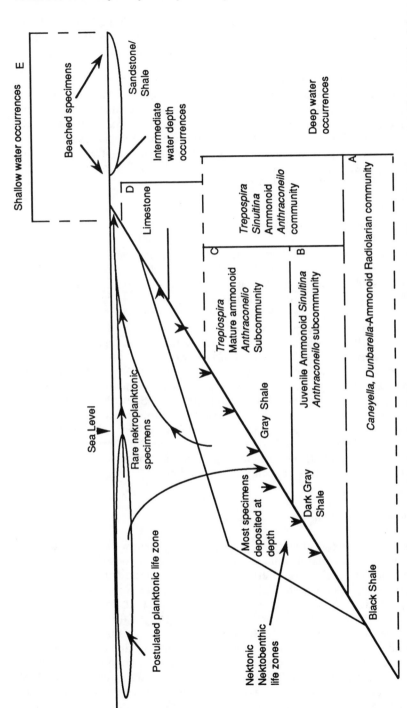

Figure 10.3 Upper Palaeozoic shoreline profile illustrating the facies-related accumulations (A–E, see also Figure 10.2) of the ectocochleate cephalopods; their life zones and arrows showing the interpreted taphonomic pathways of these fossils with the majority of the specimens deposited at depth.

their eggs (Hecht and Mapes, 1990). Hatchlings which had not had the opportunity to remove the cameral fluids while in the eggs would remain on the bottom during the mass kills postulated by Boardman *et al.* (1984).

Trepospira–mature ammonoid–*Anthraconelio* subcommunity
The *Trepospira*–mature ammonoid–*Anthraconelio* subcommunity contains a preponderance of mature ammonoids and a modest percentage of coiled and orthoconic nautiloids. The ectocochleate cephalopod fauna contains a wide variety of morphoforms covering the range of the morphologies extant during the Pennsylvanian period. The taphonomic overprint on this community includes ample evidence of predation (Boston *et al.*, 1987; Mapes *et al.*, 1987; Sims *et al.*, 1987; Mapes and Hansen, 1984).

Intermediate water-depth accumulations

The intermediate water-depth accumulations occur in a variety of shale units dominated by crinoids and brachiopods. This includes most of the fully marine gray shales of the Appalachian Basin areas of North America, such as the Brush Creek Shale interval (Boardman *et al.*, 1984; McComas *et al.*, 1986). These accumulations are most commonly straight or slightly curved shells with small siphuncles and planarspiral coiled forms of nautiloids. The most frequently encountered of these common straight forms are *Pseudorthoceras*, *Mooreoceras* and *Euloxoceras*, while the most common coiled forms are *Metacoceras*, *Tainoceras*, *Neobistrolites*, *Liroceras* and *Domatoceras*. The straight specimens appear most often to be fragmented apical ends of phragmocones containing deposits. Some specimens show slight to moderate crushing on the dorsal surface of the conch, indicating that some partially deposit-filled buoyancy chambers also occur. The cephalopod specimens are widely scattered and few retain the body chamber. The cephalopods within this taphofacies are interpreted to have been predated or those specimens that sank before reaching the strandline.

Shallow-water accumulations

Occurrences of ectocochleate cephalopods in shallow-water lithologies (shales, sandstones and limestones), particularly those deposited in shoreface or more landward areas, are interpreted to have resulted from post-mortem floating of cephalopod conchs into these environments. Supporting evidence includes the development of coquina-like deposits that contain abundant shelly ectocochleate cephalopod debris. The concentration of the shelly debris of the cephalopods in these lithologies indicates that some type of taphonomic concentration process must have occurred. The simplest of these would be the concentration of shell material in a tidal channel after nekro-planktonic drifting into tidally influenced shallow-water deposits. The best Pennsylvanian examples of these types of occurrence are those in the localities around Kansas City, Missouri. A typical shallow-water accumulation is south of Desoto, Kansas, along Kill Creek. This outcrop is part of the Farley Limestone in the area interpreted by Heckel and Cocke (1969, p. 1072) as part of an algal-mound complex. Originally collected by Newell (1936), the locality contains many relatively whole specimens with large body-chamber sections indicating mature modifications. Additionally, collection of this locality by the authors in 1988 yielded planarspiralled specimens of *Metacoceras* and *Tainoceras*; both of these taxa exhibit thick-walled ornamentation. While the presence of complete body chambers and thick-walled ornamentation should work against nekroplanktonic dispersal, these specimens were collected from an oolitic limestone of very shallow-water origin. The morphoforms recovered in 1988 from approximately 35 kg of material from this unit include at least 91 planarspiralled forms, three straight forms with small siphuncles (*Pseudotheroceras*) and three brevicone forms (?*Merceras*). It is important to note that no ammonoid or bactritoid forms were collected. A theoretical solution that explains the prevalence of the nautiloids in shallow-water occurrences lies in the calculated depth tolerance of a few tens of metres for the Palaeozoic nautiloids and a few hundreds of metres for the Palaeozoic ammonoids (Westermann, 1973). The shallow water depth at which the nautiloids would experience mortality would facilitate the attainment of nekroplanktonism, while the deeper-dwelling ammonoids would have limited probability of attaining nekroplanktonism.

CONCLUSIONS

The study of ectocochleate cephalopods is complicated by the presence of multiple taphonomic pathways caused by a variety of physical and biotic interactions that can modify the initial buoyancy parameters of a cephalopod conch. The calculations of the initial buoyancy parameters of an undamaged conch can be approximated by biometric analysis of fossil morphologies and the application of some of the conclusions derived from the study of modern *Nautilus*. The analysis of initial buoyancy parameters can establish a baseline propensity toward nekroplanktonic dispersal. However, Chamberlain *et al.* (1981) have shown that few ectocochleate cephalopod shells actually became nekroplanktonic due to water infiltration of the buoyancy chambers of the conchs. The attainment of negative buoyancy led to deposition at depth and minimal transportation of the cephalopod conch away from the habitat of the living populations.

Occasional specimens did (and do) become nekroplanktonic and drift into shallow-water environs, especially if the body-chamber material was (is) removed. However, in order fully to analyse the biostratinomic processes that have affected the post-mortem dispersal of any single ectocochleate cephalopod shell, the individual conch must be examined for evidence of modifying events, including duraphagous predation. In addition, an analysis of the lithofacies and biofacies of the strata from which the specimen was obtained must be completed in order to establish under what conditions it was deposited.

ACKNOWLEDGEMENTS

Acknowledgement is made to the donors of the Petroleum Research Fund, administered by the American Chemical Society, for support of this research (PFR no. 15821AC2 to RHM), and to Rutgers University and the Kansas Geological Survey for their support (of WBB). We also thank S.K. Donovan, J.V. Browning, A.N. Brower and S.M. Silvestri for their detailed comments on the manuscript.

REFERENCES

Boardman, D.R., II, Mapes, R.H., Yancey, T.E. and Malinky, J.M., 1984, A new model for the depth related allogenic community in successions within North American Pennsylvanian cyclothems and implications on the black shale problem. In N. Hyne (ed.), *Limestone reservoir rocks of the midcontinent, Tulsa Geological Society, Special Publication*, 2: 141–82.

Boston, W.B., McComas, G.A., Mapes, R.H. and McGhee, G.R., 1988, The fossil occurrence of epizoans on living coiled cephalopods in the Upper Paleozoic (Carboniferous), *Geological Society of America Abstracts with Programs*, 20 (1): 9.

Boston, W.B., Sims, M. and Mapes, R.H., 1987, Predation on cephalopods from the Finis Shale (Pennsylvanian–Virgilian) of Texas, *Geological Society of America Abstracts with Programs*, 19 (1): 6.

Chamberlain, J.A., Jr., 1987, Locomotion of *Nautilus*. In W.B. Saunders and N.H. Landman (eds), *Nautilus: the biology and paleobiology of a living fossil*, Plenum, New York: 489–526.

Chamberlain, J.A., Jr., Ward, P.D. and Weaver, J.S., 1981, Postmortem ascent of *Nautilus* shells: implications for cephalopod paleo-biogeography, *Paleobiology*, 7 (4): 494–509.

Donovan, S.K., 1989, Taphonomic significance of the encrustation of the dead shell of recent *Spirula spirula* (Linné) (Cephalopoda: Coleoidea) by *Lepas anatifera* Linné (Cirripedia: Thoracia), *Journal of Paleontology*, 63 (5): 698–702

Furnish, W.M. and Glenister, B.F., 1964, Paleoecology. In R.C. Moore (ed.), *Treatise on invertebrate paleontology, Part K, Mollusca 3*, Geological Society of America and University of Kansas Press, New York and Lawrence: K114–24.

Greenwald, L. and Ward, P.D., 1987, Buoyancy in *Nautilus*. In W.D. Saunders and N.H. Landman (eds), *Nautilus: the biology and paleobiology of a living fossil*, Plenum Press, New York: 547–62.

Hecht, G.D. and Mapes, R.H., 1990, Paleobiology of bactritoid cephalopods from the Pennsylvanian (Missourian) of Texas and Kansas, *Geological Society of America Abstracts with Programs*, 22 (7): 221.

Heckel, P.H., 1977, Origin of phosphatic black shale facies in Pennsylvanian cyclothems of midcontinent North America, *American Association of Petroleum Geologists Bulletin*, 61 (7): 1045–68.

Heckel, P.H. and Baesemann, J.F., 1975, Environmental interpretation of conodont distribution in Upper Pennsylvanian (Missourian) megacyclothems in eastern Kansas, *American Association of Petroleum Geologists Bulletin*, 59 (3): 486–509.

Heckel, P.H. and Cocke, J.M., 1969, Phylloid algal-mound complexes in outcropping Upper Pennsylvanian rocks of mid-continent, *American Association of Petroleum Geologists Bulletin*, 53 (5): 1058–74.

Kauffman, E.G. and Kesling, E.G., 1960, An Upper Cretaceous ammonite bitten by mosasaur, *Contributions to Museum of Paleontology University of Michigan*, **15**: 193–248.

Mapes, R.H., 1979, Carboniferous and Permian Bactritoidea (Cephalopoda) in North America, *The University of Kansas Paleontological Contributions*, **64**: 1–75.

Mapes, R.H., 1987, Upper Paleozoic cephalopod mandibles: frequency of occurrence, modes of preservation, and paleoecological implications, *Journal of Paleontology*, **61** (3): 521–38.

Mapes, R.H. and Hansen, M.C., 1984, Pennsylvanian shark–cephalopod predation: a case study, *Lethaia*, **17** (3): 175–83.

Mapes, R.H., Hansen, M.C. and Boardman, D.R., II, 1987, Evidence of predation in midcontinent Upper Paleozoic anoxic and dysaerobic marine environments, *Geological Society of America Abstracts with Programs*, **19** (1): 28.

McComas, G.A., Mapes, R.H. and Boston, W.D., 1986, Paleoecological implications of a Brush Creek (Upper Pennsylvanian) faunal succession compared to Missourian midcontinent faunal successions, *Geological Society of America Abstracts with Programs*, **18** (1): 315.

Meischner, D., 1968, Perniciöse Epökie von *Placunopsis* auf *Ceratites*, *Lethaia*, **1** (2): 156–74.

Newell, N.D., 1936, Some mid-Pennsylvanian invertebrates from Kansas and Oklahoma: III. Cephalopoda, *Journal of Paleontology*, **10** (6): 481–9.

Reyment, R.A., 1958, Some factors in the distribution of fossil cephalopods, *Stockholm Contributions to Geology*, **1** (6): 97–184.

Reyment, R.A., 1980, Floating orientation of cephalopod shell models, *Palaeontology*, **23** (4): 931–6.

Roux, M., 1990, Underwater observations of *Nautilus macromphalus* off New Caledonia, *Chambered Nautilus Newsletter*, **60** : 1.

Saunders, W.B. and Landman, N.H. (eds), 1987, *Nautilus: the biology and paleobiology of a living fossil*, Plenum, New York.

Saunders, W.B., Spinosa, C. and Davis, L.E., 1987, Predation on *Nautilus*. In W.B. Saunders and N.H. Landman (eds), *Nautilus: the biology and paleobiology of a living fossil*, Plenum, New York: 201–14.

Saunders, W.B. and Ward, P.D., 1987, Ecology, distribution, and population characteristics of *Nautilus*. In W.B. Saunders and N.H. Landman (eds), *Nautilus: the biology and paleobiology of a living fossil*, Plenum, New York: 137–62.

Scott, G., 1940, Paleoecological factors controlling distribution of Cretaceous ammonoids in Texas area, *American Association of Petroleum Geologists Bulletin*, **24** (7): 1164–203.

Sims, M.S., Boston, W.B. and Mapes, R.H., 1987, Predation on an Upper Carboniferous ammonoid *Gonioloboceras*, *Geological Society of America Abstracts with Programs*, **19** (1): 57.

Tasch, P., 1953, Causes and paleoecological significance of dwarfed fossil marine invertebrates, *Journal of Paleontology*, **27** (3): 356–444.

Teichert, C., 1964, Biostratonomy. In R.C. Moore (ed.), *Treatise on invertebrate paleontology, Part K, Mollusca 3*, Geological Society of America and University of Kansas Press, New York and Lawrence: K124–7.

Trueman, A.E., 1941, The ammonite body-chamber, with special reference to the buoyancy and mode of life of the living ammonite, *Quarterly Journal of the Geological Society of London*, **96** (4): 339–83.

Ward, P.D., 1987, *The natural history of Nautilus*, Allen & Unwin, Boston.

Ward, P.D. and Greenwald, L., 1982, Chamber refilling in *Nautilus*, *Journal of the Marine Biological Association of the United Kingdom*, **62** (2): 469–75.

Westermann, G.E.G., 1973, Strength of concave septa and depth limits of fossil cephalopods, *Lethaia*, **6** (4): 383–403.

Westermann, G.E.G., 1985, Post-mortem descent with septal implosion in Silurian nautiloids, *Paläontologische Zeitschrift*, **59** (1/2): 79–97.

Zangerl, R. and Richardson, E.S., 1963, The paleoecological history of two Pennsylvanian black shales, *Fieldiana-Geology Memoir*, **4**: 352 pp.

Chapter 11

THE TAPHONOMY OF ECHINODERMS: CALCAREOUS MULTI-ELEMENT SKELETONS IN THE MARINE ENVIRONMENT

Stephen K. Donovan

INTRODUCTION

The echinoderms occupy a unique position among marine invertebrates, possessing an endoskeleton which is composed of tens, hundreds or often thousands of separate calcite plates, called 'ossicles', which disarticulate rapidly following death. These very numerous plates have a high preservation potential. In contrast, only certain parts in the complex skeletons of arthropods (for example) are sufficiently well calcified to ensure frequent preservation (for example, the barnacle shell, the trilobite carapace and crab chelae).

The special taphonomic problems presented by the echinoderms are well illustrated by first considering how an organism with a less complex skeleton, such as a bivalve mollusc, may be preserved. The two common recognizable states in which fossil bivalves are preserved are as complete shells, with both valves conjoined, or as disarticulated valves. Even if preserved as single valves, identification of bivalves to the level of species is usually a straightforward task. There is also a diverse literature concerning the post-mortem hydrodynamic transport and sorting of bivalve shells, based on both experimental and field observations, as an aid to the interpretation of the taphonomic history of shell assemblages. Contrast this situation with the preservation of the echinoderm endoskeleton. A fossil echinoderm can be preserved in any one of a variety of modes, from articulated skeletons retaining even the most delicate ossicles to completely disarticulated and dissociated plates spread over a wide area. Few experimental data are available on the disarticulation of the echinoderm test (notable exceptions include Kidwell and Baumiller, 1989; 1990; Allison, 1990) and only on rare occasions, usually in specimens showing some exceptional feature, is it possible to determine the precise sequence of events that influenced the mode of preservation (see, for example, Maples and Archer, 1989). Further, because echinoderm taxonomy is based largely upon the features of complete or near-complete skeletons, we must make two obvious deductions: first, that taphonomic studies of disarticulated ossicles must often be incomplete because it is not currently possible to determine even the original affinities (except in a broad sense, for example, crinoid columnals) of the producing organism(s); and second, that our understanding of echinoderm palaeontology must be biased by a dependence upon preservational rarities (that is, complete specimens). The second deduction is supported by the work of Kier (1977) and Smith (1984), who demonstrated that the preservation potential of the (largely infaunal) irregular echinoids is greater than that of the epifaunal regulars, which consequently have a more imperfect fossil record; by Donovan (1989a), who showed that, although only about 50 species of crinoid have been described from the British Ordovician, at least 150 more can be inferred from the record of disarticulated stem ossicles (columnals); and by Donovan and Gale (1990), who demonstrated that, although only two poorly preserved asteroid specimens are known from the entire Triassic, the major adaptive diversification (if not radiation) of the neoasteroids probably occurred during this time interval.

Having recognized that echinoderm palaeontology is often dependent upon a biased sample showing exceptional preservation (Lasker, 1976; but see also Paul, 1982), it is apparent how important a complete understanding of their taphonomic behaviour could be to our comprehension of this group. To make a simple comparison, sedimentary events leading to the preservation of a community of crinoids are much rarer than those entombing communities of bivalves, because of the contrasting post-mortem characteristics of the two groups (see, for example, Bassler, 1913; Franzén, 1982; Brett and Eckert, 1982).

Only one previous general review of echinoderm taphonomy has been published (Lewis, 1980), although Schäfer (1972, pp. 91–105) considered the biostratinomic changes of the shallow-water eleutherozoans of the southern North Sea Basin and Smith (1984) described the post-mortem behaviour of the echinoid test. In this chapter I propose to concentrate on the processes of death, disarticulation and diagenesis that lead to the diversity of preservational styles seen in fossil echinoderms. However, it is considered essential first to discuss briefly the structure of the echinoderm test to explain how it influences preservation.

NATURE OF THE ECHINODERM TEST

Two principal features determine the robustness of the echinoderm endoskeleton: the individual calcite plates which form the test and the soft tissues that bind these ossicles together. The test plates of the echinoderms (and the closely related calcichordates; see Jefferies, 1988) are unique among invertebrates in being individual crystals of calcite which occur in a bewildering array of geometries throughout the group. These crystals are not solid chunks of calcite, however, but are porous, so that the plates have a three-dimensional meshwork structure. The calcite trabeculae (rods) that form this meshwork are called the 'stereom' (Figures 11.1E, 11.1F). The most thorough investigation of the microstructure of the echinoderm test has been made by Smith (1980; 1989; but see also Roux, 1970; 1974a; 1974b; 1975; Macurda and Meyer, 1975; Macurda *et al.*, 1978), who recognized ten different stereom architectures, each adapted to one or more particular functions and indicative of association with one or more particular types of soft tissues.

The amount of relative movement of adjacent ossicles in the echinoderm test is determined by the geometry of the articulation areas between plates and the pattern of soft tissues which bind the plates together. To illustrate the variation in test rigidity (and thus preservation potential) produced by different interplate articulations in one group of echinoderms, the structure of the echinoid corona is considered here (Figure 11.2). The echinoid corona is usually a rigid structure (Figures 11.2B–11.2D), although it is flexible in some groups (Figure 11.2A). In a flexible echinoid corona, adjacent plates are imbricate and overlap, so that one ossicle can slide over the next (indicated by arrows in Figure 11.2A; see also Lewis and Ensom, 1982, Figure 21). However, the majority of echinoids have a rigid corona, in which adjacent ossicles abut at sutures that are inflexible and perpendicular to the test surface (Figure 11.2B). The rigidity of such a structure is dependent largely upon the strength of the ligaments (see below) which bind the adjacent plates together. Such an arrangement of plates is found in the cidaroids. Soon after death the interplate ligaments rot and the cidaroid corona collapses into a mass of component ossicles. In other echinoids the test is further strengthened by having trabeculae of adjacent plates penetrating each other's stereom, resulting in a structure that remains rigid even after the ligaments have rotted (Smith, 1984, Figures 2.5b, 2.5c; Figure 11.2C herein). In some taxa interlocking peg and socket structures are developed between plates (Smith, 1984, Figure 2.5d), which are analogous to articulations developed between meres in some crinoid stems (Donovan, 1986, p. 60). Assuming that an echinoid corona with such strengthening structures is not composed from unusually thin, fragile plates, it should be capable of surviving on the sea floor for a considerable length of time following death and before final burial. It would thus be available as a hard substrate for both encrusting and boring organisms (see, for example, Joysey, 1959; Goldring and Stephenson, 1970, p. 181; Figures 11.3A, 11.3B herein). The ultimate strengthening of the echinoid corona is shown by the flattened clypeasteroids, or sand dollars, which have a pattern of internal pillars that act as rigid supports within the test and are resistant to bending stress (Smith, 1984, Figure 3.5). The genus *Clypeaster* is even more robust and secretes a second, nearly continuous calcareous layer on the inner surface of the corona (Figure 11.2D). The corona of *Clypeaster* is thus particularly strong, and both fossil and dead Recent tests frequently support a diverse epibiota. It is thus apparent that the sequence of echinoid coronal structures in Figure 11.2 represent a taphonomic gradient, from rarely preserved taxa with imbricate plating (Figure 11.2A) to the highly robust test of *Clypeaster* (Figure 11.2D). This model of increasing preservation potential is reflected by the fossil record (Kier, 1977).

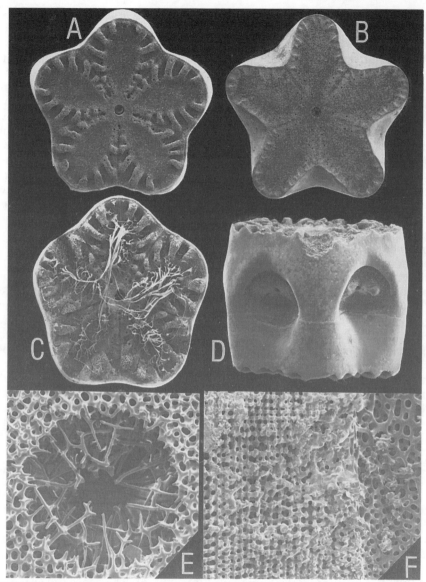

Figure 11.1 Scanning electron micrographs of columnals from the Recent isocrinid crinoid *Neocrinus decorus* Wyville Thomson. (A) Proximal articular facet of a nodal columnal of the distal column, showing the principal region of ligamentation divided into five petals with galleried stereom, with each petal surrounded by a symplectial articulation which permits moderate flexibility (after Donovan, 1984, Plate 74, Figure 2), ×13.5. (B) A cryptosymplectial articulation on the distal articular facet of a nodal, modified from a symplexy by secretion of further stereom, ×13.5. (C) Articular facet showing ligament fibres concentrated in petaloid zones (after Donovan, 1984, Plate 73, Figure 4), ×13.5.

The plates of the echinoderm test (particularly those whose interplate articulations resemble Figure 11.2A or 11.2B, or those that otherwise have flexible articulations, such as echinoid radioles or crinoid brachials) rely on collagenous ligament fibres to bind the skeleton together. These ligaments, most appropriately named 'mutable collagenous tissues' (MCTs: Wilkie and Emson, 1988; see also Wilkie, 1984; 1988; catch connective tissue of Motokawa, 1985; 1988), are distributed within the open pore spaces of certain stereom microstructures (Smith, 1980; Figures 11.1A, 11.1C, 11.1F herein). Decay of MCTs following death is often the principal factor determining the degree of disarticulation, and thus mode of preservation, in echinoderms, particularly in structures that were flexible in life and therefore were held in place primarily by ligaments, with or without the assistance of muscles (although muscles may degrade more rapidly than ligaments). That MCTs may rot quickly following death was recognized by Blyth Cain (1968, p. 192), who noted that the endoskeleton of the extant comatulid crinoid *Antedon bifida* disarticulates completely within two days of death, even under anaerobic conditions. However, it is uncertain if all MCTs rot with such rapidity. In certain fossil crinoids, for example, pluricolumnals (partially fragmented columns) were robust enough to produce tool marks during transport (Benton and Gray, 1981) or to act as hard substrates for bryozoans that encrusted both the latus and the articular facet (Donovan *et al.*, 1986).

It is well known that the fossil record of arthropods has benefited from both dead adults and their various juvenile moult stages being preserved, so that any individual may have produced more than one 'complete' fossil. Echinoderms may also make multiple contributions to the fossil record by the process of autotomy, or self-mutilation, with subsequent regeneration of lost parts. Autotomy occurs in response to adverse external stimuli (Emson and Wilkie, 1980). MCTs are mutable because they can change their tensile strength, between rigid and pliant, in response to changes in the concentrations of particular cations secreted under nervous control (Wilkie, 1983; 1984; 1988; Motokawa, 1988). During autotomy, these changes proceed irreversibly past the pliant stage and lead to MCT disintegration. Part of the skeleton may thus be lost, only to be regenerated subsequently. For example, in the stem of isocrinid crinoids (Figures 11.1B, 11.1D, 11.1E), approximately planar articular joints, called 'cryptosymplexies', are the preferred sites of autotomy. Cryptosymplexies are regularly spaced throughout all but the most proximal (youngest) stem. In response to an adverse stimulus,

---

(D) Cryptosymplectial articulation (just below centre) in lateral view (after Donovan, 1984, Plate 74, Figure 5), ×17. (E) Stereom trabeculae growing into the axial canal adjacent to a cryptosymplexy (after Donovan, 1984, Plate 74, Figure 3), ×225. (F) Longitudinal section of a columnal adjacent to the axial canal (right) (after Donovan, 1984, Plate 75, Figure 2). Stereom adjacent to the canal is rectilinear, while the microstructure of the canal wall is labyrinthic (terminology of Smith, 1980), ×100. All specimens coated with 60 per cent gold-palladium and in the collections of the Department of Invertebrate Zoology, US National Museum.

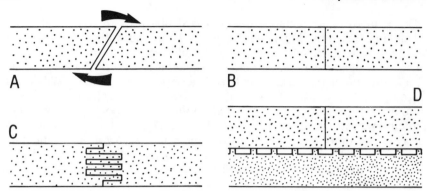

Figure 11.2   Schematic diagrams of the geometry of plate sutures within the echinoid corona. (A) Flexible test (for example, *Archaeocidaris*; see Lewis and Ensom, 1982). Arrows indicate relative movement of plates. (B) Rigid test (for example, cidaroids). (C) Rigid test with strongly interlocking trabeculae or peg and socket structures (for example, *Meoma*). (D) Rigid test of *Clypeaster* with double wall.

autotomy occurs and the distal, attached part of the stem is discarded after which the animal crawls (Messing *et al.*, 1988) to a new site. Such relocations may be a frequent occurrence in isocrinids living in high-energy environments and it is relevant to note that under such conditions *Metacrinus rotundus* has a stem growth rate of about 30–60 cm yr^{-1} (Oji, 1989). If changes of position are indeed frequent, then autotomized fragments of stem are continually being added to the surrounding sediment. The influence of local ionic environment on subsequent decay of MCTs in these autotomized remains and other dead echinoderms is unknown.

DEATH

Predation

Obtaining unequivocal evidence of predation in the fossil record is notoriously difficult (Vermeij, 1987) and usually can be done only in instances where some distinctive evidence of the predator is left on the hardparts of the prey, such as the tooth marks of a shark on a bone or a borehole in a shell made by a predatory gastropod. Predation that results in an aggregation of ossicles on the sea floor is indistinguishable from the normal post-mortem disarticulation of the echinoderm test (Figure 11.4D), although fragments broken across plate sutures may at least be suggestive (Smith, 1984, p.19, Figure 2.6.f; similar damage can be produced during storms). Predators of extant echinoderms are many and varied: they include asteroids preying on echinoids, ophiuroids, other asteroids and holothurians (Jangoux, 1982); predation of the sand dollar *Echinarachnius parma* by the green sea urchin *Strongylocentrotus droebachiensis*, which in turn falls victim to asteroids, lobsters, crabs, fishes and birds (Himmelman and Steele, 1971); comatulid crinoids being eaten by fishes (Meyer, 1985) and possibly also by crustaceans and asteroids (Mladenov, 1983; Schneider, 1988); the California sea otter feeding on echinoids (Hall and Schaller, 1964); and predation by octopus, shore birds and gastropods on a variety of Caribbean echinoids (Hendler, 1977).

Autotomy (see above) is an important defensive mechanism in some extant echinoderms, particularly ophiuroids and comatulid crinoids, which are able to lose their appendages to predators and escape. Thus, predation on echinoderms does not have to be lethal to make a contribution to the sediment budget. The ability to autotomize arms, their mobility and possible biochemical defences are among the protective adaptations of comatulids (Meyer, 1985), a group quite unlike the stalked, immobile and often heavily armoured Palaeozoic crinoids (for an excellent review of the predation of crinoids, see Meyer and Ausich, 1983, pp. 378–85). Recovery following autotomy can be fast. An amputated arm of the comatulid *Florometra serratissima* can be regrown in under nine months (Mladenov, 1983) and regeneration of the arms in some ophiuroids is even more rapid (Sides, 1987).

In a series of papers, Aronson (1987a; 1987b; 1989a; 1989b; Aronson and Harms, 1985, among others) has suggested that predation of ophiuroids has intensified since the Jurassic. Ancient examples generally lack regenerating arms (but see Bomwer and Meyer, 1987), whereas Recent ophiuroids from the Caribbean are frequently found regenerating one or more arms. Further, the density of ophiuroids in some starfish beds (see below) is much greater than is found in most modern environments. Extant ophiuroids only reach 'Palaeozoic' abundancies in environments where predation pressures are low (Aronson, 1989b; Aronson and Harms, 1985). In contrast, fossil evidence for the regeneration of skeletal elements in fossil crinoids, presumably subsequent to predation and particularly affecting the arms and anal sac, is known from as far back as the Ordovician (see, for example, Whitfield, 1904; Hattin, 1958; Strimple and Beane, 1966; Figure 11.4A herein). Lane (1984) suggested that predation on the arms of Recent crinoids is probably more intense than it was in the Palaeozoic. Comatulids have their gonads located within the arms, making them a prime target for predators, which thus ignore the dorsal cup that contains those soft tissues necessary for regeneration. The gonads were not located in the arms of the Palaeozoic crinoids and Lane has inferred that they were situated within the theca or, preferentially, in the large anal sac found in many taxa (Figure 11.4B). The latter arrangement would have had the advantage that a predator would concentrate its attentions on the anal sac (cf. comatulid arms), which could subsequently be regenerated (Lane, 1984, Figure 1).

Evidence for regeneration following non-lethal predation in fossil echinoderms is not limited to those groups with easily autotomized appendages. For example, fossil sand dollars are known with large chunks of the test broken from the circumference and which, in at least some examples, have regenerated (Seilacher, 1979, Figure 4D; Zinsmeister, 1980; Smith, 1984, p. 12). Kier and Grant (1965) noted that extant sand dollars suffer this sort of predation from fishes, but only when the echinoid is feeding epifaunally in a vertical

Figure 11.3 (A), (B). Two specimens of the Recent spatangoid *Meoma ventricosa* from the Port Royal Cays, Jamaica, both ×0.5. (A) Recently dead specimen; soft tissues and radioles are lost, but test surface is still 'fresh' and free from encrusting organisms. (B) Further exposure leads to dense encrustation by serpulid worms and (mainly) calcareous algae on the aboral surface (the adoral surface, which was resting on the sea floor, is largely free from encrustation).

position (Timko, 1976). Zinsmeister (1980) also noted that *Monophoraster darwini* is occasionally preserved with puncture wounds. Other perforations of the echinoid test in five species of irregular from the Eocene of Florida may, in at least some instances, be due to gastropod predation (Gibson and Watson, 1989), although many borings in fossil echinoids are probably post-mortem (Joysey, 1959).

Disease and parasitism

Jangoux (1987a; 1987b; 1987c; 1987d) has comprehensively reviewed those organisms that cause disease in, and which are parasitic on, extant echino-derms. The range of such disease-causing organisms is immense, and includes bacteria, fungi, cyanophytes, flagellates, amoebae, sporozoans, ciliates, algae and unidentified water borne pathogens (Jangoux, 1987a); mesozoans, sponges, cnidarians, turbellarians, trematodes, nematodes, gastropods, bi-valves, entoprocts and bryozoans (Jangoux, 1987b); and polychaetes, myzos-tomids, tardigrades, copepods, cirripedes, malacostracans, arachnids, pycnogonids, insects and fishes (Jangoux, 1987c). At least some of these associations are ultimately fatal. However, it is only in examples where infestations cause some recognizable deformity in skeletal growth, or in which the infesting organism is preserved, that any indication of disease or parasit-ism of echinoderms is likely to be recognized in the fossil record (Conway Morris, 1981). For example, the bacterial infection known as bald sea urchin disease (Maes and Jangoux, 1985; Jangoux, 1987a; Schwammer, 1989) may leave the dead echinoid test distinctly cratered and perforated. Similarly, parasites may produce distinctive galls or cysts in echinoid radioles (Warén and Moolenbeek, 1989), asteroid arms, crinoid pinnules and cirri, and ophiuroid discs (Grygier, 1988). However, other infestations are more problematic.

For example, many fossil echinoderms preserve prominent pits (Figures 11.3C, 11.4C) which are associated with growth deformities that indicate boring to have been pre-mortem (although some borings are obviously post-mortem, while other unusual perforations were produced by the echinoderm; Joysey, 1959; Franzén-Bengtson, 1983). Borings through (Kier, 1981; Gibson and Watson, 1989) and pits into (Alekseev and Endelman, 1989) fossil echinoid tests have been attributed to gastropods, both parasitic and preda-tory, depending upon the pattern of perforation and the evidence for sub-sequent growth. Pits in the endoskeleton of Palaeozoic pelmatozoans have

---

(C)–(E) Possible parasites of crinoid stems of Lower Carboniferous age from Salthill Quarry, Clitheroe, England, ×0.7. (C) Multiple borings in a pluricolum-nal. This infestation must have occurred while the crinoid was alive, as the stem has reacted by swelling. (D) The tabulate coral *Emmonsia parasitica* encrusting a pluricolumnal. (E) A broken fragment of the tabulate coral *Cladochonus*. All specimens were coated with ammonium chloride sublimate before photo-graphy. All specimens in the author's collection.

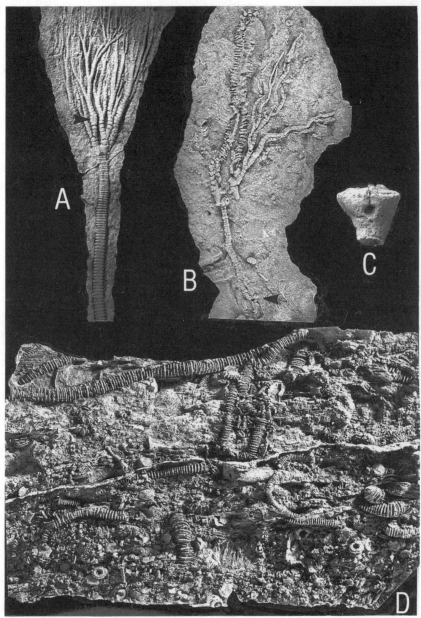

Figure 11.4. Fossil crinoids showing different styles of preservation. (A) *Caleidocrinus multiramus*, Ordovician (Caradoc), National Museum of Prague L13297 (after Donovan, 1985, Figure 3A). Well-preserved specimen showing regeneration of arm (arrowed) possibly following predation, ×1.25. (B) *Iocrinus pauli*, Ordovician (Llanvirn), private collection of Mr J.J. Savill (after Donovan, 1989b, Plate 7, Figure 6). Partially disarticulated specimen, still retaining part of

been more difficult to interpret (recent references include Franzén, 1974; 1983; Welch, 1976; Brett, 1978; 1985; Meyer and Ausich, 1983; Eckert, 1988; Donovan, 1990). Evidence for boring is most frequently found in the stem (Figure 11.3C), but is also known from the crown, particularly the dorsal cup in crinoids (Figure 11.4C). Such borings often appear to be host-specific (Brett, 1978) and frequently have resulted in an associated swelling reaction in the surrounding stereom (Figure 11.3C). These borings tend not to perforate the body cavity and instead merely form pits on the surface, particularly in crinoid stems. As the crinoid stem contains little nutrition for a potential parasite, it has been concluded (Brett, 1978; Meyer and Ausich, 1983) that many pit-forming associations represent the activities of epizoic organisms requiring a protective hard substrate (the crinoid endoskeleton) and an elevated position, possibly to facilitate filter feeding (Donovan, 1990). Nevertheless, certain infestations of the crown in some Palaeozoic crinoids are almost certainly due to myzostomid annelids (Meyer and Ausich, 1983, p. 413) and other more problematic parasites.

Palaeozoic crinoids also produced stereom overgrowths in response to encrustation by various non-boring epizoans, such as the tabulate corals *Emmonsia parasitica* (Figures 11.3D) and *Cladochonus* (Figure 11.3E: Hudson *et al.*, 1966, pp. 279–81, Figures 3–5). However, it seems probable that such associations are examples of obligate commensalism (Meyer and Ausich, 1983, pp. 401–6) and a parasitic relationship is unproven. Stereom overgrowth is a reaction of the crinoid stem to various encrusters (Le Menn, 1989), not merely to borers and/or parasites. However, not all crinoid–epizoan associations resulted in stereomic overgrowth. For example, the association between platyceratid gastropods and crinoid crowns, in which the mollusc is frequently found surmounting the anal opening of the echinoderm, produced no overgrowth (but sometimes a 'scar'), presumably because the snail remained vagile. Bowsher (1955) considered the platyceratid–crinoid association to be an obligate commensalism where the gastropod led a coprophagous existence, feeding on the echinoderm's excrement (see also Meyer and Ausich, 1983, pp. 401–3, Figure 5). However, Rollins and Brezinski (1988) have demonstrated that crinoids with platyceratid infestations tend to be smaller than those without associated gastropods, indicating that this relationship is detrimental to the host. Rather, the association may have been parasitic, with the snail possibly utilizing the detritus concentrated by the filtration fan of the crinoid, which would thus be denied a portion of its

---

the distal spiral coiled attachment structure (arrowed). Note the large anal sac, ×1.5. (C) *Synbathocrinus conicus*, Carboniferous (Chadian), Natural History Museum, London (BMNH) E71430 (after Donovan, 1990, Figure 1). Bored dorsal cup with a solitary attached columnal, ×2. (D) Indeterminate archaeocrinid, Ordovician (Caradoc), BMNH E71409 (after Donovan, 1989b, Plate 10, Figure 10). Slab preserving a hash of crinoid debris (mainly stem fragments), ×1. (A), (B) and (D) are latex casts from natural moulds. All specimens were coated with ammonium chloride sublimate before photography.

diet. Further, Baumiller (1990) has demonstrated that some platyceratids drilled into the crinoid tegman, which further suggests a parasitic association.

It may be that even mass mortalities caused by disease are unlikely to be preserved in the rock record. The Caribbean shallow-water echinoid *Diadema antillarum* was decimated by a 'water-born [sic] pathogen transported by oceanic currents' (Jangoux, 1987a, p. 159) during 1983–4 (numerous references, including Lessios *et al.*, 1984; Hughes *et al.*, 1985; Jangoux, 1987a; 1987d; Lessios, 1988). This event was geologically rapid and led to a considerable quantity of echinoid material being introduced into the sediment budget as disarticulated plates, the *Diadema* test having a notoriously poor preservation potential (Asgaard, 1985). In studies on sediment accumulations in reef environments at Bonaire and St Croix, Greenstein (1989; in press) was unable to locate a 'spike' of *Diadema* ossicles which would indicate the position in the stratigraphic record of the mass mortality, suggesting that the sedimentary record in the reef facies may not be sufficiently sensitive to preserve such an event. In contrast, Frankel (1978) and Walbran *et al.* (1989) have demonstrated previous infestations of the crown-of-thorns starfish, *Acanthaster planci*, on the Great Barrier Reef in Australia on the basis of cores from reef sediments. *A. planci* infestations are of comparable density to those of *D. antillarum*, so these contrasting results may suggest differences in mode of preservation between the Great Barrier and Caribbean reef systems, possibly related to ossicle robustness.

Environmental stress

Echinoderms are adversely effected by both desiccation and pollution. Desiccation is important only in shallow-water environments which are subjected to unusually low tides. Glynn (1968) and Hendler (1977) both observed that Caribbean reef-flat echinoids suffered peaks in mortality during subaerial exposure at times of seasonally low tides (Smith, 1984, p. 15). Glynn (1968) considered that desiccation was the principal cause of death under such conditions in reef environments in Puerto Rico, although not all taxa were equally effected. Hendler (1977) agreed with these general conclusions, while noting that in Panama exposure of reef-flat echinoids also coincided with a peak in mortality due to predation by migratory seabirds, although physical stress was nevertheless the principal control on the echinoid population. Giese and Farmanfarmaian (1963) noted a similar stress pattern in low-tide exposure of Californian *Strongylocentrotus purpuratus*. Summer low tides expose this species to high temperatures, the sun and wind; winter low tides are likely to produce reduced salinity conditions due to rain. *S. purpuratus* is able to survive only brief exposure to these environmental extremes. Certain other echinoderms can also survive exposure to low salinities (Binyon, 1966).

Pollutants, both man-made and naturally occurring, may cause either mass-mortalities of echinoderms or, in smaller concentrations, can be sublethal and insidious (Farmanfarmaian *et al.*, 1982). For example, organic matter in sewage may initially encourage rapid echinoid growth, but concentration of toxic waste in tissues thereafter inhibits development of the gonads (Delmas and Régis, 1985). Abnormally high concentrations of heavy metals can lead to developmental abnormalities (den Besten *et al.*, 1988) and can be harmful to adult organisms (Farmanfarmaian *et al.*, 1982). Abnormalities of echinoid test growth may be produced by some pollutants (Dafni, 1983). A variety of oligopygoid echinoids with growth defects are known from the Middle Eocene of Jamaica (Kier, 1967, Pl. 12). These may have been induced by increased concentrations of dissolved natural pollutants associated with the erosion of Cretaceous igneous rocks at this time.

Storms

Smith (1984, pp. 14, 15) recognized three storm-generated processes which can be fatal to shallow-water echinoids: rapid entombment of epifaunal and shallow infaunal taxa by sediment; sediment erosion and washout of (particularly) shallow infaunal irregulars which are then transported onshore (Schäfer, 1972, p. 102) or offshore (Parks, 1973, p. 1109); and crushing of regulars dwelling on rocky substrates by transported pebbles and other debris. Rapid burial is a potential hazard to all shallow-water echinoderms, and Schäfer (1972) noted that echinoids, asteroids and ophiuroids can be buried alive by thin, storm-generated sediment accumulations (for example, 60 cm of sediment can entomb the asteroid *Asterias rubens*, while only 5 cm is sufficient to immobilize and kill an ophiuroid). It is this process of rapid storm-driven sedimentation which can lead to the formation of the exquisitely preserved deposits known as 'starfish beds' (see below). However, storm sedimentation must be the agent of final burial of complete tests for such beds to be formed. For example, Schumacher (1986) noted that storm-generated currents can produce a range of preservational styles in fossil crinoids, from complete specimens to individual ossicles.

Old age

The lifespans of echinoids vary between one and 15 years (Smith, 1984, Table 3.1) and other echinoderms appear to have durations of under ten years (see, for example, Swan, 1966, p. 417). However, some deep-water species of, for example, ophiurod can live up to 20 years and may generally live longer than shallow-water taxa (Gage, 1990a; 1990b). Echinoids with shorter lifespans grow faster (Smith, 1984, p. 89) and those that only live for one year may suffer a mass-mortality following spawning, although physical factors may play a greater role than senescence in such situations. For example, Hendler (1977) recognized that echinoids which live exposed on a Caribbean reef flat reached sexual maturity in under a year, but died from physical stress due to low tides (see above); in the same environment cryptic species reproduced after two years. However, adults of many echinoderm taxa, particularly those with longer life spans, probably die of old age. Schäfer (1972) described the death through senescence of some North Sea echinoderms.

BIOSTRATINOMY: DECAY AND DISARTICULATION

Rates of decay and disarticulation

Unless buried alive or removed to an anoxic environment (see below), it is the fate of almost all echinoderms to disarticulate completely following death, as the MCTs and other soft tissues rot, and the endoskeleton is subjected to the effects of scavengers, bioturbation and physical disturbance. Only unusually sturdy tests, such as those of the sand dollar *Clypeaster*, can survive for any appreciable length of time in an agitated, oxygenated environment. In most echinoderms post-mortem disarticulation is rapid, occurring in a matter of days or weeks. Blyth Cain (1968, p. 192) reported that the comatulid crinoid *Antedon bifida* completely disarticulated within two days of dying, even under anaerobic conditions and without agitation. Liddell (1975) noted that the comatulid *Nemaster rubiginosa* completely disarticulated in two days in an agitated environment, but not so rapidly under low-energy conditions and when protected from scavengers. Lewis (1986; 1987) modelled the rates of disarticulation of Palaeozoic crinoids using Recent ophiuroids and noted a three-stage decomposition process (Table 11.1). The endoskeleton appears to be strong enough to withstand some degree of non-destructive transport in stages 1 and 2. Different ophiuroid species were found to disarticulate at different rates, which Lewis (1986) compared to the varied modes of preservation found in Palaeozoic crinoid taxa (see also Meyer *et al.*, 1990). Schäfer (1972, p. 99) noted that ophiuroid arms are slack enough to be orientated by currents immediately following death, although the arms of North Sea species start to disintegrate after only 15 hours in the natural environment. In contrast, ophiuroids investigated by Lewis (1986) took a period of weeks to disarticulate, presumably under laboratory conditions. These differences may be due at least partly to differences in the degree of agitation and also temperature, as Kidwell and Baumiller (1989) have shown that disarticulation of regular echinoids is accelerated by increased temperature.

Table 11.1 *Changes that occur during the post-mortem disarticulation of the ophiuroid endoskeleton (after Lewis, 1987)*

Stage	Effect
1	Gradual loss of colour and body fluids. Release of decomposition gases. Some stiffening of proximal arms.
2	Arms flexible, with little or no disarticulation. Dorsal integument of disc may become detached.
3	Partial decomposition, continuing to completeness.

Table 11.2 *Patterns of disarticulation in certain Recent echinoderms*

Asteroid *Asterias rubens* (after Schäfer, 1972)	Time (days)	Regular echinoid (after Smith, 1984; Schäfer, 1972)
Arms flexible, but stiffen soon after death		Radioles and pedicellariae droop; moribund
DEATH	0	DEATH
Red colouration soon fades	1	Radioles drooping
	2	Pedicellariae then radioles
Decomposition gases arch	3	detach
body and inflate carcass	4	
	5	
Dorsal integument starts to	6	
lift	7	
	8	Peristomial and periproctal
Dorsal skin becomes	9	membranes disintegrate
loosened and drifts away	10	
as pieces	11	
Decomposition has	12	Apical system largely
removed water vascular	13	detached
system, digestive system	14	
and musculature from	15	Aristotle's Lantern
ventral skeleton	16	completely disarticulated
	17	
Complete disarticulation		Corona fragments

Schäfer (1972) made detailed descriptions of the patterns of disarticulation shown by eleutherozoan echinoderms of the southern North Sea (data for asteroids and echinoids are summarized in Table 11.2 herein). The possibility exists that detailed statistical analyses can be undertaken for these and other sequences of echinoderm disintegration in order to quantify the patterns of disarticulation within the group. Similar calculations have already been made for some terrestrial tetrapods (Hill, 1979).

Transport

Of the extant classes of echinoderms, ophiuroid (Schäfer, 1972, p. 99) and crinoid (Meyer and Meyer, 1986) carcasses are not known to float, whereas asteroids (Schäfer, 1972, p. 96) and holothurians (Schäfer, 1972, p. 105) become filled with decomposition gases following death and can be rolled about on the sea floor by currents. Only the echinoids can genuinely be floated by decomposition gases (see, for example, Glynn, 1968; Reyment,

1986). This nekroplanktonic transport of dead sea urchins may be an important factor in spreading echinoid debris over a broad area, as radioles are presumably shed until the peristomial or periproctal membrane ruptures and the test sinks. The 'mummified' specimen discussed by Reyment (1986) was unusual for remaining afloat under laboratory conditions for 28 days, after it had been collected floating offshore (compare with Table 11.2). It is assumed that echinoid tests usually float for much shorter periods (if at all), although development of an algal jacket may also act to seal decomposition gases within the test (Kidwell and Baumiller, 1989), thus permitting an extended period of buoyancy.

It has been further suggested that disarticulated echinoderm (particularly crinoid) ossicles may become buoyant, due to the pore spaces within the stereom becoming filled with gases during the decomposition of the soft tissue (among others, see Müller, 1953; 1955, p. 290; Marcher, 1962, p. 831; Schwarzacher, 1963, p. 581; further references discussed by Meyer and Meyer, 1986, pp. 298–9). However, there is little, if any, evidence to support this speculation. Blyth Cain (1968, pp. 196–8, Figure 2) monitored the change in specific gravity of the remains of the comatulid *Antedon bifida* following death and recognized an overall increase in effective density. Blyth Cain also commented that the decomposition gases produced by a crinoid are probably soluble in water, thus eliminating any likelihood of buoyant transport. Meyer and Meyer (1986) noted that reef-dwelling comatulids die and then rapidly disarticulate *in situ*, without buoyant transport.

Kidwell and Baumiller (1989) have examined the effect of transport on the regular echinoid test using tumbling experiments. Freshly killed specimens proved to be remarkably sturdy; for example, *Strongylocentrotus purpuratus* still retained radioles after eight hours of tumbling without grit. However, tests that have been allowed to decay for a period of weeks before tumbling disintegrated far more rapidly. Transport immediately following death may thus have little effect on the excellence of preservation of a regular echinoid.

However, repeated reworking is probably an important factor in disarticulating the echinoderm endoskeleton. Many fragments of fossil echinoderms appear to have survived at least some transport, suggesting that multiple episodes of reworking may be necessary to disintegrate certain skeletal structures completely. For example, pluricolumnals of Palaeozoic crinoids were sturdy enough to form tool marks (Benton and Gray, 1981), to become orientated as current accumulations (Schwarzacher, 1963) and to survive impact following downhill saltation with only limited fragmentation (Broadhurst and Simpson, 1973).

Perhaps the most important aspect of the biostratinomy of echinoderms is that they 'are potentially sensitive indicators of post-mortem depositional processes because differential disarticulation can be correlated with exposure time and degree of transportation prior to final burial' (Meyer *et al.*, 1990). For example, Schumacher (1986) recognized three principal modes of preservation shown by Upper Ordovician crinoid-rich storm deposits: autochthonous crinoid preservation, with spectacular accumulation of near-complete specimens which are randomly orientated and parallel to bedding; autochthonous-winnowed crinoid preservation, with specimens from crowns to solitary ossicles showing a preferred orientation; and allochthonous crinoid preservation, with lateral gradation from more to less complete specimens. These modes of preservation are associated with mud-supported carbonates produced by: thick, rapidly deposited beds (poorly sorted); lenticular, vertically-graded beds with erosional bases and parallel to hummocky cross-stratification (moderately well sorted); and lenticular, laterally-graded deposits with erosional bases (moderately to well sorted). More detailed studies by Meyer *et al.* (1989; 1990; see also Ausich and Meyer, 1990) on the Lower Carboniferous of Kentucky have demonstrated a close link between lithofacies and style of preservation shown by pelmatozoans.

DIAGENESIS

While diagenesis of carbonate skeletons is discussed by Tucker (this volume), it is considered essential to emphasize the principal processes of post-burial change that affect the echinoderm endoskeleton.

In Recent echinoderms the endoskeleton is composed of high magnesium calcite (HMC), normally containing 10–20 mol% $MgCO_3$. Ancient echinoderms have a low magnesium calcite (LMC) test which has been diagenetically altered, as can be demonstrated by comparing Recent and Upper Pleistocene tests of the same species (see, for example, MacQueen *et al.*, 1974). This change occurs because HMC is metastable, so stabilization is inevitable (Land, 1967). The transformation from HMC to LMC apparently occurs by a solid-state process of incongruent dissolution (Land, 1967; Bathurst, 1971, pp. 335–8), resulting in ossicles that contain 1.5–4 mol% $MgCO_3$ (Scoffin, 1987, p. 111). This transformation may require the preliminary formation of a thin layer of syntaxial cement in optical continuity with the ossicle (Bathurst, 1971, p. 50). The exsolved Mg^{2+} cations may be incorporated in rhombs of microdolomite within the LMC lattice (see, for example, MacQueen and Ghent, 1970; MacQueen *et al.*, 1974; Lohmann and Meyers, 1977; Leutloff and Meyers, 1984).

Growth of syntaxial cements on echinoderm ossicles commences with infilling of the pore space (Evamy and Shearman, 1965; Bathurst, 1971), unless the stereom has already been infilled by clays, which act to preserve the microstructure against precipitation of the spar (Roux, 1975; Figure 11.5 herein). The stereom meshwork may be destroyed in muddy deep-sea environments by the growth of framboidal pyrite which bursts the meshwork (Roux, 1975; 1987) and is biocorroded by fungal and bacterial activity (Roux, 1987). Roux (1975) considered that the initial overgrowth of some crinoid ossicles is by aragonite needles which eventually coalesce to form a crust. At an early stage this aragonite is transformed into calcite. Calcitic overgrowths develop as a cluster of pyramidal crystals, initially on those faces of the host ossicle that are orientated in the direction of the crystallographic C-axis (Evamy and Shearman, 1965; 1969; Bathurst, 1971) and crystal growth is always fastest in this direction. Overgrowth masks the original shape of an ossicle. Calcite overgrowths are in optical continuity with the ossicle and usually coalesce as growth proceeds (but see Braithwaite and Heath, 1989).

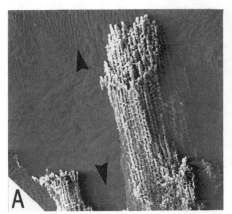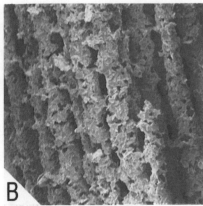

Figure 11.5. Mud infilling galleried stereom of a crinoid columnal from the Ordovician (Ashgill) Rhiwlas Limestone of north Wales. The columnal has been dissolved away with 10% hydrochloric acid, leaving the mud infill as a 'ghost' microstructure. (A) Infill of stereom adjacent to axial canal (arrowed; compare with Figure 11.1F), ×115. (B) Enlargement of the central part of (A), ×850.

Overgrowth on monocrystalline echinoderm ossicles is faster than on polycrystalline bioclasts. Crystal growth does not occur where the grain has been micritized and overgrowths in micritic sediments have been shown to fill voids (Görür, 1979; Walkden and Berry, 1984).

EXCEPTIONAL PRESERVATION

Preservation of soft tissues

Soft tissues of echinoderms are rarely retained by the fossil record. However, tube feet are known from a few exceptional specimens. For example, the holotype of the Middle Cambrian (Burgess Shale) crinoid(?) *Echmatocrinus brachiatus* bears a number of stout tube feet (Sprinkle, 1973). This specimen is crushed flat and the endoskeleton has been pyritized, with the tube feet 'composed of an extremely thin sheet of pyrite-like material containing organic carbonaceous specks' (Sprinkle, 1973, p. 183). In contrast, the Upper Ordovician (Caradocian) asteroid *Siluraster caractaci* is preserved as an external mould which retains simple, unsuckered tube feet in three dimensions (Gale, 1987, Figure 7F). Preservation of the plated tube feet of some primitive echinoderms cannot be regarded as exceptional (for example, the ophiocistioid *Gillocystis*: see, Jell, 1983).

   Haugh (1975) identified various species of Lower Carboniferous camerate crinoid whose thecae are either silicified and hollow, or silicified chert steinkerns. The silicification extends to internal structures that have been interpreted as visceral organs. Haugh considered that preservation may have been aided by these organs originally having calcitic supporting spicules.

Holothurians

A further example of soft tissue preservation in echinoderms is shown by the extremely rare occurrences of fossilized holothurians. The holothurian body lacks an articulated test, unlike other echinoderms, and instead comprises a leathery body wall which includes microscopic calcareous sclerites (Frizzell *et al.*, 1966). Because of this lack of an articulated skeleton, the fossil record of 'complete' holothurians is extremely poor, although isolated sclerites are well known as microfossils. Pawson (1980) considered only four records of complete fossil holothurians to be reliable, but further specimens have subsequently been discovered (for example, Sroka, 1988; Smith and Gallemí, 1991). Preservation of holothurians and other echinoderm soft tissues was undoubtedly a diagenetically rapid event. This is well demonstrated by the mode of preservation of the new species (Sroka, 1988) from the Upper Carboniferous Essex Fauna of Illinois. Specimens are generally preserved as compression fossils within diagenetically early siderite nodules. Preservation is often spectacular, with, for example, retention of the tube feet and gut traces, but other specimens have been decomposed or scavenged (Baird *et al.*, 1986).

Echinodermen-Lagerstätten

The majority of exceptionally well-preserved echinoderm faunas do not, however, retain soft tissues. Rather, such deposits preserve very complete echinoderm tests, often in great numbers and/or variety. For example, fossil echinoid tests are usually preserved in a 'denuded' state, having lost lantern elements (if originally present), radioles, pedicellariae, peristomial plating and, in many regulars, periproctal plating (although a Recent specimen, the spatangoid in Figure 11.3A shows features typical of a fossil echinoid test). In contrast, Aslin (1968) recorded exceptionally well-preserved echinoids from the Jurassic of Northamptonshire which retain radioles, the apical disc and the Aristotle's Lantern. Such preservation is unusually good and is correctly termed an 'Echinodermen-lagerstätte' or echinoderm 'mother lode' (Seilacher *et al.*, 1985).

Two principal processes lead to the preservation of exceptional echinoderm deposits. At least some fossil echinoderms have been preserved following transport into, or introduction of, anoxic environmental conditions, which had the effect of discouraging bioturbation and scavengers while retarding the rate of decay of soft tissues. Such environments are well known for their association with pseudoplanktonic crinoids, which were introduced into such hostile conditions after their floating substrate (usually a log) became waterlogged and sank to the sea floor (see, for example, McIntosh, 1978; Simms, 1986; Wignall and Simms, 1990). However, well-preserved accumulations of fossil echinoderms are more usually associated with rapid sedimentation events in a shallow-water environment (see, for example, Aslin, 1968). Such pulses of sedimentation bury echinoderms alive (see discussion of death during storms, above) and, providing there is no subsequent reworking

(Schumacher, 1986), can lead to truly exquisite preservation. It is thus apparent that the best preservation of echinoderms is associated with rapid death and immediate introduction into an environment that favours preservation.

A particular group of exceptional echinoderm deposits are called 'starfish beds'. These are usually well-preserved accumulations of asteroids and ophiuroids, two groups with an otherwise poor fossil record. Goldring and Stephenson (1972) examined the depositional environments of three starfish beds in Britain and concluded that smothering by sediment was the principal factor in preservation (also see Meyer, 1984). Goldring and Stephenson considered the preservation of echinoderms in the Lady Burn Starfish Bed (Upper Ordovician) of Scotland to be due to rapid local resedimentation (due to storm action?) of a shallow-water fauna living on a sandy substrate (although Ingham, 1978, p. 173, attributed preservation to deposition by downslope currents and turbulence on an unstable submarine fan). In contrast, echinoderms in the Leintwardine Starfish Bed (Upper Silurian) were preserved in fine-grained turbidites within channel deposits following transport, while the Middle Lias (Lower Jurassic) starfish bed of Dorset represents *in situ* burial of ophiuroids by rapid deposition of a layer of sediment too thick for subsequent bioturbation to penetrate and disturb the echinoderm horizon. A similar range of burial conditions was probably responsible for producing exceptionally well-preserved crinoid horizons (Brett and Eckert, 1982; Franzén, 1982).

SUMMARY

The taxonomy and palaeobiology of fossil echinoderms is almost entirely dependent upon the record of 'complete' specimens, yet the structure of the endoskeleton in this group does not favour excellent preservation. It is the fate of the majority of echinoderms to disarticulate into a myriad separate ossicles following death. Even those ossicles that can be identified to a high taxonomic level can rarely be used to reconstruct the producing organism because of the great variation found in echinoderm test architecture. However, it is this skeletal complexity and the potential that echinoderms may have for 'timing' the interval between death and final burial that makes their processes of fossilization so fascinating. From the cidaroid that retains its pedicellariae and Aristotle's Lantern to the solitary crinoid columnal, all echinoderm fossils have a potential unique among marine invertebrates to tell a tale.

ACKNOWLEDGEMENTS

This chapter was written during the period of National Geographic Society grant 4264–90, which I gratefully acknowledge. Photographs not taken by the author were supplied by Cornelis J. Veltkamp (Figures 11.1, 11.5), D. George Solomon (Figure 11.3) and the photographic unit of the Natural

History Museum, London (Figures 11.4C, 11.4D). Harold L. Dixon printed the photographs in Figure 11.5. I thank Dr William I. Ausich for his help in supplying certain references. Drs Christopher R.C. Paul, Richard B. Aronson, Michael J. Simms and Professor George D. Sevastopulo made constructive review comments on an early draft of this chapter.

REFERENCES

Alekseev, A.S. and Endelman, L.G., 1989, Assotsiatsiya ehktoparaziticheskikh perednezhabernykh gastropod s lozdnemelovymi morskimi ezhami *Galerites, Problemy Izucheniya Iskopaemykh i Sovremennykh Iglokozhikh, Akademiya Nauk Ehstonskoj SSR*: 166–74 [in Russian].

Allison, P.A., 1990, Variation in rates of decay and disarticulation of echinodermata: implications for the application of actualistic data, *Palaios*, **5**.

Aronson, R.B., 1987a, Predation on fossil and Recent ophiuroids, *Paleobiology*, **13** (2): 187–92.

Aronson, R.B., 1987b, A murder mystery from the Mesozoic, *New Scientist* **116** (1581): 56–9.

Aronson, R.B., 1989a, A community-level test of the Mesozoic marine revolution theory, *Paleobiology*, **15** (1): 20–5.

Aronson, R.B., 1989b, Brittlestar beds: low-predation anachronisms in the British Isles, *Ecology*, **70** (4): 856–65.

Aronson, R.B. and Harms, C.A., 1985, Ophiuroids in a Bahamian saltwater lake: the ecology of a Paleozoic-like community, *Ecology*, **66** (5): 1472–83.

Asgaard, U., 1985, Hvorfor er *Diadema* så succesfuld?, *Dansk Geologisk Forening*, 1984: 109–10.

Aslin, C.J., 1968, Echinoid preservation in the Upper Estuarine Limestone of Blisworth, Northamptonshire, *Geological Magazine*, **105** (6): 506–18.

Ausich, W.I. and Meyer, D.L., 1990, Origin and composition of carbonate buildups and associated facies in the Fort Payne Formation (Lower Mississippian, south-central Kentucky): an integrated sedimentologic and paleoecologic analysis, *Geological Society of America Bulletin*, **102** (1): 129–46.

Baird, G.C., Sroka, S.D., Shabica, C.W. and Kuecher, G.J., 1986, Taphonomy of Middle Pennsylvanian Mazon Creek area fossil localities, northeast Illinois: significance of exceptional fossil preservation in syngenetic concretions, *Palaios*, **1** (3): 271–85.

Bassler, R.S., 1913, Notes on an unusually fine slab of fossil crinoids, *Proceedings of the United States National Museum*, **46** (2009): 57–9.

Bathurst, R.G.C., 1971, *Carbonate sediments and their diagenesis*, Elsevier, Amsterdam.

Baumiller, T.K., 1990, Non-predatory drilling on Mississippian crinoids by platyceratid gastropods, *Palaeontology*, **33** (3): 743–8.

Benton, M.J. and Gray, D.I., 1981, Lower Silurian distal shelf storm-induced turbidites in the Welsh borders: sediments, tool marks and trace fossils, *Journal of the Geological Society*, **138** (6): 675–94.

Binyon, J., 1966, Salinity tolerance and ionic regulation. In R.A. Boolootian (ed.), *Physiology of Echinodermata*, Wiley, New York: 359–77.

Blyth Cain, J.D., 1968, Aspects of the depositional environment and palaeoecology of crinoidal limestones, *Scottish Journal of Geology*, **4** (3): 191–208.

Bomwer, T. and Meyer, C.A., 1987, Ophiurous oddities—deviation from pentamerous symmetry in non-fissiparous ophiuroids: a comparison of Recent and fossil species, *Eclogae Geologicae Helvetiae*, **80** (3): 897–905.

Bowsher, A.L., 1955, Origin and adaptation of platyceratid gastropods, *University of Kansas Paleontological Contributions, Mollusca Article* **5**: 1–11.

Braithwaite, C.J.R. and Heath, R.A., 1989, Inhibition, distortion and dissolution of overgrowth cements on pelmatozoan fragments, *Journal of Sedimentary Petrology*, **59** (2): 267–71.

Brett, C.E., 1978, Host-specific pit-forming epizoans on Silurian crinoids, *Lethaia*, **11** (3): 217–32.

Brett, C.E., 1985, *Tremichnus*: a new ichnogenus of circular-parabolic pits in fossil echinoderms, *Journal of Paleontology*, **59** (3): 625–35.

Brett, C.E. and Eckert, J.D., 1982, Palaeoecology of a well-preserved crinoid colony from the Silurian Rochester Shale of Ontario, *Life Sciences Contributions, Royal Ontario Museum*, **131**: 1–20.

Broadhurst, F.M. and Simpson, I.M., 1973, Bathymetry on a Carboniferous reef, *Lethaia*, **6** (4): 367–81.

Conway Morris, S., 1981, Parasites and the fossil record, *Parasitology*, **82** (3): 489–509.

Dafni, J., 1983, Aboral depressions in the tests of the sea urchin *Tripneustes* cf. *gratilla* (L.) in the Gulf of Eilat, Red Sea, *Journal of Experimental Marine Biology and Ecology*, **67** (1): 1–15.

Delmas, P. and Régis, M.B., 1985, Influence of the domestic pollution generated by the discharge of the Marseilles–Cortiou sewage outlet on the distribution and morphometry of *Paracentrotus lividus* (Lmk.). In B.F. Keegan and B.D.S. O'Connor (eds), *Echinodermata: Proceedings of the Fifth International Echinoderm Conference, Galway, 24–29 September, 1984*, Balkema, Rotterdam: 381.

den Besten, P.J., Zandee, D.I. and Voogt, P.A., 1988, Effects of cadmium on the development of the embryos of the sea star *Asterias rubens* L. In R.D. Burke, P.V. Mladenov, P. Lambert and R.L. Parsley (eds), *Echinoderm Biology: Proceedings of the Sixth International Echinoderm Conference, Victoria, British Columbia, 23–28 August, 1987*, Balkema, Rotterdam: 788.

Donovan, S.K., 1984, Stem morphology of the Recent crinoid *Chladocrinus (Neocrinus) decorus*, *Palaeontology*, **27** (4): 825–41.

Donovan, S.K., 1985, The Ordovician crinoid genus *Caleidocrinus* Waagen and Jahn, 1899, *Geological Journal*, **20** (2): 109–21.

Donovan, S.K., 1986, Pelmatozoan columnals from the Ordovician of the British Isles, part 1, *Palaeontographical Society Monograph*, London, **138** (568): 1–68.

Donovan, S.K., 1989a, The significance of the British Ordovician crinoid fauna, *Modern Geology*, **13** (3/4): 243–55.

Donovan, S.K., 1989b, Pelmatozoan columnals from the Ordovician of the British Isles, part 2, *Palaeontographical Society Monograph*, London, **142** (579): 69–114.

Donovan, S.K., 1990, Site selectivity of a Lower Carboniferous boring organism infesting a crinoid, *Geological Journal*, **25**.

Donovan, S.K. and Gale, A.S., 1990, Predatory asteroids and the decline of the articulate brachiopod, *Lethaia*, **23** (2): 77–86.

Donovan, S.K., Hollingworth, N.T.J. and Veltkamp, C.J., 1986, The British Permian crinoid '*Cyathocrinites*' *ramosus* (Schlotheim), *Palaeontology*, **29** (4): 809–25.

Eckert, J.D., 1988, The ichnogenus *Tremichnus* in the Lower Silurian of western New

York, *Lethaia*, **21** (3): 281–3.

Emson, R.H. and Wilkie, I.C., 1980, Fission and autotomy in echinoderms, *Oceanography and Marine Biology Annual Review*, **18**: 155–250.

Evamy, B.D. and Shearman, D.J., 1965, The development of overgrowths from echinoderm fragments, *Sedimentology*, **5** (3): 211–33.

Evamy, B.D. and Shearman, D.J., 1969, Early stages in development of overgrowths on echinoderm fragments in limestones, *Sedimentology*, **12** (3/4): 317–22.

Farmanfarmaian, A., Socci, R. and Iannacconne, V., 1982, Interaction of heavy metals with intestinal transport mechanisms in echinoderms. In J.M. Lawrence (ed.), *Echinoderms: Proceedings of the International Conference, Tampa Bay, 14–17 September, 1981*, Balkema, Rotterdam: 339–44.

Frankel, E., 1978, Evidence from the Great Barrier Reef of ancient *Acanthaster* aggregations, *Atoll Research Bulletin*, **220**: 75–93.

Franzén, C., 1974, Epizoans on Silurian–Devonian crinoids, *Lethaia*, **7** (4): 287–301.

Franzén, C., 1982, A Silurian crinoid thanatotope from Gotland, *Geologiska Föreningens i Stockholm Förhandlingar*, **103** (4): 469–90.

Franzén, C., 1983, Ecology and taxonomy of Silurian crinoids from Gotland, *Acta Universitatis Upsaliensis*, **665**: 1–31.

Franzén-Bengtson, C., 1983, Radial perforations in crinoid stems from the Silurian of Gotland, *Lethaia*, **16** (4): 291–302.

Frizzell, D.L., Exline, H. and Pawson, D.L., 1966, Holothurians. In R.C. Moore (ed.), *Treatise on invertebrate paleontology, Part U, Echinodermata* 3 (2), Geological Society of America and University of Kansas Press, New York and Lawrence: U641–72.

Gage, J.D., 1990a, Skeletal growth markers in the deep-sea brittle stars *Ophiura ljungmani* and *Ophiomusium lymani*, *Marine Biology*, **104** (3): 427–35.

Gage, J.D., 1990b, Skeletal growth bands in brittle stars: microstructure and significance as age markers, *Journal of the Marine Biological Association of the United Kingdom*, **70** (1): 209–24.

Gale, A.S., 1987, Phylogeny and classification of the Asteroidea (Echinodermata), *Zoological Journal of the Linnean Society*, **89** (2): 107–32.

Gibson, M.A. and Watson, J.B., 1989, Predatory and non-predatory borings in echinoids from the upper Ocala Formation (Eocene), north-central Florida, U.S.A., *Palaeogeography, Palaeoclimatology, Palaeoecology*, **71** (1–3): 309–21.

Giese, A.C. and Farmanfarmaian, A., 1963, Resistance of the purple sea urchin to osmotic stress, *Biological Bulletin*, **124** (2): 182–92.

Glynn, P.W., 1968, Mass mortalities of echinoids and other reef flat organisms coincident with midday, low water exposures in Puerto Rico, *Marine Biology*, **1** (3): 226–43.

Goldring, R. and Stephenson, D.G., 1970, Did *Micraster* burrow? In T.P. Crimes and J.C. Harper (eds), *Trace fossils*, Seel House Press, Liverpool: 179–84.

Goldring, R. and Stephenson, D.C., 1972, The depositional environment of three starfish beds, *Neues Jahrbuch für Geologie und Paläontologie Monatshefte*, **1972** (10): 611–24.

Görür, N., 1979, Downward development of overgrowths from echinoderm fragments in a submarine environment, *Sedimentology*, **26** (4): 605–8.

Greenstein, B.J., 1989, Mass mortality of the West-Indian echinoid *Diadema antillarum* (Echinodermata: Echinoidea): a natural experiment in taphonomy, *Palaios*, **4** (5): 487–92.

Greenstein, B.J., in press, Taphonomic biasing of subfossil echinoid populations adjacent to St. Croix, U.S.V.I., *Proceedings of the Twelfth Caribbean Geological Conference, St. Croix, August 7–11, 1989.*

Grygier, M.J., 1988, Unusual and mostly cysticolous crustacean, molluscan, and myzostomidan associates of echinoderms. In R.D. Burke, P.V. Mladenov, P. Lambert and R.L. Parsley (eds), *Echinoderm Biology: Proceedings of the Sixth International Echinoderm Conference, Victoria, British Columbia, 23–28 August, 1987*, Balkema, Rotterdam: 775–84.

Hall, K.R.L. and Schaller, G.B., 1964, Tool-using behaviour of the California sea otter, *Journal of Mammalogy*, **45** (2): 287–98.

Hattin, D.E., 1958, Regeneration in a Pennsylvanian crinoid spine, *Journal of Paleontology*, **32** (4): 701–2.

Haugh, B.N., 1975, Digestive and coelomic systems of Mississippian camerate crinoids, *Journal of Paleontology*, **49** (3): 472–93.

Hendler, G., 1977, The differential effects of seasonal stress and predation on the stability of reef-flat echinoid populations, *Proceedings of the Third International Coral Reef Symposium, Miami*: 217–23.

Hill, A., 1979, Disarticulation and scattering of mammal skeletons, *Paleobiology*, **5** (3): 261–74.

Himmelmann, J.H. and Steele, D.H., 1971, Foods and predators of the green sea urchin *Strongylocentrotus droebachiensis* in Newfoundland waters, *Marine Biology*, **9** (4): 315–22.

Hudson, R.G.S., Clark, M.J. and Sevastopulo, G.D., 1966, The palaeoecology of a Lower Viséan crinoid fauna from Feltrim, Co. Dublin, *Scientific Proceedings of the Royal Dublin Society, Series A*, **2** (17): 273–86.

Hughes, T.P., Keller, B.D., Jackson, J.B.C. and Boyle, M.J., 1985, Mass mortality of the echinoid *Diadema antillarum* Phillipi in Jamaica, *Bulletin of Marine Science*, **36** (2): 377–84.

Ingham, J.K., 1978, Geology of a continental margin 2: Middle and Late Ordovician transgression, Girvan. In D.R. Bowes and B.E. Leake (eds), *Crustal evolution in northwestern Britain and adjacent regions*, Seel House Press, Liverpool: 163–76.

Jangoux, M., 1982, Food and feeding mechanisms: Asteroidea. In M. Jangoux and J.M. Lawrence (eds) *Echinoderm nutrition*, Balkema, Rotterdam: 117–59.

Jangoux, M., 1987a, Diseases of Echinodermata. I. Agents microorganisms and protistans, *Diseases of Aquatic Organisms*, **2** (2): 147–62.

Jangoux, M., 1987b, Diseases of Echinodermata. II. Agents metazoans (Mesozoa to Bryozoa), *Diseases of Aquatic Organisms*, **2** (3): 205–34.

Jangoux, M., 1987c, Diseases of Echinodermata. III. Agents metazoans (Annelida to Pisces), *Diseases of Aquatic Organisms*, **3** (1): 59–83.

Jangoux, M., 1987d, Diseases of Echinodermata. IV. Structural abnormalities and general considerations on biotic diseases, *Diseases of Aquatic Organisms*, **3** (3): 221–9.

Jefferies, R.P.S., 1988, How to characterize the Echinodermata—some implications of the sister-group relationship between echinoderms and chordates. In C.R.C. Paul and A.B. Smith (eds), *Echinoderm phylogeny and evolutionary biology*, Clarendon Press, Oxford: 3–12.

Jell, P.A., 1983, Early Devonian echinoderms from Victoria (Rhombifera, Blastoidea and Ophiocistioidea), *Memoirs of the Association of Australasian Palaeontologists*, **1**: 209–35.

Joysey, K.A., 1959, Probable cirripede, phoronid, and echiuroid burrows within a Cretaceous echinoid test, *Palaeontology*, **1** (4): 397–400.

Kidwell, S.M. and Baumiller, T.K., 1989, Post-mortem disintegration of echinoids: effects of temperature, oxygenation, tumbling, and algal coats, *Abstracts, 28th International Geological Congress, Washington, D.C., 9–19 July*, **2**: 2.188–9.

Kidwell, S.M. and Baumiller, T.K., 1990, Experimental disintegration of regular echinoids: roles of temperature, oxygen and decay thresholds, *Paleobiology*, **16**: 247–71.

Kier, P.M., 1967, Revision of the oligopygoid echinoids, *Smithsonian Miscellaneous Collections*, **152** (2): 149 pp.

Kier, P.M., 1977, The poor fossil record of the regular echinoid, *Paleobiology*, **3** (2): 168–74.

Kier, P.M., 1981, A bored Cretaceous echinoid, *Journal of Paleontology*, **55** (3): 656–9.

Kier, P.M. and Grant, R.E., 1965, Echinoid distribution and habitats, Key Largo Coral Reef Preserve, Florida, *Smithsonian Miscellaneous Collections*, **149** (6): 68 pp.

Land, L.S., 1967, Diagenesis of skeletal carbonates, *Journal of Sedimentary Petrology*, **37** (3): 914–30.

Lane, N.G., 1984, Predation and survival among inadunate crinoids, *Paleobiology*, **10** (4): 453–8.

Lasker, H., 1976, Effects of differential preservation on the measurement of taxonomic diversity, *Paleobiology*, **2** (1): 84–93.

Le Menn, J., 1989, Contrôle de la sécrétion stéréomique dans les stolons d'un Crinoïde nouveau du Dévonien inferieur d'Algérie, *Lethaia*, **22** (4): 395–404.

Lessios, H.A., 1988, Mass mortality of *Diadema antillarum* in the Caribbean: what have we learned?, *Annual Review of Ecology and Systematics*, **19**: 371–93.

Lessios, H.A., Robertson, D.R. and Cubit, J.D., 1984, Spread of *Diadema* mass mortality through the Caribbean, *Science*, **226** (4672): 335–7.

Leutloff, A.H. and Meyers, W.J., 1984, Regional distribution of microdolomite inclusions in Mississippian echinoderms from southwestern New Mexico, *Journal of Sedimentary Petrology*, **54** (2): 423–46.

Lewis, D.N. and Ensom, P.C., 1982, *Archaeocidaris whatleyensis* sp. nov. (Echinoidea) from the Carboniferous Limestone of Somerset, and notes on echinoid phylogeny, *Bulletin of the British Museum (Natural History)*, Geology, **36** (2): 77–104.

Lewis, R.D., 1980, Taphonomy. In T.W. Broadhead and J.A. Waters (eds), *Echinoderms: notes for a short course, University of Tennessee Studies in Geology*, **3**: 27–39.

Lewis, R.D., 1986, Relative rates of skeletal disarticulation in modern ophiuroids and Paleozoic crinoids, *Geological Society of America Abstracts with Programs*, **18** (6): 672.

Lewis, R.D., 1987, Post-mortem decomposition of ophiuroids from the Mississippi Sound, *Geological Society of America Abstracts with Programs*, **19** (2): 94–5.

Liddell, W.D., 1975, Recent crinoid biostratinomy, *Geological Society of America Abstracts with Programs*, **7** (7): 1169.

Lohmann, K.C. and Meyers, W.J., 1977, Microdolomite inclusions in cloudy prismatic calcites: a proposed criterion for former high-magnesium calcites, *Journal of Sedimentary Petrology*, **47** (3): 1078–88.

MacQueen, R.W. and Ghent, E.D., 1970, Electron microprobe study of magnesium distribution in some Mississippian echinoderm limestones from western Canada, *Canadian Journal of Earth Sciences*, **7** (5): 1308–16.

MacQueen, R.W., Ghent, E.D. and Davies, G.R., 1974, Magnesium distribution in living and fossil specimens of the echinoid *Peronella lesueuri* Agassiz, Shark Bay, Western Australia, *Journal of Sedimentary Petrology*, **44** (1): 60–9.

Macurda, D.B., Jr and Meyer, D.L., 1975, The microstructure of the crinoid endoskeleton, *University of Kansas Paleontological Contributions, Paper* **74**: 1–22.

Macurda, D.B., Jr, Meyer, D.L. and Roux, M., 1978, The crinoid stereom. In R.C. Moore and C. Teichert (eds), *Treatise on invertebrate paleontology, Part T, Echinodermata* 2 (1), Geological Society of America and University of Kansas Press, New York and Lawrence: T217–28, T230, T232.

Maes, P. and Jangoux, M., 1985, The bald-sea-urchin disease: a bacterial infection. In B.F. Keegan and B.D.S. O'Connor (eds), *Echinodermata: Proceedings of the Fifth International Echinoderm Conference, Galway, 24–29 September, 1984*, Balkema, Rotterdam: 313–14.

Maples, C.G. and Archer, A.W., 1989, Paleoecological and sedimentological significance of bioturbated crinoid calyces, *Palaios*, **4** (4): 379–83.

Marcher, M.V., 1962, Petrography of Mississippian limestones and cherts from the northwestern Highland Rim, Tennessee, *Journal of Sedimentary Petrology*, **32** (4): 819–32.

McIntosh, G.C., 1978, Pseudoplanktonic crinoid colonies attached to Upper Devonian (Frasnian) logs, *Geological Society of America Abstracts with Programs*, **10** (7): 453.

Messing, C.G., RoseSmyth, M.C., Mailer, S.R. and Miller, J.E., 1988, Relocation movement in a stalked crinoid (Echinodermata), *Bulletin of Marine Science*, **42** (3): 480–7.

Meyer, C.A., 1984, Palökologie und Sedimentologie der Echinodermen lagerstätte Scholgraben (mittleres Oxfordian, Weissenstein, Kt. Solothurn), *Eclogae Geologicae Helvetiae*, **77** (3): 649–73.

Meyer, D.L., 1985, Evolutionary implications of predation on Recent comatulid crinoids from the Great Barrier Reef, *Paleobiology*, **11** (2): 154–64.

Meyer, D.L. and Ausich, W.I., 1983, Biotic interactions among Recent and among fossil crinoids. In M.J.S. Tevesz and P.L. McCall (eds), *Biotic interactions in Recent and fossil benthic communities*, Plenum, New York: 377–427.

Meyer, D.L., Ausich, W.I. and Terry, R.E., 1989, Interpretation of Paleozoic crinoid taphonomy: limitations of Recent models, *Abstracts, 28th International Geological Congress, Washington, D.C., 9–19 July*, **2**: 2.419.

Meyer, D.L., Ausich, W.I. and Terry, R.E., 1990, Comparative taphonomy of echinoderms in carbonate facies: Fort Payne Formation (Lower Mississippian) of Kentucky and Tennessee, *Palaios*, **4** (6): 533–52.

Meyer, D.L. and Meyer, K.B., 1986, Biostratinomy of Recent crinoids (Echinodermata) at Lizard Island, Great Barrier Reef, Australia, *Palaios*, **1** (3): 294–302.

Mladenov, P.V., 1983, Rate of arm regeneration and potential causes of arm loss in the feather star *Florometra serratissima* (Echinodermata: Crinoidea), *Canadian Journal of Zoology*, **61** (12): 2873–9.

Motokawa, T., 1985, Catch connective tissue: the connective tissue with adjustable mechanical properties. In B.F. Keegan and B.D.S. O'Connor (eds), *Echinodermata: Proceedings of the Fifth International Echinoderm Conference, Galway, 24–29 September, 1984*, Balkema, Rotterdam: 69–73.

Motokawa, T., 1988, Catch connective tissue: a key character for echinoderms' success. In R.D. Burke, P.V. Mladenov, P. Lambert and R.L. Parsley (eds), *Echinoderm Biology: Proceedings of the Sixth International Echinoderm*

Conference, Victoria, British Columbia, 23–28 August, 1987, Balkema, Rotterdam: 39–54.

Müller, A.H., 1953, Bemerkungen zur Stratigraphie und Stratonomie der obserseno-nen Schreibkreide von Rügen, *Geologie*, **2** (1): 23–34 [not seen].

Müller, A.H., 1955, Beiträge zur Stratonomie und Ökologie des germanischen Muschelkalkes, *Geologie*, **4** (3): 285–97.

Oji, T., 1989, Growth rate of stalk of *Metacrinus rotundus* (Echinodermata: Crinoidea) and its functional significance, *Journal of the Faculty of Science, University of Tokyo*, **22** (1): 39–51.

Parks, N.B., 1973, Distribution and abundance of the sand dollar, *Dendraster excentricus*, off the coast of Oregon and Washington, *Fisheries Bulletin of the National Oceanic and Atmospheric Administration of the United States*, **71**: 1105–9.

Paul, C.R.C., 1982, The completeness of the echinoderm fossil record. In J.M. Lawrence (ed.), *Echinoderms: Proceedings of the International Conference, Tampa Bay, 14–17 September, 1981*, Balkema, Rotterdam: 89.

Pawson, D.L., 1980, Holothuroidea. In T.W. Broadhead and J.A. Waters (eds), *Echinoderms: notes for a short course, University of Tennessee Studies in Geology*, **3**: 175–89.

Reyment, R.A., 1986, Nekroplanktonic dispersal of echinoid tests, *Palaeogeography, Palaeoclimatology, Palaeoecology*, **52** (3/4): 347–9.

Rollins, H.B. and Brezinski, D.K., 1988, Reinterpretation of crinoid–platyceratid interaction, *Lethaia*, **21** (3): 207–17.

Roux, M., 1970, Introduction à l'étude des microstructures des tiges de crinoïdes, *Geobios*, **3** (3): 79–98.

Roux, M., 1974a, Les principaux modes d'articulation des ossicules du squelette des Crinoïdes pédonculés actuels. Observations microstructurales et conséquences pour l'interprétation des fossiles, *Compte Rendu de l'Academie des Sciences*, Paris, **278**: 2015–18.

Roux, M., 1974b, Observations au microscope électronique à balayage de quelques articulations entre les ossicules de squelette des Crinoïdes pédonculés actuels (Bathycrinidae et Isocrinina), *Travaux du Laboratoire de Paléontologie*, Orsay: 10 pp.

Roux, M., 1975, Microstructural analysis of the crinoid stem, *University of Kansas Paleontological Contributions, Paper* **75**: 1–7.

Roux, M., 1987, Biocorrosion et micritisation des ossicules d'echinodermes en milieu bathyal au large de la Nouvelle-Calédonie, *Compte Rendu de l'Academie des Sciences*, Paris, **305**: 701–5.

Schäfer, W., 1972, *Ecology and palaeoecology of marine environments*, University of Chicago Press, Chicago.

Schneider, J.A., 1988, Frequency of arm regeneration of comatulid crinoids in relation to life habit. In R.D. Burke, P.V. Mladenov, P. Lambert and R.L. Parsley (eds), *Echinoderm Biology: Proceedings of the Sixth International Echinoderm Conference, Victoria, British Columbia, 23–28 August, 1987*, Balkema, Rotterdam: 531–8.

Schumacher, G.A., 1986, Storm processes and crinoid preservation, *Abstracts, Fourth North American Paleontological Convention, Boulder, 12–15 August*: A41.

Schwammer, H.M., 1989, Bald-sea-urchin disease: record of incidence in irregular echinoids—*Spatangus purpureus*, from the SW-coast of Krk (Croatia-Jugoslavia), *Zoologischer Anzeiger*, **223** (1/2): 100–6.

Schwarzacher, W., 1963, Orientation of crinoids by current action, *Journal of*

Sedimentary Petrology, **33** (3): 580–6.

Scoffin, T.P., 1987, *An introduction to carbonate sediments and rocks*, Blackie, Glasgow.

Seilacher, A., 1979, Constructional morphology of sand dollars, *Paleobiology*, **5** (3): 191–221.

Seilacher, A., Reif, W.E. and Westphal, F., 1985, Sedimentological, ecological and temporal patterns of fossil Lagerstätten, *Philosophical Transactions of the Royal Society of London*, **B311**: 5–24.

Sides, E.M., 1987, An experimental study of the use of arm regeneration in estimating rates of sublethal injury on brittle-stars, *Journal of Experimental Marine Biology and Ecology*, **106** (1): 1–16.

Simms, M.J., 1986, Contrasting lifestyles in Lower Jurassic crinoids: a comparison of benthic and pseudopelagic Isocrinida, *Palaeontology*, **29**, (3): 475–93.

Smith, A.B., 1980, Stereom microstructure of the echinoid test, *Special Papers in Palaeontology*, **25**: 1–81.

Smith, A.B., 1984, *Echinoid palaeobiology*, George Allen & Unwin, London.

Smith, A.B., 1989, Biomineralization in echinoderms. In J.G. Carter (ed.), *Skeletal biomineralization: patterns, processes and evolutionary trends*, American Geophysical Union Short Course in Geology, **5** (2): 117–47.

Smith, A.B. and Gallemí, J., 1991, Middle Triassic holothurians from northern Spain. *Palaeontology*, **34** (1).

Sprinkle, J., 1973, *Morphology and evolution of blastozoan echinoderms*, Museum of Comparative Zoology, Harvard.

Sroka, S.D., 1988, Preliminary studies on a complete fossil holothurian from the Middle Pennsylvanian Francis Creek Shale of Illinois. In R.D. Burke, P.V. Mladenov, P. Lambert and R.L. Parsley (eds), *Echinoderm Biology: Proceedings of the Sixth International Echinoderm Conference, Victoria, British Columbia, 23–28 August, 1987*, Balkema, Rotterdam: 159–60.

Strimple, H.L. and Beane, B.H., 1966, Reproduction of lost arms on a crinoid from Le Grand, Iowa, *Oklahoma Geology Notes*, **26** (1): 35–7.

Swan, E.F., 1966, Growth, autotomy and regeneration. In R.A. Boolootian (ed.), *Physiology of echinodermata*, Wiley, New York: 397–434.

Timko, P.L., 1976, Sand dollars as suspension feeders: a new description of feeding in *Dendraster excentricus, Biological Bulletin*, **151** (1): 247–59.

Vermeij, G.J., 1987, *Evolution and escalation: an ecological history of life*, Princeton University Press, Princeton, New Jersey.

Walbran, P.D., Henderson, R.A., Jull, A.J.T. and Head, M.J., 1989, Evidence from sediments of long-term *Acanthaster planci* predation on corals of the Great Barrier Reef, *Science*, **245** (4920): 847–50.

Walkden, G.M. and Berry, J.R., 1984, Syntaxial overgrowths in muddy crinoidal limestones: cathodoluminescence sheds new light on an old problem, *Sedimentology*, **31** (2): 251–67.

Warén, A. and Moolenbeek, R., 1989, A new eulimid gastropod, *Trochostilifer eucidaricola*, parasitic on the pencil urchin *Eucidaris tribuloides* from the southern Caribbean, *Proceedings of the Biological Society of Washington*, **102** (1): 169–75.

Welch, J.R., 1976, *Phosphannulus* on Paleozoic crinoid stems, *Journal of Paleontology*, **50** (2): 218–25.

Whitfield, R.P., 1904, Notice of a remarkable case of reproduction of lost parts shown on a fossil crinoid, *Bulletin of the American Museum of Natural History*, **20** (37): 471–2.

Wignall, P.B. and Simms, M.J., 1990, Pseudoplankton, *Palaeontology*, **33** (2): 359–78.

Wilkie, I.C., 1983, Nervously mediated change in the mechanical properties of the cirral ligaments of a crinoid, *Marine Behavioural Physiology*, **9** (3): 229–48.

Wilkie, I.C., 1984, Variable tensility in echinoderm collagenous tissues: a review, *Marine Behavioural Physiology*, **11** (1): 1–34.

Wilkie, I.C., 1988, Design for disaster: the ophiuroid invertebral ligament as a typical mutable collagenous structure. In R.D. Burke, P.V. Mladenov, P. Lambert and R.L. Parsley (eds), *Echinoderm Biology: Proceedings of the Sixth International Echinoderm Conference, Victoria, British Columbia, 23–28 August, 1987*, Balkema, Rotterdam: 25–38.

Wilkie, I.C. and Emson, R.H., 1988, Mutable collagenous tissues and their significance for echinoderm palaeontology and phylogeny. In C.R.C. Paul and A.B. Smith (eds), *Echinoderm phylogeny and evolutionary biology*, Clarendon Press, Oxford: 311–30.

Zinsmeister, W.J., 1980, Observations on the predation of the clypeasteroid echinoid, *Monophoraster darwini*, from the Upper Miocene Entrerrios Formation, Patagonia, Argentina, *Journal of Paleontology*, **54** (5): 910–12.

Chapter 12

BONES AS STONES: THE CONTRIBUTION OF VERTEBRATE REMAINS TO THE LITHOLOGIC RECORD

David M. Martill

INTRODUCTION

The abundance of vertebrate skeletal elements in sediments is highly variable. In the majority of sedimentary sequences vertebrates are exceedingly rare or absent. Some horizons are known to yield vertebrates at regular intervals, but their remains are not obvious in outcrop. Deposits are usually only referred to as 'bone beds' when the rock contains an appreciable quantity of vertebrate elements such that each sample contains some or many elements. Historically, bone accumulations have been the cause of some debate between catastrophists and gradualists. The former saw them as evidence of mass-mortalities, while gradualists recognized the importance of sedimentary reworking and condensation in their genesis.

The genesis of bone beds is of great concern to vertebrate palaeontologists and, to a lesser extent, to sedimentologists. To the former, bone beds are a rich source of fossils, while to the latter they offer insights into the processes of sedimentary reworking, and size and density sorting. In general, accumulations of vertebrate remains are of little economic importance, except when they occur in sufficient quantities to be a suitable source of phosphate for fertilizers.

This chapter deals largely with the fate of vertebrate remains and the way their hard tissues become incorporated into the lithologic record. The passage of vertebrate remains into the fossil record can be divided into a number of interdependent and frequently overlapping stages: the ecology and biology of the original organisms; their mode of death; the taphonomy, transport, burial and diagenesis of their remains. Because they are a particular research interest of the author, some emphasis is placed on the accumulations of

Jurassic marine reptiles in clay sequences of the northwest European province.

VERTEBRATE MINERALIZED TISSUES

The origins of vertebrate mineralized tissues have been discussed in great detail (see Halstead, 1969) and need not be considered here. It is, however, relevant to discuss briefly the various vertebrate hard tissue types and their compositions.

Vertebrates are characterized by internal, multi-component skeletons. Vertebrate hard tissues are made of a variety of types, ranging from almost wholly collagenous cartilage to various mineralized types. Most common is bone, of which there are a number of recognized sub-varieties according to microstructure. Other hard tissues of similar composition to bone include aspidin, dentine and enamel, the last largely restricted to surface coatings of teeth and scales. Cartilage may also be mineralized, particularly in a number of elasmobranchs, where small prisms of apatite may form a structure superficially similar to bone, but with a highly distinct microstructure.

A number of biominerals have been recognized within vertebrate hard tissues, but by far the most common is calcium hydroxy apatite with the general composition ($Ca_{10}(PO_4)_6.2OH$), although some substitution of hydroxyl by fluorine may occur, especially in teeth (Prevot and Lucas, 1989), and PO_4 may be exchanged for CO_3.

In contrast, within the auditory capsule of many vertebrates occur small grains of calcium carbonate ($CaCO_3$), generally known as 'otoliths', which perform a sensory role. This is usually related to balance and orientation, and is particularly important in fishes. Several polymorphs of calcium carbonate have been identified including calcite, aragonite and vaterite (Nolf, 1985). These elements are not major constituents of the rock record, but they can be abundant in micropalaeontological residues. Enhanced concentrations of otoliths relative to fish bone are thought to be attributable to the ability of otoliths to resist acid digestion in fish guts, where bone readily dissolves. Otoliths have not been reported as rock producers.

Bone contains living cells within its structure and is continually modified during ontogeny. The vertebrate skeleton can therefore be considered as a living tissue. Bone tissue consists of its organic matrix and mineral component, within which occur numerous elongate cavities, or lacunae, which in life are occupied by cells, called 'osteocytes', that are responsible for bone manufacture and resorption. The cells are interconnected by numerous small canals, the cannaliculi, through which essential nutrients can be exchanged and transported.

The ultrastructure of bone is also highly complex. The organic and inorganic phases are intimately connected, rather than being segregated into discreet layers. The organic matrix is largely fibrillar collagen, in the form of a tropocollagen molecule (Miller, 1984). Within this molecule exist sites where nucleation of apatite can take place (Glimcher, 1984). These sites are extremely small and the individual crystallites of apatite remain small. The

C-axis of the crystallite is parallel to the length of the collagen molecule (Miller, 1984). However, the collagen molecules are frequently arranged in twisted rope-like configurations (Figure 12.1). In crossed polarized light, thin sections of bones show a shadowy extinction that travels along the length of the collagen fabric (Figure 12.2).

The ultra microscopic size of the bone mineral phase (individual crystallites are less than 500 mm) has made it difficult to study, especially with regard to the diagenetic processes that may have affected the original structure. This contrasts strongly with the relatively massive size of the crystallites in, for example, mollusc shells. However, the structure of bone prevents fracturing along the large cleavage surfaces which make many mollusc shells so fragile.

The different types of tissues in bones have distinct mechanical properties. In life some are load bearers, whereas others are protective shells. The resistance to stress even varies within a single bony element according to where that stress is applied (for example, our own bones readily break when we apply stress in the wrong place). These differences in mechanical properties influence the potential of a bone to enter into the fossil record. For example, the light hollow bones of pterosaur wings were resistant to torsion and were extremely light. They were not resistant to any type of impact or compression. Thus, pterosaurs are rare as fossils, and, when found, are frequently crushed. On the other hand, the auditory bones of large cetaceans are very dense, a requirement for the transmission of low-frequency sound. As a consequence they are resistant to physical abrasion, dissolution and compaction. This results in an enhanced abundance of these strong, resistant bony elements compared to light, delicate bones. This selective preservation of more resistant elements is most vividly demonstrated by the great abundance of teeth over bone in reworked bone beds.

There are a number of non-mineralized skeletal components present in some vertebrates which may occasionally occur in the rock record. These include the more resistant macromolecules (such as keratin) which form structures such as hair, claws and horn. They are generally rare as fossils, however, and occur only under conditions of exceptional preservation.

THE TAPHONOMY OF VERTEBRATES

The taphonomy of terrestrial vertebrates has been extensively discussed in recent years (see, for example, Behrensmeyer and Hill, 1980; Franzen, 1985), especially with regard to faunas associated with hominid remains in Africa. A number of authors have extended taphonomic studies into the Mesozoic to examine the accumulations of terrestrial (mainly dinosaur) vertebrates (Wood *et al.*, 1988) and marine reptiles (Martill, 1985), and a few recent studies have examined Upper Palaeozoic bone-bearing sequences with respect to taphonomy and sedimentology (Martin-Sander, 1989; Parrish, 1978).

The term 'taphonomy' is used in a number of different ways. Many palaeontologists use it as a lumping term for all post-mortem processes, including diagenesis. Sedimentologists, on the other hand, tend to restrict the term to those processes that take place before burial, and consider the post-burial

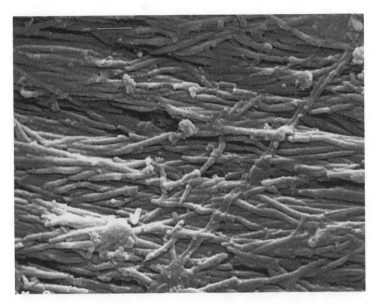

Figure 12.1. Scanning electron micrograph of plesiosaur centrum showing original mineralogy of bone tissue. Note that the large mineralized collagen fibrils are woven in bundles. Specimen from Lower Oxford Clay (Middle Callovian, Middle Jurassic) of Peterborough, England. Approximately × 10 000.

environment in terms of diagenesis. The two overlap, however, as bacterial breakdown begins pre-burial and in marine environments is a major component of post-burial diagenetic pathways.

The taphonomic processes affecting vertebrates are highly varied, and too numerous to discuss them all here, but there are a few that can be summarized briefly and which are of particular importance. Scavenging of carcasses is frequently disruptive, often resulting in skeleton disarticulation and scattering of skeletal elements. The eating of bones by predators and scavengers generally results in their total destruction, although some elements may be passed out in faecal pellets or regurgitated as in the case of pellets from raptor birds. Microbial breakdown in ubiquitous, but it may be slowed down or temporarily stopped by high- and low-pH environments, and by highly arid environments in both arctic and hot regimes. Microbial decay is not stopped by anoxia alone (Allison, 1988).

Death

It is exceedingly difficult to ascertain the cause of death in palaeontological material, except in the rare specimens that exhibit teeth marks or other unequivocal evidence of predation, and in cases where sedimentary structures give clues to catastrophic events such as rapid burial or dessication.

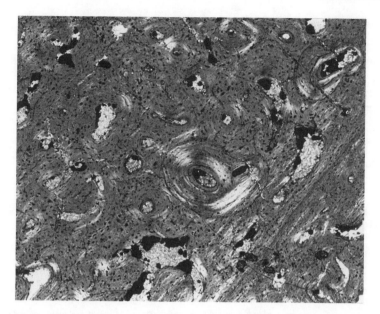

Figure 12.2. Thin section of plesiosaur bone under crossed polars. Note the shadowy extinction of the circular haversian structure in the centre of the photograph. Specimen from Lower Oxford Clay of Peterborough, England. Approximately × 100.

Mass-mortality events are frequently recorded in the rock record, but again their cause can be difficult to establish. They may involve a few tens of individuals, and be of limited areal distribution (Turnbull and Martill, 1988), or, in the case of some fish mass-mortalities, may contain millions of individuals spread over a wide area (Figure 12.3). It is most likely that many mass-mortality events go unrecognized due to taphonomic filtering and extensive sedimentary reworking.

Transport

Between death and incorporation into the fossil record, most carcasses undergo some transport. This may be minimal in the case of vertebrates living within marine or lacustrine sedimentary basins (for example, fishes may simply sink to the sea floor). On the other hand, carcasses may be transported considerable distances by the agencies of predators, scavengers, fluvial systems (Behrensmeyer, 1988), and by marine currents (in the case of floating cadavers). Predators and scavengers that transport carcasses into caves are frequently responsible for large accumulations of bones, especially when caves are occupied as lairs for several successive generations.

Post-mortem drifting of floating carcasses may be prolonged (Schäfer, 1962), especially in the case of large vertebrates such as baleen whales (Figure

Figure 12.3. Mass-mortality of small herring (*Gosiutichthys parvus* Grande) from the Green River Formation of Green River Basin, Eighteen Mile Canyon, near Rock Springs, Wyoming. These herrings occur on a single bedding plane. They form a bone bed of articulated skeletons that is less than 1 mm thick. Approximately × 0.5.

12.4). It is possible that carcasses may travel tens or even hundreds of kilometres in ocean currents before descending to the sea floor. Post-mortem drifting has been elegantly demonstrated in the Lower Oxford Clay (Callovian, Middle Jurassic) of the UK, where several dinosaur skeletons have been reported from fully marine sediments. The nearest landmass to the discovery sites is thought to have been at least 80 km distant. Most of the dinosaur skeletons are incomplete, suggesting that considerable scavenging took place during transportation.

Skeletal disarticulation

Although the most spectacular vertebrate skeletons are those that are complete and display true bone-to-bone relationships, they are infrequent compared to the large number of disarticulated remains. There is a broad

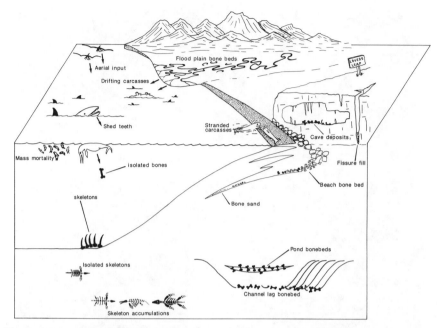

Figure 12.4.   Diagram illustrating some of the transportational and depositional systems affecting vertebrate carcasses.

spectrum of preservational styles between fully articulated and three-dimensional, as in the case of fish skeletons from the Santana Formation of Brazil (?Albian, Lower Cretaceous; Martill, 1988), to totally disarticulated, but complete, as is often the case for marine reptile skeletons in shell beds from the Lower Oxford Clay (Callovian, Middle Jurassic) of the UK. The degree of scatter can be variable, as are the agencies that cause disarticulation (Figure 12.5).

Bone breakage

A majority of bones encountered in the fossil record have suffered damage of one sort or another. The cause of breakage and its implications have been discussed at length by Hill (1980). Damage to bone may occur at several stages during the taphonomic and diagenetic history of a bone, and it is convenient to divide these into the following categories.

- *Pathological and traumatic*: Damage sustained during life, including diseased bone and natural fractures.
- *Post-mortem damage*: This might be due to predation, scavenging, trampling by elephants (Behrensmeyer *et al.*, 1979), transport (Hill, 1980) or even due to gnawing by porcupines to keep their teeth worn down (Brain, 1980; Figure 12.6A herein).

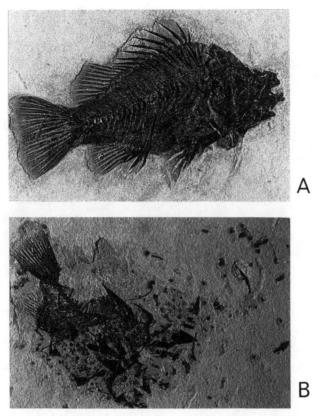

Figure 12.5. A spectrum of articulation may be observed in a single deposit. Here two specimens of the actinopterygian *Priscacara* cf. *liops* from the Green River Formation of Fossil Basin, near Kemmerer, Wyoming, show varying degrees of articulation and disarticulation. (A) Complete specimen, fully articulated, though laterally flattened due to compaction, × 0.5. (B) Specimen in which anterior portion of body and skull are disarticulated; approximately × 0.25. This may be attributable to the rapid escape of gases from the decomposing carcass. Both specimens were obtained from a well-laminated micrite approximately 45 cm thick.

- *Post-burial damage*: Sustained due to compaction, cracking in septarian concretions or dissolution (Figure 12.6B).

It is important to be able to recognize damaged from undamaged bone and to be able to determine how the damage was sustained. Damage sustained during life may show evidence of healing, including new bone growth. Damage sustained postmortem will not show new bone growth and it may be difficult to distinguish from damage sustained during compaction. It may be necessary to examine thin-sections through fractures to distinguish post-mortem from post-burial damage.

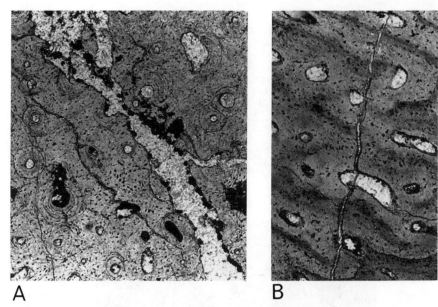

A   B

Figure 12.6. Fracturing in a plesiosaur bone. (A) Pre-burial fracturing. The fracture is lined with early diagenetic pyrite and calcite; approximately × 100. (B) Bone and diagenetic infills fractured. This fracture probably accompanied compaction due to high overburden pressure; approximately × 150. Specimen from Lower Oxford Clay (Middle Callovian, Middle Jurassic) of Peterborough, England.

REMOVAL OF BONE FROM THE SEDIMENTARY CYCLE

Despite the robust nature of dense bone and tooth enamel, many sediments contain little or no vertebrate hard tissues, but were nevertheless deposited in environments similar to those which today contain many vertebrates. The explanation that vertebrates were absent from the fauna is not usually tenable in Mesozoic or Cainozoic sediments. Often simple dilution of vertebrate remains in sediment is an insufficient explanation and we need to discover how bone has been removed from the system. Explanations can be found in both biological and physico-chemical processes, and bone may be removed before burial, during transportation and even post burial. Thus, bone removal may be either a taphonomic or a diagenetic phenomenon. Sediments apparently lacking vertebrates in hand specimen may frequently indicate the presence of microvertebrates by simple washing and screening.

Biological removal of bone

Phosphate is an essential nutrient, but a rare commodity in many environments. Bones may thus represent an important resource to be utilized by other organisms. This is not simply restricted to predators crunching bones and eating them along with meat. A number of invertebrates and vertebrates are known to feed actively on bone material that lacks soft tissue. For example, archaeogastropods have been reported feeding on large whale skeletons (Haszprunar, 1988; Marshall, 1987), while some echinoids may feed almost entirely on fish bones. The organic component of bone may also be a rich source of nutrients for a number of organisms. In many so-called 'bone feeders' it might be the organic phase rather than the mineral phase that is being sought, but the effect is the same: the destruction of the bone. Much bone material, therefore, does not become incorporated into the sedimentary record, but is biologically recycled.

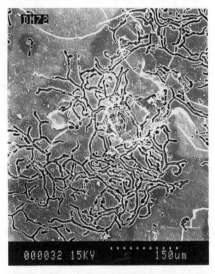

Figure 12.7. Biological recycling and weakening of bone. Scanning electron micrograph of fish scale from *Notelops* sp., Santana Formation, northeast Brazil. The microbored surface has been much weakened due to intense activity of fungi or ?bacteria.

Microbial activity also plays an important role in the destruction of bone material (Figure 12.7). Bacteria and possibly fungi are known to attack bone and teeth (Goujet and Locquin, 1979), some being restricted to specific mineralized tissues (Martill, 1989), for example, dental caries. It is not apparent how much bone material is removed by microbial processes. Such attack may not remove bone entirely, but may weaken it, making it highly susceptible to any subsequent physical abrasion.

Physical and chemical break-up of bone

The bones of terrestrial vertebrates may be exposed to bright sunlight for some time. Such bones are rapidly bleached (organic component oxidized).

This aids the destruction of the organic matrix and may result in complete break-up of bone almost to a white powder.

Physical transport in sedimentary environments is a major cause of damage to bones, but may not always be as severe as previously thought. Bones introduced into river systems shortly after death may still contain appreciable quantities of organic material, and be able to resist abrasion due to their slightly plastic/elastic surface. Many bones are light due to their highly porous nature and may float at the surface or mid-way in the water column. They may travel significant distances in this way, rarely coming into contact with abrasive sediment. However, when bones reach environments in which they are subjected to continued attrition by sediment, they will ultimately be severely damaged or even destroyed. The rate of attrition depends on the current velocity, and the amount, size and composition (= hardness) of the sedimentary particles. In the case of small plesiosaur remains from the Hugh Miller Reptile Bed (Middle Jurassic) of Scotland (Hudson, 1966), the sediment is composed largely of calcitic shell fragments. Here vertebrae are frequently highly polished, but otherwise complete. In fine-grained sands of the Kellaways Beds of Peterborough, England, large bones of marine reptiles are worn, and on occasions the inner trabecular bone may be exposed. Continued abrasion in coarser-grained sediments usually results in total destruction.

Chemical removal from the rock record

Chemical alteration, and even complete removal of bone from the sedimentary cycle, may take place both pre- and post-burial. Post-burial removal of bone is frequently attributable to low pH in soils and ground water (Williams, 1987). This is often the case with exceptionally well-preserved human remains in peat deposits (Glob, 1969), where soft tissues have been preserved, but skeletal elements have been completely dissolved.

Complete dissolution of apatite by pore waters with low pH is probably a common occurrence, but is difficult to positively demonstrate due to an absence of fossils. Only when soft tissue outlines are preserved or when well-consolidated rock contains external moulds can dissolution be adequately demonstrated. A number of fossil localities are known in which bones occur as external moulds (Milner, 1980), from which skeletons may be reconstructed using latex replicas.

VERTEBRATE REMAINS AS SEDIMENTARY PARTICLES

Surprisingly little work has examined the sedimentology of bone beds (Heldt *et al.*, 1947), although individual and well-known horizons have received intermittent attention for over 150 years (for example, Murchison, 1856; for a review see Antia, 1979). The great emphasis has been on the study of

Siluro-Devonian bone beds in the UK (Antia, 1981; Antia and Whitaker, 1979), and Triassic bone beds in Germany and the United Kingdom (Reif, 1976; 1982).

The contribution to the lithic record of vertebrate remains is small compared to contributions of skeletal carbonate made by calcareous shelled invertebrates. Nevertheless, vertebrate hard tissues, most notably bones and teeth, can be important locally and are rarely even of economic significance. Processes resulting in the continued accumulation of large quantities of vertebrate remains are varied, but include macro-, meso- and microscale phenomenon. On the macro scale, ocean upwelling zones several hundred kilometres in length and rich in nutrients may offer opportunities for population explosions of marine vertebrates, usually fishes, and also provide mechanisms for mass-mortalities (Brongersma-Sanders, 1948). On the meso-scale, sedimentary processes such as reworking and density sorting may concentrate the remains of vertebrates by several orders of magnitude. At the micro-scale, predation by organisms that habitually transport their prey to lairs may concentrate large quantities of small bones, such as the activities of owls producing pellets and cave-dwelling mammals. Additionally, diagenetic processes may concentrate vertebrate remains with respect to other skeletal elements. For example, small decreases in pH of pore water may remove aragonite- and even calcite-shelled invertebrates, while the slightly more resistant bone may remain unaffected.

By themselves, many of these processes may not result in large-scale accumulations, but often the interaction between two or more processes may produce accumulations of skeletal elements in sufficient concentrations that some sediments may comprise more than 50 per cent vertebrate remains. Such deposits are widely referred to as 'bone beds'. However, there is as yet no clear agreement as to what constitutes a bone bed (see Antia, 1979). Bone-rich sediments are of great value as source rocks for vertebrate palaeontologists. Consequently, much work on bone accumulations has concentrated on their palaeontological significance, rather than on their genesis. Their unusual and somewhat rare occurrence has resulted in bone bed being considered something of an oddity.

A notable feature of many bone beds is their regularity of grain size and consistency of grain composition. In assemblages composed largely of fish scales or teeth from a single taxon, this may reflect hydrodynamic sorting, but it is important to consider that in monospecific assemblages the original contribution may have been highly size-specific. Sorting of different skeletal elements is, however, an important process. Fluvial and marginal marine vertebrate accumulations in the Permian Dokum Group of north Texas frequently contain an abundance of palaeoniscoid fishes, acanthodians and, less commonly, therapsid reptiles. In a 20 m section I counted more than five individual bone beds. One bone bed contained only the spines of acanthodians, while another contained only their dermal denticles. Another bone bed was dominated by imbricated palaeoniscoid scales, while a fine sandstone contained scattered scales and some larger palaeoniscoid skull bones. The fifth bone bed contained teeth and bones of *Dimetrodon* and cf. *Ophiacodon*, as well as elements present in all the other bone beds. Clearly a great deal of hydrodynamic sorting, probably associated with other taphonomic filtering, had taken place.

PHYSICAL ABRASION OF BONE

Bone apatite is one of the hardest of the common biominerals. However, beneath the seemingly solid surface of a bone, the interior may be very delicate and brittle. In some cases (such as pterosaur wing bones) this is taken to extremes, with bones being thin hollow tubes supported by struts less than 1 mm thick. Once the outer surface of the bone is worn away, continued attrition rapidly destroys the bone. However, modern studies have shown that some bones may persist for prolonged periods in fluvial systems charged with abrasive particles such as sand and grit. This may be attributable to the slightly elastic surface of bone which retains it organic matrix, thus absorbing some of the shock of impacting sand grains.

BONES AS BENTHIC ISLANDS: ENCRUSTED BONES

In many sedimentary environments, highly mobile or very soft sediment, or anoxic bottom water, may prevent colonization of the sea floor by epifauna. Vertebrate skeletons can provide refuges (benthic islands) for encrusting plants and animals in such environments (Seilacher, 1982; Martill, 1987). This is particularly noticeable in basins dominated by clay sedimentation, where later physical abrasion of bone does not usually occur and epifauna remain attached, and in oxygen-depleted environments, where chemosynthetic communities may find bones suitable substrates for colonization (Smith *et al.*, 1989). In the Jurassic clay sequences of the UK, the skeletons of marine reptiles are frequently encrusted on the upper surfaces of the bones (Figure 12.8). Exposed skeletons help increase faunal diversity in impoverished or restricted communities dominated by deposit-feeding burrowers.

As the vertebrate skeleton is wholly internal, encrustation of bones can only take place after the death of the individual, and after exposure of the skeleton. Encrusted skeletons may therefore demonstrate that bottom waters were oxygenated. In contrast, encrusted externally shelled organisms, such as ammonites, where encrustation may have occurred during life as well as post-mortem, do not give unequivocal evidence as to bottom-water conditions.

GEOGRAPHICAL AND TEMPORAL DISTRIBUTION OF BONE BEDS

Living vertebrates are distributed globally. In some lacustrine, marginal marine and coastal environments they may occur in super-abundance. This has not always been the case. The modern abundance and success of vertebrates is attributable to a number of adaptive radiations that have given them a competitive edge, especially in terrestrial and aerial environments. This is reflected in the temporal and spatial distribution of vertebrate-bearing

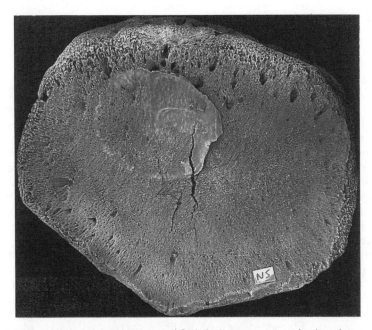

Figure 12.8. Ulna of ichthyosaur (*Ophthalmosaurus* sp.) showing surface encrustation by oyster. The oyster has only laid down a very thin coating of carbonate at the margins of the shell. Note that the crack in the bone also cuts the oyster shell, demonstrating the post-mortem and post-encrustation timing of the crack. Specimen from the Lower Oxford Clay of Milton Keynes, England. Approximately life size.

sediments, and in the composition of bone beds. The earliest unequivocal vertebrates with preservable hardparts appeared in the Ordovician (for example, see Gagnier *et al.*, 1986), although records of earlier possible vertebrates exist (Repetski, 1978). These remains are of early jawless fish (agnathans) in which the body was armoured with large plates, or numerous small denticles composed largely of dentine in the earliest forms. They sometimes accumulated in such abundance as to be the major dense particle in fine gritstones. Although such accumulations are not abundant in the Ordovician, there are numerous occurrences of Silurian age (Antia, 1981). By the end of the Devonian a series of adaptive radiations had resulted in the appearance of the major fish clades and the tetrapods. Innovations that arose sometime during the Ordovician to Devonian included the ossification of cartilaginous skeletons and the development of jaws and teeth. The development of bone tissue was perhaps the single greatest innovation. Most evolutionary events after that largely affected the form and function of the various skeletal elements. As the vertebrate skeleton and its components became more complex, so did the composition of bone beds.

The earliest bone beds (commonly of Late Silurian and Early Devonian age) are dominated by thelodonts, and more rarely 'fin' spines. Later

Devonian vertebrate accumulations frequently contain an abundance of strong, enamel-coated scales of actinopterygian fish. By the Carboniferous vertebrates had become highly diversified, with marine accumulations often containing large numbers of elasmobranch teeth and denticles, as well as smaller numbers of fine spines. Mesozoic bone accumulations are highly varied in composition. Marine bone beds of the Upper Triassic contain large numbers of hybodont shark teeth as well as the enamel scales of palaeoniscoid and other related fishes. By the end of the Jurassic fishes with hard enamel scales were of little significance in the biota, and, as a consequence, their remains usually form only a small fraction of vertebrate accumulations. Mesozoic terrestrial vertebrate accumulations are dominated by reptile teeth, often in vast quantities. But after the Cretaceous–Tertiary boundary event, the reptile component of bone beds dropped and mammalian teeth became dominant. Quasi-marine vertebrate accumulations at the end of the Jurassic, especially in the UK (Lulworth Formation), show a large increase in the quantity of strong turtle carapace bones. These resistant elements are often important in non-marine vertebrate accumulations throughout the rest of the Mesozoic and Tertiary.

At certain times during the Upper Palaeozoic to Recent, different vertebrates with resistant skeletal elements have been abundance for short intervals. Their remains can be used to construct broad biochrons. A notable example are the placodonts, whose highly resistant and readily identifiable teeth are a common component of Middle Triassic bone accumulations in the Tethyan region. Similarly, the large marine varanid lizard *Mososaurus globidens* from the Upper Cretaceous had a highly resistent globident dentition. Its teeth form a common component of Upper Cretaceous bone accumulations in North America.

There is some evidence that marine bone bed distribution is controlled in part by transgressive and regressive cycles. Reif (1982) has demonstrated that storm activity, or a lowering of wave base during regressive cycles, is a contributory factor in bone bed genesis. However, it might also be the case that bone beds are produced where reworking of fluvial plain sediments occur at the onset of transgression. Marine bone beds should be dominated by resistant elements of shelf sea vertebrates, whereas freshwater and terrestrial vertebrates dominate fluvial bone-rich deposits.

Organic-rich marine mudrocks, such as the Lower Lias, the Upper Lias, the Oxford Clay and the Kimmeridge Clay (all Jurassic of the UK), are frequently rich in vertebrates in their lower parts (Martill, 1985). There appears to be a correlation with an increased abundance of vertebrates and elevated organic carbon content of the sediment, possibly a reflection of high levels of surface productivity rather than anoxia with slow sedimentation. These relatively thick clay sequences are characteristic of widespread marine transgressions, and, although bone beds are infrequent, when they do occur, they too appear to be better developed during the early stages of the transgression.

Geographically, vertebrate accumulations are distributed globally, but there are certain restrictions between differing types of accumulation. Fissure fills and cave deposits are restricted largely to areas of widespread carbonate rocks that have been exposed to dissolution. In the case of the Lower Carboniferous limestones of south Wales, two episodes of dissolution have resulted in vertebrate accumulations: a Late Triassic to Lower Jurassic phase (Fraser, 1988) and a Pleistocene phase (Stuart, 1982). Bone accumulations from mass-mortalities associated with the dessication of the landlocked seas and large lakes are generally restricted to lower latitudes, as might be the case of the Green River Formation (Eocene) of Wyoming, and the Santa Formation (Lower Cretaceous) of Brazil. Other accumulations are possibly controlled by the distribution of deep-ocean currents and their associated upwelling zones (Antia, 1979). At the present time this is dominantly on the western margins of the South American and African continents in the southern hemisphere.

The combination of geological, biological and temporal constraints which control the distribution of vertebrate accumulations may allow for the prediction of their occurrence. This could be a useful procedure for vertebrate palaeontologists in their search for new specimens and could provide a test for hypotheses concerning bone-bed genesis.

DIAGENESIS OF BONES

Very few workers have been concerned with the diagenesis of vertebrate remains. This is surprising, as vertebrate skeletons are frequently the only robust uncrushed elements in highly compacted mudrocks and also offer void space for the accumulation of diagenetic minerals.

The diagenesis of vertebrate hard tissues can be exceedingly complex. In order to understand the processes involved it is convenient to consider three diagenetic environments: the diagenesis of the bone tissue itself; the diagenesis in pore spaces and cavities within the bones; and the diagenesis in the sediment surrounding the bone. Each diagenetic environment may produce a distinct suite of mineral phases and petrofabrics. I will consider each of these in turn.

Diagenesis of bone tissue

In a large majority of cases the only diagenetic alteration of original bone fabric to take place is ionic substitution between OH and F, or to a lesser degree Cl, in the bone apatite, and between PO_4 and CO_3. Recrystallization of the original bone tissue can occur, as well as overgrowth of cryptocrystalline calcium phosphate as an early diagenetic phase. However, both occur only rarely.

In some environments bone apatite may be replaced by diagenetic phases, most notably pyrite (Figure 12.9). Bones occurring in acidic groundwater with an adequate supply of iron may exhibit efflorescences of the bright blue iron phosphate vivianite. It is not clear whether the phosphorous here is derived from the bone or is liberated from the decomposition of organic material and is simply nucleating on the bone surface.

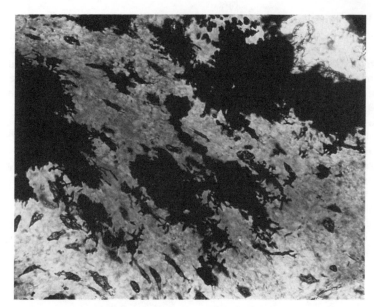

Figure 12.9. Thin-section through an ichthyosaur bone showing pyrite (black) infilling of large voids, lacunae and canaliculae, and partially replacing the bone itself. Specimen from Lower Oxford Clay of Milton Keynes, England.

Diagenetic environment within pore spaces in bone

Two pore-space environments can be considered to occur within bones. The first is the spongy space within trabecular or cancellous bone. This includes the hollow of bones with high marrow content. This void space is frequently connected to the exterior environment by fenestrae on the bone surface formerly perforated by blood vessels. The second pore-space environment includes the lacunae and linking canaliculi of the bone itself (Figure 12.9). These pores are numerous, but usually constitute a much smaller volume than the former. They are not well connected to the exterior environment except when bones are damaged and the internal structure is exposed.

The spongy cavity of bone is frequently filled with diagenetic minerals. In the case of bones from marine reptiles of the Oxford Clay there can be as many as seven distinct mineral phases filling these voids, including two distinct phases of pyrite (FeS_2), two calcites ($CaCO_3$), sphalerite (ZnS), baryte ($BaSO_4$), and very rarely celestine ($SrSo_4$). Bones that have been in the weathering zone may also contain selenite ($CaSO_4.2H_2O$) as a by-product of pyrite breakdown.

Diagenesis of the host sediment

This subject is far too large to enter into in detail here, but it is necessary to comment that the diagenetic environment within the sediment surrounding bones is of significance. First, it directly affects the outer surface of the bone. Second, it affects the ability of the host sediment to transport pore waters and may influence the composition of those waters. Third, the ability of the host sediment to resist compaction is at least partly controlled by the presence of diagenetic cements. Early diagenesis of the surrounding sediment frequently produced concretions which may have prevented the bones from being compacted.

GEOCHEMISTRY OF BONE

Vertebrate remains can be used for stable isotope studies (oxygen and phosphorus). The results can be compared with data obtained from carbonate skeletons, and offer a test for possible vital, mineralogical and diagenetic isotopic fractionation effects. The PO_4 portion of bone apatite is particularly stable and unlikely to alter due to diagenesis. Thus, palaeotemperatures obtained using data obtained from $^{16}O/^{18}O$ ratios derived from phosphate are thought to be more accurate than those obtained from most invertebrate skeletons (Kolodny *et al.*, 1983; Luz and Kolodny, 1985). However, PO_4 may undergo some substitution form CO_3, from which $^{16}O/^{18}O$ values can also be obtained, the latter probably being in isotopic equilibrium with diagenetic pore waters rather than ancient marine or meteoric waters. Similarly, the ability of bone apatites to take up trace elements, including rare earths, renders them suitable for determining redox potentials of ancient seawaters (Wright *et al.*, 1987), although caution is necessary in determining when and where trace elements became incorporated.

It is also becoming possible to examine the small fraction of organic material that remains trapped within well-preserved bone from ever older rocks, both for molecular fossil and stable isotope studies, especially sulphur, carbon and nitrogen. Such data can be used to indicate maturation levels within sediments and to identify trophic levels for specific organisms.

COMPACTION OF VERTEBRATE TISSUES

Vertebrate hard tissues are subjected to compaction just like any other fossils. However, their resistance to compaction depends on a number of factors, including internal composition, the thickness of cortical bone and orientation within the sediment. This resistance is greatly enhanced by the infilling of pore spaces within bones by early diagenetic minerals (Figure 12.10).

The teeth of marine reptiles (pliosaurs, etc.) are characterized by robust, dense crowns, supported by large, elongate roots. The roots are generally hollow and are frequently compacted, but the crown usually remains uncrushed. The vertebrae of marine reptiles are highly varied in shape and internal structure. In the Lower Oxford Clay it is possible to find highly compacted and uncompacted vertebral centra on the same specimen. The controls on the compaction are varied. Small caudal vertebrae are often uncrushed due to the higher ratio of dense cortical bone to trabecular bone

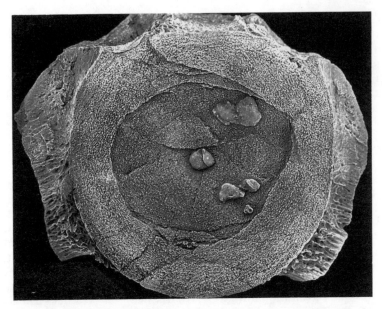

Figure 12.10.   Differential compaction of a pliosaur cervical vertebra. The relatively dense outer surface of the bone has resisted compaction, whereas the highly vascular inner part of the centrum has collapsed. This is due to the inability of the bone to withstand the high overburden pressures which were in place before diagenetic infills had formed. Specimen from Lower Oxford Clay of Peterborough, England. Life size.

internally. However, larger dorsal vertebrae may be compacted in excess of 50 per cent volume reduction (Figure 12.11), while the neighbouring centrum is uncrushed. In most cases this can be attributed to the filling of the uncrushed vertebral centrum by early diagenetic minerals such as pyrite and calcite.

PREFOSSILIZATION

As discussed above, a number of bone beds clearly contain evidence of reworking, in the form of worn and damaged bones and teeth, and in the mixing of distinct faunal assemblages. Reif (1976) proposed that many of the elements contained within such deposits had been 'fossilized' prior to reworking. Indeed, it is unlikely that density sorting of bone (not teeth) could take place unless void spaces within the bone had been infilled with minerals, as the bulk density of bones is highly variable between elements even of the same type. Reif proposed the term 'prefossilization' for such early diagenesis. Recognition of prefossilized vertebrate remains uses criteria similar to those used to determine the presence of derived fossils in any assemblage. Of particular value are the diagenetic infills which may have a marked contrast to the diagenetic suite characteristic of the rock as a whole.

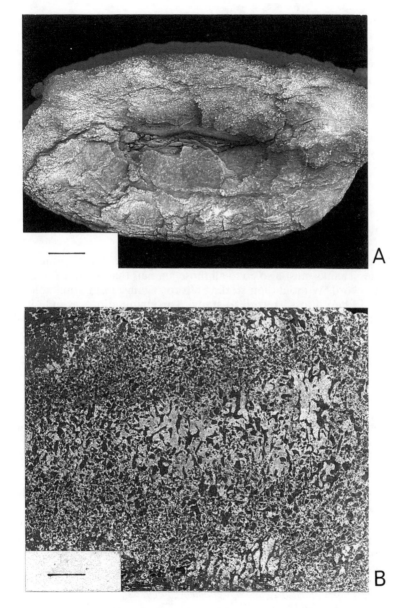

Figure 12.11. (A) Ichthyosaur centrum showing collapse due to overburden pressure. Scale bar represents 10 mm. (B) The same centrum in thin-section, showing complete brecciation of the bone internally. Scale bar represents 1 mm. The specimen is held intact by late ferroan calcites. If subjected to acid preparation it would collapse into a hash of angular bone shards. Specimen from Lower Oxford Clay of Milton Keynes, England.

SUMMARY

Clearly, vertebrate remains can make an important contribution to the lithologic record. The processes by which vertebrate tissues become incorporated into the rock record may be complex, involving repeated episodes of burial and reworking, with varying degrees of disarticulation and damage. The post-burial environment is also highly variable and vertebrate remains may be affected to varying degrees by diagenesis and compaction. Nevertheless, vertebrates may be common as fossils and exhibit a broad range of preservational styles from highly abraded bone pebbles to perfectly articulated skeletons, which very rarely may even show preservation of soft tissues and stomach contents.

ACKNOWLEDGEMENTS

I would particularly like to thank my mum and dad for bringing me up as a vertebrate. Thanks must also go to all those animals that bothered to get into the fossil record. Without their sterling efforts, against insurmountable odds, there would be no fossil record to study. This manuscript was only written because I had let so many other people down and was beginning to feel guilty. I freely acknowledge that NERC in no way financed this work. Steve Donovan phoned while I was eating my dinner.

REFERENCES

Allison, P., 1988, *Konservat-Lagerstätten*: cause and classification, *Paleobiology*, **14** (4): 331–44.

Antia, D.D.J., 1979, Bone-beds: a review of their classification, occurrence, genesis, diagenesis, palaeoecology, weathering and microbiotas, *Mercian Geologist*, **7** (2): 93–174.

Antia, D.D.J., 1981, The Temeside Bone-Bed and associated sediments from Wales and the Welsh borderland, *Mercian Geologist*, **8** (3): 163–215.

Antia, D.D.J. and Whitaker, J.H.McD., 1979, A scanning electron microscope study of the genesis of the Upper Silurian Ludlow Bone Bed. In W.B. Whalley (ed.), *Scanning electron microscopy in the study of sediments*, GeoAbstracts, Norwich: 119–36.

Behrensmeyer, A.K., 1988, Vertebrate preservation in fluvial channels, *Palaeogeography, Palaeoclimatology, Palaeoecology*, **63** (2): 183–99.

Behrensmeyer, A.K. and Hill, A.P. (eds), 1980, *Fossils in the making*, University of Chicago Press, Chicago.

Behrensmeyer, A.K., Western, D. and Dechant Boaz, D.E., 1979, New perspectives in vertebrate paleoecology from a recent bone assemblage, *Paleobiology*, **5** (1): 12–21.

Brain, C.K., 1980, Some criteria for the recognition of bone collecting agencies in African caves. In A.K. Behrensmeyer and A.P. Hill (eds), *Fossils in the making*, University of Chicago Press, Chicago: 107–30.

Brongersma-Sanders, M., 1957, Mass mortality in the sea. In J.W. Hedgepath (ed.), *Treatise on marine ecology and palaeoecology, 1, Ecology, Geological Society of America Memoir*, **67**: 941–1010.

Franzen, J.L., 1985, Exceptional preservation of Eocene vertebrates in the lake deposits of Grube Messel (West Germany), *Philosophical Transactions of the Royal Society of London*, **B311**: 181–6.

Fraser, N.C., 1988, Rare tetrapod remains from the Late Triassic fissure infillings of Cromhall Quarry, Avon, *Palaeontology*, **31** (3): 567–76.

Gagnier, P.Y., Blieck, A.R.M. and Rodrigo, S., 1986, First Ordovician vertebrates from South America, *Geobios*, **19** (5): 629–34.

Glimcher, M.J., 1984, Recent studies of the mineral phase in bone and its possible linkage to the organic matrix by protein-bound phosphate bonds, *Philosophical Transactions of the Royal Society of London*, **B304**: 479–507.

Glob, P.V., 1969, *The bog people: iron age man preserved*, Faber and Faber, London.

Goujet, D. and Locquin, M.V., 1979, Découverte de spores fongiques dans les écailles de poissons et d'agnathes paléozoiques: *Mycobystrovia lepidophaga* gen. et sp. nov, *Compte Rendu, 104e Congrèss de la Société des Savants, Bordeaux 1979*, **1**: 87–99.

Halstead, L.B., 1969, *The pattern of vertebrate evolution*, Oliver and Boyd, Edinburgh.

Haszprunar, G., 1988, Anatomy and relationships of the bone-feeding limpets, *Cocculinella minutissima* (Smith) and *Osteopelta mirabilis* Marshall (Archaeogastropoda), *Journal of Molluscan Studies*, **54**: 1–20.

Heldt, M., Riviere, A. and Bellair, P., 1947, Origines possibles des bonebeds, *Compte Rendu de l'Academie des Sciences, Hebd.*, **225**: 882–3.

Hill, A.P., 1980, Early postmortem damage to the remains of some contemporary East African mammals. In A.K. Behrensmeyer and A.P. Hill (eds), *Fossils in the making*, University of Chicago Press, Chicago: 131–52.

Hudson, J.D., 1966, Hugh Miller's Reptile Bed and the Mytillus Shales, Middle Jurassic, Isle of Eigg, Scotland, *Scottish Journal of Geology*, **2** (3): 265–81.

Kolodny, J., Luz, B. and Navon, O., 1983, Oxygen isotope variations in phosphate of biogenic apatites, I. Fish bone apatites—rechecking the rules of the game, *Earth and Planetary Science Letters*, **64** (4): 398–404.

Luz, B. and Kolodny, J., 1985, Oxygen isotope variations in phosphate of biogenic apatites, IV. Mammal teeth and bones, *Earth and Planetary Science Letters*, **75** (1): 29–36.

Marshall, B.A., 1987, Osteopeltidae (Mollusca: Gastropoda): a new family of limpets associated with whale bone in the deep-sea, *Journal of Molluscan Studies*, **53** (2): 121–7.

Martill, D.M., 1985, The preservation of marine vertebrates in the Lower Oxford Clay (Jurassic) of central England, *Philosophical Transactions of the Royal Society of London*, **B311**: 155–65.

Martill, D.M., 1987, A taphonomic case study of a partially articulated ichthyosaur, *Palaeontology*, **30** (3): 543–56.

Martill, D.M., 1988, Preservation of fish in the Cretaceous Santana Formation of Brazil, *Palaeontology*, **31** (1): 1–18.

Martill, D.M., 1989, Fungal borings in neoselachian teeth from the Lower Oxford Clay of Peterborough, *Mercian Geologist*, **12** (1): 1–4.

Martin Sander, P., 1989, Early Permian depositional environments and pond bonebeds in central Archer County, Texas, *Palaeogeography, Palaeoclimatology,*

Palaeoecology, **69** (1): 1–21.

Miller, A., 1984, Collagen: the organic matrix of bone, *Philosophical Transactions of the Royal Society of London*, **B304**: 455–77.

Milner, A., 1980, The tetrapod assemblage from Nyrany, Czechoslovakia. In A.L. Panchen (ed.), *The terrestrial environment and the origin of land vertebrates*, Academic Press, London: 439–96.

Murchison, R.I., 1856, On the bone beds of the Upper Ludlow Rock and the base of the Old Red Sandstone, *Report of the British Association for the Advancement of Science for 1856*: 70–1.

Nolf, D., 1985, *Otolithi piscium, Handbook of paleoichthyology*, Gustav Fischer, Stuttgart.

Parrish, W.C., 1978, Paleoenvironmental analysis of a Lower Permian bonebed and adjacent sediments, Wichita County, Texas, *Palaeogeography, Palaeoclimatology, Palaeoecology*, **24** (3): 209–37.

Prevot, L. and Lucas, J., 1989, Phosphate. In D.E.G. Briggs and P.R. Crowther (eds), *Palaeobiology: a synthesis*, Blackwell Scientific, Oxford: 256–7.

Reif, W.-E., 1976, Sedimentologie und Genese von Bonebeds, *Zentralblatt für Geologie und Pälaontologie*, **2** (5/6): 252–5.

Reif, W.-E., 1982, Muschelkalk/Keuper bone beds (Middle Triassic, SW-Germany) —storm concentration in a regressive cycle. In G. Einsele and A. Seilacher (eds), *Cyclic and event stratification*, Springer-Verlag, Berlin: 299–325.

Repetski, J.E., 1978, A fish from the Upper Cambrian of North America, *Science*, **200** (4341): 529–31.

Schäfer, W., 1962, *Aktuo-Paläontologie, nach Studien in der Nordsee*, Verlag W. Kramer, Frankfurt am Main.

Seilacher, A., 1982, Posidonia shales (Toarcian, S. Germany)—stagnant basin model revalidated. In E. Montanaro-Gallitelli (ed.), *Palaeontology, essential of historical geology*, STEM Mucchi, Moderna Press: 25–55.

Smith, C.R., Kukert, H., Wheatcroft, R.A., Jumars, P.A. and Deming, J.W., 1989, Vent fauna on whale remains, *Nature*, **341** (6237): 27–8.

Stuart, A.J., 1982, *Pleistocene vertebrates in the British Isles*, Longman, London.

Turnbull, W.D. and Martill, D.M., 1988, Taphonomy and preservation of a mono-specific titanothere assemblage from the Washakie Formation (Late Eocene), southern Wyoming: an ecological accident in the fossil record, *Palaeogeography, Palaeoclimatology, Palaeoecology*, **63** (1): 91–108.

Williams, C.T., 1987, Alteration of chemical composition of fossil bones by soil processes and groundwater. In G. Grupe *et al.* (eds), *Trace elements in environmental history*, Springer-Verlag, Heidelberg: 27–40.

Wood, J.M., Thomas, R.G. and Visser, J., 1988, Fluvial processes and vertebrate taphonomy: the Upper Cretaceous Judith River Formation, south-central Dinosaur Provincial Park, Alberta, Canada, *Palaeogeography, Palaeoclimatology, Palaeoecology*, **66** (2): 127–43.

Wright, J., Schrader, H. and Holser, W.T., 1987, Paleoredox variations in ancient oceans recorded by rare elements in fossil apatites, *Geochimica et Cosmochimica Acta*, **51** (3): 631–44.

SYSTEMATIC INDEX

SUBJECT INDEX

DATE DUE

JUL 1 7 1993		
FEB 1 2 1997		
NOV 13 '94		